FOUNDATIONS OF NON-COOPERA

Foundations of Non-Cooperative Game Theory

KLAUS RITZBERGER

OXFORD
UNIVERSITY PRESS

OXFORD

UNIVERSITY PRESS

Great Clarendon Street, Oxford OX2 6DP

Oxford University Press is a department of the University of Oxford.
It furthers the University's objective of excellence in research, scholarship,
and education by publishing worldwide in

Oxford New York

Auckland Bangkok Buenos Aires Cape Town Chennai
Dar es Salaam Delhi Hong Kong Istanbul Karachi Kolkata
Kuala Lumpur Madrid Melbourne Mexico City Mumbai Nairobi
São Paulo Shanghai Singapore Taipei Tokyo Toronto

with an associated company in Berlin

Oxford is a registered trade mark of Oxford University Press
in the UK and in certain other countries

Published in the United States
by Oxford University Press Inc., New York

© Klaus Ritzberger, 2002

The moral rights of the author have been asserted
Database right Oxford University Press (maker)

First published 2002

British Library Cataloguing in Publication Data

Data available

Library of Congress Cataloging in Publication Data

Data available

ISBN 0-19-924785-4 (hbk.)
ISBN 0-19-924786-2 (pbk.)

1 3 5 7 9 10 8 6 4 2

Typeset by the author
Printed in Great Britain
on acid-free paper by
Biddles Ltd., *www.biddles.co.uk*

for Victoria

Preface

This book has grown out of my lecture notes on the various game theory courses that I have taught over the years. Courses have ranged from applied game theory and primer courses to advanced and topics courses in game theory. Yet, the course I taught most frequently was a first-year graduate course in game theory. It is this course that has shaped the collection of material in this book.

Somewhat surprisingly, I had never used a single textbook as a basis of my courses, but rather material from a number of books, papers, and even unpublished work. One reason for this eclectic approach was that most textbooks leave out some material which I consider important. For instance, a proof of Kuhn's theorem is usually not given in textbooks, nor is there a discussion of the Thompson transformations.

Moreover, many texts start from the "bottom up" with solutions of games, before going on to cover what is to be solved. Though this has certain pedagogical merits, it can be confusing for students, who desire a solid understanding of the theory.

Therefore, this book brings together most of the basic material in non-cooperative game theory that has been studied since its early days. This includes some material that has lain almost dormant since the 1950s. In particular, the book covers perfect recall, Kuhn's theorem, and the Thompson transformations, but also more recent material, like normal form information sets, strategic stability, and index theory. To the best of my knowledge, this is the *only* book that offers a textbook-level treatment of all these topics.

The book addresses first-year graduate students or advanced undergraduates, as well as researchers, who desire a consistent overview of the field. The required mathematical background is comparable to a solid microeconomics course; more advanced material at times requires somewhat more advanced mathematical machinery. But I have made an effort to include explanations and definitions of these formal tools, as much as possible. For readers who need it, even an appendix (to Section 5.3) on fixed point theory is included. For most of the book, a certain maturity in dealing with abstract reasoning is more important than advanced mathematical training.

Also, I have tried to avoid giving the impression that game theory is a "closed" field, where most open issues have been settled. Having been raised on Luce and Raiffa's [105] wonderful book, I still view game theory as an exciting ground for intellectual adventure. Thus, I have tried to point to inherent problems and unsettled questions wherever the flow of the argument has allowed it.

An ample number of examples is included in every section. Most of these are drawn from economics, because they originate from courses on applications of game theory. But theoretical concepts too are systematically motivated by examples.

How to Use this Book   The organization of the material is strictly logical. It starts out with decision theory (Part I), goes on to game representations (Part II), and finally turns to solutions and refinements thereof (Part III). But nothing forces the user to stick to that order.

A number of different courses can be drawn up with the material in this book. The following table offers three combinations which I have successfully tried out. (Numbers in the list refer to sections.)

Game Theory I	Game Theory II	Advanced GT
$1.1 - 3$ $2.1 - 3$ $3.1 - 3$ 4.1 $5.1 - 5$	$2.2 - 5$ $3.2 - 5$ 4.1 $5.1 - 5$	$3.2 + 4$ $4.2 - 4$ 5.3 $6.1 - 5$

These were full semester courses with two to three hours per week. Shorter courses can be based on single chapters. For instance, Chapter 6 on equilibrium refinements together with Section 5.3 constitutes a shorter course of its own— which, however, requires prior knowledge of game theory. A course on game representations can be devised with Chapters 3 and 4.

The book's organization makes it easy to use for reference. Every section consists of a main body, followed by a brief summary, often an appendix, exercises on the current material, and finally "Notes on the Literature". The latter collect the references for the material presented in the section, because in the main text citations are avoided.

In those sections that include an appendix, this presents the proofs for results in the text. Only proofs that either students "must see" or that help to advance the argument are given in the text.

Wherever appropriate, the main text contains an intuitive explanation for the proofs gathered in the appendix. Systematically gathering proofs in an appendix has, in my view, the advantage in that the flow of the argument in the text is not interrupted by technicalities. At times, however, the appendix is also used for more technical extensions of results (Sections 4.2 and 6.4), or for explaining the mathematical background (Section 5.3). In teaching, an appendix may or may not be used, depending on how much rigor is desired.

Acknowledgments My deepest intellectual indebtedness is to my students and the Institute for Advanced Studies in Vienna. Students have provided suggestions for improvements in and criticisms of my courses, most of which have left their mark on this book. But more importantly, their questions and persistence have aided my own understanding of the material.

The Institute for Advanced Studies in Vienna has provided an ideal environment for my intellectual development for many years. Here Anatol Rappaport first taught me game theory. And here I taught my first courses. But also, I have learned from the numerous prominent visitors to the Institute, and from my colleagues. The contact with these scholars has kept my interest alive and up-to-date.

Even beyond the Institute, I have greatly benefited from the environment in Vienna. Egbert Dierker has provided lasting inspiration and intellectual support. Carlos Alos-Ferrer had a more direct bearing on this book; material from a paper that we are about to write jointly has been incorporated into Section 3.1. Georg Kirchsteiger has been a co-author with me of a paper that has inspired the discussion in Section 4.5.

Many other friends and co-authors have influenced my thinking: Jürgen Eichberger, David Kelsey, and Frank Milne during my time in Canberra; Jörgen Weibull and Karl Wärneryd in Stockholm; Helmut Bester (now in Berlin) and Eric van Damme in Tilburg; Fernando Vega-Redondo, Carmen Herrero, and Luis Corchón (now in Madrid) in Alicante; Werner Güth and Wolfgang Leininger in Germany, to name but a few. During my teaching period in Bonn I have learned from Reinhard Selten. Certain insights that are now incorporated in Sections 4.3 and 5.1 I owe to Larry Samuelson (Wisconsin). On some material in Section 6.5 I have collaborated with Stefano Demichelis from Pavia. None of these, of course, is to blame for any errors and mistakes in this book.

The two books that have shaped my views most are Luce and Raiffa's [105] textbook and Eric van Damme's [37] treatise on Nash refinements. James Friedman's [62] book has strongly influenced my interests in applications of the theory.

Andrew Schuller from Oxford University Press persuaded me to write this book and kept me going. Isabella Andrej and Harald Hutter helped me finish the manuscript and do the layout.

Contents

Contents

Part III. **Solutions of Games**

List of Figures

Part I

Games and Decisions

1

Introducing Games

Game theory has become an integral part of economics and, indeed, of other sciences, like biology. Since it originates, to a large extent, from mathematics, it requires extensive formal machinery. This will be developed in later chapters. This one is devoted to some of game theory's intuitive content which often transcends received economic "wisdom".

1.1 Playing Games

Three Examples In the movie *Last Year at Marienbad* directed by Alain Resnais (1961) a character, whom the script calls M, repeatedly persuades another, called X, to play a simple game. He arranges cards on the table in a triangular shape, one in the first row, three in the second, five in the third, and seven cards in the fourth and last row. The player who makes the first move may pick as many cards as he likes from one, but only one, row. Then the other player picks as many cards as he likes from one row. This goes on until one of the players has to pick the last card, and loses. Politely, the aristocratic M always lets X make the first move. But, alas, M always wins the game.

The game being played is a "nim" game, a class of games probably originating from China. The name is thought to derive from the German word *nimm* which means "take". The first European mention of nim apparently dates from the 15th century. Nim games have long attracted the attention of mathematicians and philosophers. And they have been an inspiration for a young discipline, emerging in the 20th century, called *game theory*. One of the early findings from this discipline is that for nim games there exists an optimal strategy. That is, one of the two players of a nim game can always force a win or, at least, a draw. Have you tried it out?

Scene-change. With the ending of the Cold War, radio spectrum became available in the United States which had previously been reserved for military use. It was decided to designate it for newly invented personal communications services, like pocket telephones, portable fax machines, and wireless computer networks. Historically, spectrum rights had been assigned by administrative procedures. This cumbersome method had produced a big backlog of unassigned licenses. So, in the 1980s it was replaced by lotteries. But the prospects of windfall gains attracted large numbers of applicants who were not seriously intending to use the rights. The value of cellular licenses that the government gave away through lotteries was estimated to total US$46 billion. Hence, in the 1990s Congress turned to a new method.

New Zealand, the United Kingdom, and Australia had previously held spectrum *auctions*—with varying degrees of success. Consequently, in 1993 the US Congress gave the Federal Communications Commission (FCC) the authority to auction spectrum licenses. In the light of experience, the FCC decided to draw on available know-how and it held a large consulting process, out of which the precise rules of the auction were drawn up. This marked the start of a case study in the policy application of economic theory. In the process of designing one of the largest auctions in history—the spectrum on offer was estimated to be worth US$10.6 billion—the government, the FCC, and the major companies involved all relied on the advice of the academics. One of the main consultants for the FCC writes:

> The analysis of how auctions work is one of the successes of modern mathematical economics. Developed to try out new ideas in game theory, auction theory has turned out to have considerable practical content. ... When the theorists met the policy-makers, concepts like Bayes-Nash equilibrium, incentive-compatibility constraints, and order-statistic theorems came to be discussed in the corridors of power. ([112], p. 146)

By the late 20th century, then, game theory had come "down to earth". With its practical applications, an entirely new perspective on economics, and on other disciplines, had emerged.

This new perspective has served to change economic theory. In economics the neoclassical paradigm had portrayed economies as an assembly of many "Robinson Crusoes". Consumers and firms were price-takers, unaware of any influence that their action might have on the market outcome, and prices were determined by the abstract consistency condition that all markets have to clear simultaneously.

The innovation that game theory introduced was that individuals could now be aware of their influence on the behavior of others—the "strategic dimension". As a simple example, compare the decisions of a lumberjack and a general. A lumberjack when chopping wood does not expect the wood to fight back; i.e., his environment is neutral. The general, on the other hand, has to overcome and anticipate resistance from the enemy. Indeed, he has to expect that his enemy will try to outguess his plans. Neoclassical economics portrayed lumberjacks, while game theory studies generals.

The key idea which enabled such an analysis to be made was the use of parlor games as models of what may happen in more serious circumstances. This has given the discipline its nickname.

Let us see how this works in an indeed serious, though fictitious, situation. (The following is a classical example from the *Theory of Games and Economic Behavior* [134].)

Sherlock Holmes boards a train from London to Dover to escape his arch-enemy, Professor Moriarty. Just when the train leaves, Holmes catches sight of Moriarty on the platform, and Moriarty has spotted him. Holmes knows that right after his train a faster train will leave which will reach Dover sooner. There is only a single stop before Dover—in Canterbury—for both trains. If Holmes can make it to Dover without meeting Moriarty, he will be able to escape to the Continent. But if he runs into Moriarty, either in Canterbury or in Dover, he will most likely be killed.

What should Holmes do? If he figures that Moriarty expects him to go straight to Dover, he should get off at Canterbury; otherwise he will experience a fatal meeting in Dover. But Moriarty is known to be a genius. So he will certainly be able to reproduce Holmes' reasoning, and thus will wait in Canterbury. Hence Holmes had better go straight to Dover. Yet, even this reasoning Moriarty will anticipate.

Moriarty, on the other hand, is in the same situation. If he banks on Holmes going to Dover, Holmes will anticipate this and get off at Canterbury. So, should Moriarty wait in Canterbury? If he expects Holmes to get off at Canterbury, Holmes will understand that and will go on to Dover. The two seem to be caught in circular reasoning.

The point of this example is not what they should do, but how important it is for each of them to hide the decision from the other. If Moriarty can find out where Holmes will get off the train, he will be able to kill him. If Holmes can figure out where Moriarty will wait, he will escape.

One way to hide a decision is not to take it, for example by leaving it to a random device, like a coin toss. This says that in the "game" between Holmes and Moriarty each one has a 50% chance that his most preferred outcome will obtain.

The parlor game associated with Holmes' and Moriarty's problem is a children's game known as "Matching Pennies". Two players each secretly put a coin on the table. Once they have done so, they reveal their choices. If both coins are heads up or tails up, player 1 pays a dollar to player 2; if the up sides do not match, player 2 pays a dollar to player 1.

Again, the most important aspect here is to hide the decision. Though a monetary payment is incomparable to being killed, the substance of the two strategic situations is the same. The conflict between the two parties makes hiding the decision vital.

Conflicts of Interest   To see why parlor games serve as good models for situations in which conflicts of interest arise, consider what a parlor game consists of. There are three ingredients,

- interacting *parties* who participate in the game, called the "players",

- *rules of the game*, which specify *who* can *when* do *what*, technically called the "game form", and

- *outcomes*, according to what was done in the course of the game, e.g. win and lose, payments made between the players, scores earned, etc.

Most interactive social situations consist of the same three ingredients. For example, consider standard economic analysis. An economy is made up of (a) agents, like consumers and firms, and (b) constraints, like prices, endowments (for consumers), and technology sets (for firms). And the interaction of the first under the constraints specified by the second give rise to (c) an allocation of goods and services.

The first two items constitute the *data* of the situation at hand, technically treated as exogenous. Adopting the assumption that each agent will do what is best for her- or himself ("maximizing agents"), the agents' interaction and the data determine a result, such as an allocation for an economy or, more generally, some outcome.

Yet, when economic agents simply maximize under constraints and at given prices, there is something missing from the picture. In particular, the "strategic dimension" is missing, the awareness that each agent's decision has an impact on the decisions of others. It is unlikely that General Motors will design its model policy without giving a thought to how Ford or the Japanese car producers will react.

The strategic dimension can be brought into the picture by being precise about what agents can do, i.e., by a precise specification of the rules of the game. And this is where the parlor game model comes in. In such games, like chess, the rules *are* very precise. Thus, it is common for chess players to anticipate the reaction of their partners according to the rules. Much like General Motors, they do not expect their partners to react to different opening moves always in the same way. Rather, opening moves are chosen in view of the anticipated variable reaction.

Being precise about the rules, of course, also imposes a restriction. In many relevant social situations the rules are not as explicit as they are in parlor games. There are general constraints, such as the legal system, rules of good conduct, and the like. But often there is nothing as precise as the rules of a chess game.

The way theory copes with this is to allow as many different ways of behavior as can reasonably be expected to conform to broad rules. Indeed, *how* precisely the rules of the game are specified makes for a distinction between two basic branches of game theory.

Cooperative and Non-Cooperative Theory One branch of game theory deals with games in which the rules are very broadly defined. In fact, in this approach the rules of the game are kept *implicit* in the formal specification of the game. This is known as the "cooperative" approach to games.

Since rules are defined only broadly, the individual decision problems of players cannot be directly analyzed. Cooperative game theory circumvents this difficulty by emphasizing *coalitions* of players. The study of coalitions implicitly assumes that such coalitions can be formed and will be maintained. Hence, behind the emphasis on coalitions is a presumption that players can *commit* themselves, either by explicitly signing enforceable contracts or by transferring their decision-making powers.

Hence, cooperative game theory is characterized by three features.

- Rules are kept implicit.

- The emphasis is on coalitions.

- Commitments are available.

There are many social situations where the rules are so vague or complex that the best approach to analysis is the cooperative. Still, this text will not deal with cooperative game theory at all. This is for three reasons.

First, cooperative game theory has not experienced such a rapid development as non-cooperative theory has, over the past fifty years. Hence, there is not so much material in cooperative theory that needs to be brought together. Second, modern applications of game theory, in particular to economics, use mainly non-cooperative theory. So readers interested in a solid background in applied modern economics will benefit more from learning non-cooperative theory. Third, cooperative theory is guided largely by axioms on group behavior (i.e. what coalitions can do), rather than on individual behavior. While this is certainly a valid approach, it does not fit directly with the individualistic view of social phenomena prevalent in modern economics.

This text will deal with "non-cooperative" game theory. In contrast to the other branch, this insists on a precise specification of the rules of the game. In fact, non-cooperative game theory can be defined as the theory of *games with complete rules*. The latter enables an analysis of individual decisions by the players and does not have to rely on commitment devices. Hence, non-cooperative game theory can be described by the following three features.

- Rules are complete.

- The ultimate decision units are the individual players.

- Commitments are not available, unless allowed for by the rules of the game.

The results of the two approaches sometimes differ. Consider, for example, the analysis of markets. A cooperative approach would impose Pareto efficiency as an axiom on the predicted outcome, arguing that anything that is Pareto-inefficient can be improved upon by forming suitable coalitions. A number of non-cooperative models of markets, on the other hand, have argued that market outcomes can be inefficient. So, the self-interest of market participants may not support what is socially desirable.

Hence, there is a tension between the predictions emerging from the two sub-disciplines. A reconciliation of this is probably one of the great challenges for future research.

Summary

- Game theory studies social interaction when individuals are aware of their impact on the decisions of others.

- There are three main ingredients to a game: players, rules, and outcomes.

- Game theory has two branches: cooperative and non-cooperative theory.

- The latter is the theory of games with complete rules.

Exercises

Exercise 1.1 Can you explain why large software companies often make their programs incompatible with the products of other producers?

Exercise 1.2 Suppose a company announces that it will always sell its product at the lowest price charged by any competing company. Will this drive prices down or will it help to maintain the going price level?

Exercise 1.3 Do you think that it could be a good idea for a commanding general of an army to toss a coin to decide whether to attack today or tomorrow?

Exercise 1.4 If the Marienbad game were to be played with only two rows (of 1 and 3 cards, respectively), can you figure out who will win? Can you also do it with three rows (of 1, 3, and 5 cards, respectively)?

Exercise 1.5 Is it possible that a member of a committee finds it optimal to vote for the alternative he or she likes least? (Hint: Consider a committee with three members who vote in two stages, first on A versus B, and then on the winner of the first round versus C, by majority voting. Suppose the three members rank the alternatives in the following order. The first A, C, B, the second C, A, B, and the third B, A, C.)

Notes on the Literature: The foundation book on game theory was by von Neumann and Morgenstern [134]. It still focussed on cooperative theory. The Sherlock Holmes tale is taken from this source. The Marienbad game is also discussed by Moulin ([125], p. 30). The story on the genesis of the FCC auction follows McMillan [112]. For an ex post evaluation of the spectrum auction, see e.g. Cramton [34] or Milgrom [118]. Introductions to cooperative game theory are available in the textbooks by Moulin [125], Myerson [128], Osborne and Rubinstein [138], and Owen [139].

1.2   Simple Examples

To develop some feel for what is to come, a few very simple examples will be discussed now. As yet there is no intention to be precise. Rather, much of what follows will involve considerable hand-waving.

Prisoners' Dilemma   Consider first two students who have handed in two good, but suspiciously similar, exams and are therefore suspected of cheating. They are being separately interrogated by the Professor, without a chance to coordinate their statements. The professor offers each of them separately the following deal.[1] If one confesses and the other does not, the confessor will get a B, but the other will fail the exam. If both confess, they will both pass, but with the worst possible grade, say, D. If both hold out, they will both pass, but with a C.

The decision problem faced by each of the two students may be represented by the following table, which gives the grades earned as a function of the student's own action and the decision of the other student. The student's own options are charted vertically and the other student's choices horizontally.

| | *Other student* | |
Student	confesses	holds out
confesses	D	B
holds out	fail	C

What should a student do? If the other student confesses, then she will fail the exam if she holds out. Hence, if the other is going to confess, it is better also to confess. If the other student holds out, then holding out will result in a C. But this can be improved upon by confessing, which results in a B *if* the other holds out. It follows, that *whatever the other student does*, it is better to confess. If students are interested in improving grades, therefore, this deal will result in both confessing—even if they have not cheated at all!

The situation described above appears somewhat artificial. But it has more important applications. Think of two institutions that independently produce economic forecasts. It is often claimed by observers that economic forecasters have a tendency to predict "no change" which is not justified by the available data. There is a strategic reason for that.

If the public reacts adversely to forecasts that predict change, each forecasting institution faces the following situation. If the other predicts change,

---

[1]   The deal offered by the Professor is *not* recommended. As will shortly be seen, it will result in both students scoring a D, even if they are innocent.

but there is a good chance that no change will occur, then it pays to predict no change. This is because the public will be comforted in the short run, even if the other institution will possibly be praised for having been right half a year later. If, on the other hand, the other predicts no change, then predicting change runs a risk of causing an embarrassment. If you are wrong, you will be blamed for creating panic. Thus, whatever the other does, predicting no change is the more attractive option.

The structure described by these simple games is one of the best known and is traditionally referred to as the *prisoners' dilemma*. Despite its simplicity, it carries important insights. Going back to the two students, it is easily seen that if they could coordinate what they say, they might both get away with a C. Lacking the means to communicate, they will both score D's. Hence, from a "social" point of view, the "socially efficient solution" would be for them both to hold out. But this is not what will happen under the deal offered by the mean Professor.

In other words, self-interested behavior can result in socially undesirable results. This is, in fact, the first lesson that can be learned from this simple example. *Individually* rational behavior need not generate *socially* efficient outcomes. And this becomes evident only when the "strategic dimension" is accounted for. The insight, obviously, is in stark contrast to standard micro-economic theory with its (two) "welfare theorems".

Coordination The prisoners' dilemma is somewhat special, because what is best for each party does *not* depend on what the other does. But in general, it will.

Suppose you are on the station platform, ready to board the train, and you meet an old friend who has reserved a seat in a different car from yours. You agree to meet in the diner. After you board the train a steward comes through making reservations, and you discover that there is a first-class diner and a second-class buffet car. You would rather eat in first class, but you suspect that your friend prefers the buffet car. You would prefer to make a reservation that coincides with hers. Do you choose the buffet car or the diner?

friend you	1st class	2nd class
1st class	3 2	1 1
2nd class	0 0	2 3

To represent the decision problems of you and your friend, a tabulation can be used, as before. Chart vertically your own choices and horizontally the decisions of your friend (see the table above). Moreover, express preferences by assigning numbers; e.g., missing your friend is worst, so "pays" 0, meeting pays two extra units, and for each of you eating in your preferred environment adds another unit. Enter these numbers in the table such that the upper left number represents your "payoff" and the lower right number represents your friend's "payoff".

Here, what is best for you crucially depends on what your friend does. If she chooses the diner, you will be best off by also choosing first class. If she chooses the buffet car, the best you can do is to settle for second class. Hence, you always want to coordinate, but would prefer to coordinate on first class.

In contrast to the prisoners' dilemma, this "coordination game" allows for (at least) two "solutions": both choose first class *and* both choose second class. The meaning of the word "solution" here is that it is a combination of choices that none of the "players" can improve upon by choosing something different. Given that your friend chooses first (second) class, the best you can do is to also choose first (second) class. And for your friend, precisely the same argument applies.

Simple as it is, the example still makes an important point. In particular, it shows that self-interested rational behavior does *not* always imply a *unique* solution. The interaction of rational individuals may well be resolved by *multiple* solutions. As the example shows, such multiplicity does not require complicated structures of interaction. It can happen in extremely simple situations.

Now consider a variant of the story. Suppose you know that your friend is in a car that the steward will reach after yours and you can accurately describe your friend, so the steward will identify her. Then you may send a message to your friend, letting her know where you have reserved *two* seats. Though this seems like a minor wiggle in the story, it has striking consequences.

To represent the new situation, use a "flow chart" representation. The "game" starts at the leftmost node in Figure 1.1, where you (player 1) can make the reservation and tell the steward to inform your friend. Whatever you choose, your friend will learn of it (the two middle nodes labeled "player 2"). She can then make her choice conditional on yours—which leads to the terminal nodes at the far right. Each combination of choices ultimately yields a payoff to each of you. Those are written as vectors with the first entry as your payoff and the second your friend's. (Such a game representation will in Chapter 3 be called an *extensive form*.)

The flow chart representation can be used to solve the game "backwards", i.e. starting at the end. If your friend learns that you chose the buffet car, she will also do so. If she learns that you will dine first class, she will be best off by also choosing first class. Given the behavior of your friend, your decision problem reduces to the choice between first and second class. And you do prefer first class. Hence, your having a commitment opportunity will not only help to coordinate, but will induce your friend to dine where *you* prefer.

Once again, this carries an important message. While often one would think that more flexibility is better, this simple example reveals that in strat-

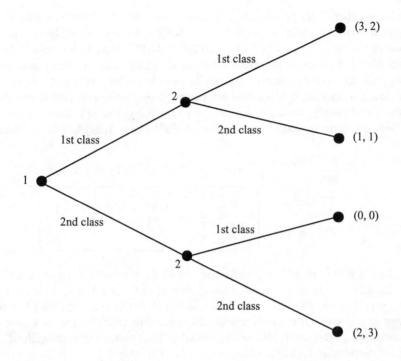

Figure 1.1: Coordination with one-way communication

egic situations giving up flexibility may help. Your commitment, in a sense, forces the other player to accept your preferred solution. The message, therefore, is that *commitment* often *helps*.

What your capacity to send a one-way message has changed is the informational basis for your friend's decision. She can decide conditional on your reservation. Therefore, she effectively has more options (to be called *strategies* later on) available than in the first variant of the story: she may unconditionally choose the diner regardless of what you chose ("$1_1 1_2$"); she may unconditionally choose the buffet car regardless of whether you chose the diner or the buffet ("$2_1 2_2$"); she can choose whatever you chose ("$1_1 2_2$"), or she can choose the opposite of what you chose ("$2_1 1_2$").

Since there are more options for your friend, the tabular representation (to be called a *normal form* in Chapter 4) of the game changes. (Again, in the table below in each cell the upper left entry represents your payoff and the lower right, your friend's.) Since your friend can now use information on what you will do, she has more options. In other words, with a change in the *information* provided to the players, the *game* changes.

Yet, there is a peculiarity here. In particular, looking for strategy com-

binations that form "solutions", in the sense that no player can improve by
deviating (to be called an *equilibrium* in Chapter 5), not only do you find
your preferred solutions (1st class, $1_1 1_2$) and (1st class, $1_1 2_2$): in addition, if
your friend stubbornly ignores your message, she can force you to dine second
class. Given that you expect her to choose the buffet car regardless of what
she hears from you, it is optimal for you to make the reservation there. And if
your friend anticipates that this is how you will react to her stubbornness, she
is best off ignoring your message. Hence, (2nd class, $2_1 2_2$) is still a solution.

	friend $1_1 1_2$	$2_1 2_2$	$1_1 2_2$	$2_1 1_2$
you 1st class	3 2	1 1	3 2	1 1
2nd class	0 0	2 3	2 3	0 0

But it is one in which your friend would do something she does not wish
to do, albeit only in circumstances that will not materialize. It is true that
you would make the reservation in the buffet car if you expected her to go
there in any case. But in the counterfactual event that she learns about your
reservation in the diner, she would probably have second thoughts. Hence, not
all solutions may be alike. (More on this in Chapter 6.)

On the other hand, your friend being stubborn acts much like your choice
of reservation (communicated to her via the steward). If you *know* that your
friend—say, for budgetary reasons—always eats second class, you are forced
to give in. And this is why (2nd class, $2_1 2_2$) remains a solution.

The example may again appear somewhat artificial. But think of firms
in an industry that have to develop an industry standard for their products. .
Each of them may have some preference for what the standard should be,
because, say, they have invested in research and development. But it is most
important for them to coordinate. The firms' problem is of the same type as
you meeting a friend on the train.

Battle of the Sexes Consider two (risk-neutral) market makers who compete in buying and selling a risky asset by setting bid and ask prices. Customers are not concerned about the return to the asset, but trade only for
reasons of liquidity. Yet, they will always buy at the lowest ask price and sell
at the highest bid price.

Prior to making the market, the two market makers can decide to acquire
(precise) information about the risky return to the asset at some fixed cost
$c > 0$. Hence, once the floor opens they can be in one of three types of
situation: they can both be informed about the risky return, or they can both
be uninformed, or one can be informed and the other not.

If both are informed, then price competition will drive both their profits from trading to zero. The same holds (in expected terms) if both are uninformed: since they have identical knowledge, price competition will drive both their (expected) profits to zero. Yet, when one is informed and the other not, only the uninformed market maker's (expected) profit will be driven to zero.

The reason for this is simple. First, the uninformed trader will never expect losses, because if she did, she would stay out of the market in the first place. But, second, the uninformed cannot expect to make a profit either: for, if she did, then for some values of the asset's return she would have to profitably capture at least one side of the market, say, the buy (bid) side, so that her bid would fall short of the asset's (true) return. But then the informed market maker can bid a price above the uninformed market maker's, but below the true return. This will steal the buy side of the market away from the uniformed trader and improve the informed market maker's profit. Since a symmetric argument holds on the sell (ask) side of the market, the uniformed trader has to expect precisely zero profit.

For the informed market maker, on the other hand, it is reasonable to assume that she makes a positive profit $\Pi > 0$ when she faces an uninformed competitor. After all, the informed market maker has all the options that the uninformed one has, and, moreover, she can condition her pricing on information about the asset's true return.

Now turn to the decisions on whether or not to acquire information at the cost $c > 0$. While once on the floor the two traders will know who is informed and who not, their decisions on information acquisition must be taken independently and in ignorance about what the other does. The problem can again be represented by a table in which the upper left entry represents trader 1's profit and the lower right entry represents trader 2's profit (net of information costs).

trader 2 ⟍ trader 1	informed	uninformed
informed	$-c$ $\quad -c$	$\Pi - c$ $\quad 0$
uninformed	0 $\quad \Pi - c$	0 $\quad 0$

Suppose now that the cost of information acquisition is sufficiently small; in particular, $c < \Pi$. Would it make sense if both paid for information? No, because they would not earn any profit from trading, but would have to bear the cost of information. Hence, if one acquires information the other is better off not buying information. On the other hand, if both are uninformed, this is not a stable constellation either. For, if the opponent is uninformed, then it pays to become informed and to capture the market at a net profit of $\Pi - c > 0$.

Thus, we are led to consider asymmetric situations. Indeed, when one trader is informed and the other not, then neither of them can do better. If the uninformed trader were to buy information, trading profits would be squeezed to zero, but the information cost would have to be born. If the informed trader would rather not spend money on information acquisition, she would lose her positive profit, because then two uninformed traders would be competing in prices. Hence, the asymmetric constellations do qualify as solutions.

Simple as it is, this example nevertheless conveys an important lesson: even when the underlying situation is symmetric, in a strategic interaction the resolution may be *asymmetric*. Strategic behavior does *not* necessarily translate symmetric situations into symmetric solutions (though it may).[2]

Matching Pennies   Finally, return to Sherlock Holmes and Professor Moriarty. Despite the apparent complexity of their decision problem, representing it as a game is rather simple. Each has two options: to get off either in Canterbury, or in Dover. Associate some payoff numbers with each combination of those which reflect how the two opponents may feel about it. Say, being killed amounts to −10 units for Sherlock Holmes and 5 units of "pleasure" for Moriarty. If they do not meet, they both receive zero units. The corresponding normal form game is given below. Once again, in each cell Holmes' payoff is in the upper left and Moriarty's at the lower right. Can you find a solution, i.e. a strategy combination, such that neither of the two can profitably deviate?

	Moriarty Dover	Canterbury
Holmes		

Dover	−10
	    5
		0
		   0

Canterbury	0
	  0
		−10
		     5

Clearly, wherever Holmes gets off, it is best for Moriarty to get off at the *same* station. On the other hand, wherever Moriarty gets off, it is best for Holmes to get off at the *other* station. Hence, there does not seem to be a solution there.

This is not quite true. Let us take a somewhat different approach. Suppose Holmes (resp. Moriarty) estimates that the chances that Moriarty (resp. Holmes) gets off in Dover are higher than for Moriarty (resp. Holmes) to get off in Canterbury. Then Holmes (resp. Moriarty) should clearly get off in Canterbury (resp. Dover). But then Moriarty (resp. Holmes) will anticipate this and get off in Canterbury. But this contradicts Holmes' (resp. Moriarty's)

---

[2]   The story presented here is obviously *not* the one that motivates the name "Battle of the Sexes".

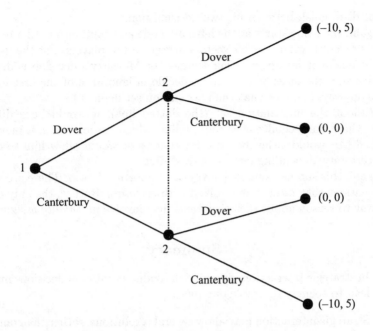

Figure 1.2: Holmes versus Moriarty

original hypothesis that Moriarty (resp. Holmes) is more likely to get off in Dover than in Canterbury. Hence, this constitutes inconsistent reasoning.

Try the hypothesis that Moriarty (resp. Holmes) is more likely to get off in Canterbury than in Dover. An analogous reasoning chain reveals that this is also inconsistent—with the presumption that both are geniuses. So, how should Holmes (resp. Moriarty) then reason about Moriarty's (resp. Holmes') choices? The only possible conclusion is that Holmes (resp. Moriarty) has to expect Moriarty (resp. Holmes) to get off at the two stops with *equal* chances.

This is *as if* both would flip a (fair) coin to determine where to get off. If they had coins available, this would then indeed constitute a solution. If Holmes (resp. Moriarty) flips a coin, Moriarty (resp. Holmes) is indifferent as to where to get off. Since he is indifferent, he might as well flip a coin. In terms of payoffs, this says that Holmes can expect −5 and Moriarty can expect 5/2 from this situation. Neither will get his most preferred outcome with certainty.

One might wonder whether this situation can also be represented by an extensive form (a "flow chart"). And indeed it can. But it requires an extra graphical device to depict that no player learns what the other chooses. Hence, in Figure 1.2 a dotted line connects the two decision points of Moriarty (labeled "2"). This is known as an "information set" and indicates that Moriarty cannot distinguish between his two decision points.

Again, the game starts at the leftmost node and continues until a terminal node is reached, where payoff vectors accrue to the players. For the game at hand it does not matter whether Holmes' or Moriarty's decision is depicted first, because the other player has to decide in ignorance of the first mover's choice anyway. (This is what the information set depicts.)

Without the information set (the dotted line), it would be a different game, where the second mover could tell at which station the first mover got off. And this would either be fatal for Holmes or would allow him to escape with certainty, depending on who moved first.

Again, this simple example conveys an important lesson. There are strategic situations that *cannot* be resolved with *certainty*. Rather, the only consistent way to reason about such situations may involve *probability* assignments.

Summary

- In strategic interaction, pursuing individually rational decisions may *not* lead to socially desirable outcomes.

- Strategic interaction may allow several resolutions, rather than only one.

- The informational basis for the players' decisions determines what the game is.

- Symmetric situations may translate into asymmetric solutions.

- Commitments can help in strategic interaction.

- Consistent reasoning about strategic interaction may require probabilistic beliefs.

Exercises

Exercise 1.6 Try a variant of the "coordination game" in which two friends have to meet either in the buffet car or the diner of a train (and the steward does not relay messages). But now assume that meeting is "worth" 1 unit (rather than 2) for each, and eating in the preferred environment adds 2 units (rather than 1). Will they meet? And which other example does your result resemble?

Exercise 1.7 Take the original version of the two friends on the train, but now relabel the strategies by (horizontally) first listing your friend's preferred choice (the buffet car), keeping your strategies fixed. What has changed, and of which other example does the result remind you?

Exercise 1.8 Delete the information set in Figure 1.2 and solve the game between Sherlock Holmes and Professor Moriarty when Moriarty can tell where Holmes is going to get off. Also represent this new game in normal form by writing down a new table.

Exercise 1.9 Why do expensive soccer players often score fewer goals than their value would suggest? And why may they still be valuable to the team?

Exercise 1.10 In sailboat racing the key is to sail where the winds are most favorable. Does this imply that the trailing boats will follow the leader or that the leader will imitate what the trailing boats do?

Notes on the Literature: Where the prisoners' dilemma originates is somewhat unclear. It first made it to a textbook in Luce and Raiffa [105]. The example of two friends meeting on the train, illustrating a "2 × 2 coordination game", is from Schelling ([160], p. 213). The story about the two market makers illustrating the "Battle of the Sexes" is inspired by the "Grossman-Stiglitz paradox" (see [72]), for which a resolution is offered by Bester and Ritzberger [20]. The interpretation of the "matching pennies" example, in terms of probabilistic beliefs about what the opponent does, draws on Aumann [7].

1.3 The Cornerstones

Three Building Blocks Recall that the three inputs to a game or, more generally, any social interaction are the interacting parties ("players"), *W*hen can *W*ho do *W*hat ("rules"), and the possible results ("outcomes"). Accordingly, non-cooperative game theory consists of three building blocks:

- decision theory (see Chapter 2),

- representation theory (see Chapters 3 and 4),

- solution theory (see Chapters 5 and 6).

The first item, decision theory, is not as simple as it is in, say, standard microeconomics. The reason is illustrated by the Holmes-Moriarty example. Games may involve random events. And this requires a theory of decision making under *uncertainty*. Therefore, one of the earliest contributions of game theory to economics was an axiomatic theory of decisions under uncertainty.

This theory is known as *von Neumann-Morgenstern* decision theory. Its core contribution is a method of representing preferences over lotteries on arbitrary alternatives, known as the "expected utility theorem". This has made its way into many other fields of economics, outside the realm of game theory, e.g. the standard microeconomic theory of general equilibrium with incomplete markets, or agency theory.

The second item concerns formal machinery for the representation of arbitrary rules of a game. Simple examples for the two most basic representations were given in the previous section. The two most basic representations are the *extensive form* and the *normal form* (sometimes also called the "strategic form" in the older literature) representation.

The first, the extensive form, is like a flow chart representation of what can happen during a playing of the game and what the informational constraints are. Its primitives are a "tree", "choices" over which various players have control, and—possibly—a specification of explicit random moves. That such an extensive form representation is possible follows from the definition of a non-cooperative game, i.e., a game with complete rules.

The second, the normal form, takes as one of its primitives complete plans of players for all possible contingencies—plans known as "strategies". The other primitive specifying a normal form is a function which assigns payoffs for all players to any combination of strategies. These payoff functions are derived using the extensive form representation combined with von Neumann-Morgenstern decision theory.

The third item on the list corresponds to the ultimate goal of game theory. It deals with how to assign solutions to games. Its core concept emphasizes that all players simultaneously will do whatever is best for them individually. Hence, it requires that no player can improve by deviating from the solution as long as she expects all other players to adhere to the solution. Such a strategy combination is known as a *Nash equilibrium*.[3]

In the presence of multiple solutions, a Nash equilibrium requires some coordination of the players' expectations. Therefore, game theory also deals with weaker solution concepts, which in games like the prisoners' dilemma still give good predictions (see Chapter 5). On the other hand, the observation that equilibria can have different properties (see the modified version of the two friends on the train) motivates inquiries into how to distinguish them (see Chapter 6).

A general point about solutions is worth making. In principle, there are two possible interpretations of a solution concept. One is that a solution prescribes how perfectly rational players "should" behave in a game; i.e., it constitutes a *normative* recommendation. The other is that a solution is what the analyst expects to happen when the game is played by real-world people. The latter is a *positive* interpretation, which takes the underlying assumptions as being at least approximately fulfilled in reality. Such an approach is often used when a certain given phenomenon is to be explained from strategic considerations. Which interpretation is adopted, ultimately, is a matter of the application.

Assumptions No solution theory can do without a few basic assumptions on how players will take decisions. Non-cooperative game theory, as most other fields in economics, here rests on a "rationality" assumption. But in the presence of strategic interaction "rationality" becomes a more intricate concept than when decision makers simply optimize with respect to their preferences. It consists of two parts.

The first part of game theory's rationality assumption is indeed that (all) players will do whatever is best for them individually. Call this the assumption of *optimizing behavior*. It means that each player will decide so as to obtain the best possible outcome with respect to her individual preferences. Hence, each player is assumed to decide consciously in view of all information available to her. This is not like in everyday life, where many "decisions" are guided by routine and do not really deserve to be called "decisions".

This assumption also is taken to mean that no player faces any bounds on her computational abilities, nor on her capacity to reproduce the reasoning process of other players. It is as if every player had a super-computer at hand

---

[3] John Nash was the first to provide an existence proof for equilibria and, thereby, to set game theory on its track. The concept itself, however, seems to have been around as folklore even before his seminal contribution. The name "NASH" also provides an acronym for four other main contributors to modern game theory: John von *N*eumann, Robert *A*umann, Reinhard *S*elten, and John *H*arsanyi.

which allowed her to solve the most complicated optimization problems. In this sense Sherlock Holmes and Professor Moriarty are the perfect example of players in a game: they are both men of genius and each considers the opponent to be a genius. The latter point is particularly important to allow for the "strategic dimension", the awareness that what is best for you depends on what the others do.

The second part of game theory's rationality assumption concerns the players' information about the game. It is assumed that each player knows

(a) the rules of the game, and

(b) the preferences for each player, and that

(c) items (a) and (b) are common knowledge.

Item (a) means that each player is fully informed about the relevant representation of the game and understands what it formalizes. Item (b) means that each player is fully aware of every other player's preferences and, of course, her own. And she knows that every player will act according to those individual preferences.

Item (c) is the most involved. It means that each player not only knows (a) and (b), but also knows that every other player knows (a) and (b); she also knows that every player knows that every player knows (a) and (b), that every player knows that every player knows that every player knows (a) and (b); and so on ad infinitum. Together with the assumption of optimizing behavior, this "common knowledge" assumption constitutes the model for the strategic dimension. It is necessary to enable each player to reconstruct the reasoning process of her opponents, without resorting to ad hoc assumptions about what others may believe.

If a player attempts to reconstruct how others reason about the game, she has to know what others know, and she has to be aware that the others are in a comparable situation. Otherwise she will be forced to make assumptions about how the others view the game, and those assumptions may be inconsistent across players. Such inconsistent beliefs would then lead to a theory that is based on ill-founded beliefs by the various players. But this is *not* what game theory is after. The goal of game theory is to understand what may happen *in the absence of* rules of thumb, religion, or prejudice. This is to be understood as it is in physics, where, for instance, the law of gravity also abstracts from frictions and turbulences.

The assumption that rules and preferences are common knowledge among the players is called the assumption of *complete information*. The assumptions of optimizing behavior and complete information together imply that what players believe about the game and the behavior of others has to be consistent with the data of the game. It forces players to make only logically consistent inferences from information they may receive in the course of playing.

Hence, the rationality assumption of game theory combines the usual optimizing behavior with complete information. In particular, the complete information assumption seems stringent at first sight. But one of the achievements of game theory is that this assumption can be mitigated without truly giving it up. Thus, a non-cooperative game can model violations of the complete information assumption, though at the cost of specifying a larger "meta"-game (see Section 3.5). Still, for all that follows, complete information is an assumption maintained throughout.

Summary

- Non-cooperative game theory consists of a decision theory under uncertainty, representation theory, and solution theory.

- The basic decision theory is expected utility.

- The basic representations are extensive forms and normal forms.

- There are several solution concepts, but the core one is Nash equilibrium.

- The rationality assumption imposed on players in a non-cooperative game consists of optimizing behavior and complete information.

Exercises

Exercise 1.11 Try representing the Marienbad game by an extensive form (a "flow chart" representation). Use dots ("nodes") to indicate where a player has to take a decision, and edges (which connect the nodes) to represent what will happen for each decision.

Exercise 1.12 Consider a parlor game and discuss which aspects of it the participating players may be uncertain about.

Exercise 1.13 Consider the following "goofspiel" between two players. A deck of cards is divided into suits, one of which (say, clubs) is discarded, and a second (say, spades) is shuffled and placed face down on the table. Each of the two players has, as her hand, a complete suit. In the course of play the spades are turned over one by one and each is captured by one of the players. They are valued: ace = 1, numbered cards their numerical value, jack = 11, queen = 12, king = 13. The player capturing the larger total value of spades wins. The procedure is as follows. The first spade is turned over so that both players can see it. Each player then selects whatever cards she wishes from her hand, and these are shown simultaneously. The one with the higher value takes the spade that was showing. If the selected cards are of equal value, the victor of the next round takes both spades. This process is then repeated, except that each hand is now depleted by the card used on the first spade. The procedure continues until the spades are exhausted.

For simplicity, assume that each suit consists of only three cards (with numerical values $1, 2,$ and 3). Try an extensive form representation of this game. And try to get an idea of the reasoning process required to win.

Exercise 1.14 Reconsider Exercise 1.13 and discuss which aspects of this game the players may be uncertain about. When a player in the "goofspiel" reasons about which card to select, what must she assume about her opponent?

Exercise 1.15 The rules of parlor games are often simple enough, so that playing them would be boring for rational players. What, do you think, are the main violations of game theory's rationality assumptions that make playing simple parlor games challenging for humans?

Notes on the Literature: The theory of decisions under uncertainty was introduced by von Neumann and Morgenstern [134] and Savage [159]. The basic representations were also developed by the earlier two authors, drawing on contributions by Borel [29] and von Neumann [133]. Kuhn [104] has developed the modern formulation of an extensive form representation. The idea of Nash equilibrium goes back at least to Cournot [33]. Nash ([131] and [132]) has provided the first existence proof and, thereby, popularized the equilibrium notion. Exercise 1.13 is from Luce and Raiffa [105].

2

Decisions under Uncertainty

Game theory, in a sense, constructs "virtual laboratories", called "models", of interactions among individuals. The goal is to perform "virtual experiments" (*Gedankenexperiment*) about what will happen in such interactions, under certain assumptions. This requires that decisions about actions are left exclusively to the participants in the game, the "players". Ideally, there should be no room for decisions other than those by the players, except for random events. This, of course, requires a theory of individual decision making in the first place.

The basic elements of such a theory are developed in the present section. It starts out with decisions among abstract *alternatives*. Those are then adapted to account for possible uncertainty. More precisely, the case of uncertainty generated from random events, or *risky* alternatives, is dealt with here: *strategic* uncertainty, which stems from how a decision maker reasons about others, who reason about her, is the topic of the theory of solutions for games and dealt with later.

2.1 Preferences

Throughout, decision making will be considered in an abstract setting. An individual decision maker is given a set A of alternatives $a \in A$ from which she is supposed to choose.

The set of alternatives being given is the only constraint on decision making. In particular, there are no time and no memory constraints, so the decision maker can consider all alternatives in the set, and in any combination. This is certainly unrealistic, but consistent with the presumption of a rational decision maker. Most importantly, it is assumed that what the decision maker likes or dislikes is *independent* of the particular (sub)menu of alternatives presented to her.

Preference Relations This abstract setting enables an understanding of the guiding principles of individual choice. It may seem surprising that something as private and concealed as the decision process of an individual can be subjected to analysis at all. After all, tastes, needs, and motivations of different individuals vary a great deal. And indeed, there is little hope of predicting from pure theory what a particular individual will decide in particular circumstances.

But this is not the goal of decision theory. Rather, it attempts to unravel the *guiding principles* behind the concrete decisions taken in particular cases. This requires a theory that is applicable to all, or almost all, tastes, needs, or preferences.

Example 2.1 *A typical example from economics is the decision problem of a consumer. The alternatives $a \in A$ are bundles of commodities. Such bundles comprise various quantities of different goods which the consumer more or less desires. This is usually formalized by representing consumption bundles as vectors in the non-negative orthant of some Euclidean space, $A = \mathbb{R}^l_+$. The consumer is then confronted with various commodity bundles and considers whether she likes one better than another or not.*

Quite obviously, theory cannot predict which of a given number of consumption bundles a particular consumer will actually choose. But what it can do is to identify how this decision will depend on the constraints of the consumer, such as prices and income. This identification, however, relies on the assumption that the consumer's tastes are independent of prices and income. And it relies on a machinery to represent any consumer's preferences by identifying what is common among the otherwise diverse tastes and needs.

The clue to a theory of choice is what happens when a person makes a choice: she will *compare* various alternatives before she actually picks one. Such hypothetical comparisons lead the decision maker to judge one alternative as better or worse than another, or to the conclusion that no such judgement is possible. Imagine the decision maker considering two alternatives, $a_1, a_2 \in A$. She may conclude that "a_2 is better than a_1" or that "a_1 is better than a_2", she may feel that one or both of those statements are false, or she may feel unable to decide. In any case, the decision maker can assign to any *ordered* pair of alternatives $(a_1, a_2) \in A \times A$ precisely one of the following two statements:

"a_2 is truly better than a_1", denoted $a_1 \prec a_2$, or

"a_2 is *not* truly better than a_1", denoted $a_1 \not\prec a_2$.

An assignment of one of those two statements to any ordered pair $(a_1, a_2) \in A \times A$ is known as a (binary) *relation* on the set A.

In general, a binary relation R on a set A assigns to every ordered pair $(a_1, a_2) \in A \times A$ precisely one of the statements "$a_1 \, R \, a_2$" (read "a_1 is in relation R to a_2") or "$a_1 \, \not{R} \, a_2$" (read "a_1 is not in relation R to a_2"). Since all pairs $(a_1, a_2) \in A \times A$ which satisfy $a_1 \, R \, a_2$ constitute a subset of the Cartesian product $A \times A$, the relation R on A can be identified with the set of all such pairs. Conversely, any subset of $A \times A$ induces a relation on A. Hence, viewing R as a subset of $A \times A$, $a_1 \, R \, a_2$ if and only if $(a_1, a_2) \in R \subseteq A \times A$.

The particular relation "$\prec$" describes the decision maker's (strict) *preferences* over alternatives $a \in A$. This interpretation suggests that $\prec$ should have further properties beyond simply being a relation on A. In the technical jargon of mathematics a few possible properties that a general relation R on A may have are as follows. The relation R may be:

- *reflexive*; i.e., $a \, R \, a$ for all $a \in A$;

- *irreflexive*; i.e., there is no $a \in A$ such that $a \, R \, a$;

- *symmetric*; i.e., if $a_1 \, R \, a_2$, then $a_2 \, R \, a_1$ for all $a_1, a_2 \in A$;

- *asymmetric*; i.e., there are no $a_1, a_2 \in A$ such that $a_1 \, R \, a_2$ and $a_2 \, R \, a_1$;

- *transitive*; i.e., if $a_1 \, R \, a_2$ and $a_2 \, R \, a_3$, then $a_1 \, R \, a_3$ for all $a_1, a_2, a_3 \in A$;

- *complete*; i.e., for all $a_1, a_2 \in A$ either $a_1 \, R \, a_2$, or $a_2 \, R \, a_1$, or both.

Which of these applies to a particular relation depends on what it models. For example, if $a_1 \prec a_2$ is meant to be "a_2 is better than a_1", then certainly the relation $\prec$ should not be symmetric. If it were, then "a_2 is better than a_1" would imply that also "a_1 is better than a_2", which is contradictory. Thus, for the interpretation at hand, the relation $\prec$ should rather be *asymmetric*.

Two types of relation are fundamental and will be used throughout. They are characterized by the following properties.

Definition 2.1 *A relation is called a (strict)* **partial ordering** *if it is irreflexive and transitive. A relation is called an* **equivalence** *relation if it is reflexive, symmetric, and transitive.*

We have seen that a (strict) *preference relation* ("better than") should *not* be an equivalence relation, because it should be asymmetric (rather than symmetric). On the other hand, to require a preference relation to be a partial ordering seems natural. No alternative should be judged better than itself; i.e., a preference relation should be *irreflexive*. Moreover, if "going to the movies" is better than "an evening at home", and "attending a party" is better than "going to the movies", we expect that "attending a party" is better than "an evening at home". Hence, a preference relation should be *transitive*. Moreover, a partial ordering is always asymmetric. For, if there would be $a_1, a_2 \in A$ such that $a_1 \prec a_2$ and $a_2 \prec a_1$, then by transitivity $a_1 \prec a_1$ yields a contradiction to irreflexivity.

Transitivity is key for a rational decision maker's preference relation. Without it the decision maker may judge $a_1 \prec a_2$ and $a_2 \prec a_3$, but then also $a_3 \prec a_1$ which keeps her locked into a cycle. In such a case the decision maker will never decide. But a rational decision maker *will* eventually make a decision. To that end, transitivity seems a minimal requirement. When mere mortals (as opposed to rational decision makers) evaluate alternatives far from common experience, however, transitivity may be hard to satisfy. Almost imperceptible differences between alternatives may lead to violations, as may certain presentation or framing effects.

Example 2.2 *The relation "a_1 is shorter than a_2" on a set of roads is a partial ordering. No road can be shorter than itself (irreflexivity). And, if road a_1 is shorter than road a_2 and a_2 is shorter than road a_3, then a_1 must be shorter than a_3 (transitivity).*

If $A = \mathbb{R}^l$, the relation $a_1 \ll a_2$, defined by "every component of the vector a_1 is strictly smaller than the corresponding component of the vector a_2", is a partial ordering. (Convince yourself.)

The relation on the set A of books defined by "book a_1 has the same number of pages as book a_2" is an equivalence relation. Every book has the same number of pages as itself (reflexivity). If a_1 has the same number of pages as a_2, then a_2 has the same number of pages as a_1 (symmetry). And if a_1 has the same number of pages as a_2, and a_2 the same number of pages as a_3, then a_1 has the same number of pages as a_3 (transitivity).

The relation "a_1 is a sibling of a_2" (by having at least one parent in common) on the set of humans is neither a partial ordering nor an equivalence relation, because it is not necessarily transitive: a_1 and a_2 may have the same father but different mothers, while a_2 may have the same mother as a_3, but a_3 may have a different father from a_1 and a_2.

While it seems compelling that a preference relation should be a partial ordering, other properties are less obvious. For instance, can the preference relation of a rational decision maker be such that there are three alternatives with $a_1 \prec a_2$, $a_3 \nprec a_2$, and $a_1 \nprec a_3$? If this were the case and the decision maker held a_2, she would be willing to exchange a_2 for a_3 at no cost—after all, $a_3 \nprec a_2$ means that "a_2 is not better than a_3". By the same token, she should be willing to exchange a_3 for a_1, because $a_1 \nprec a_3$ ("a_3 is not better than a_1"). But at this point, when she holds a_1, she will be ready to pay for exchanging a_1 for a_2, thus returning to her initial position at a loss.

A rational decision maker presumably cannot be exploited by offering her this chain of exchanges. Hence, her preferences will be such that there are no three alternatives which allow such trades.

Definition 2.2 *A **rational** (strict) **preference relation** on a set A of alternatives is a (strict) partial ordering $\prec$ on A such that*

$$\text{if } a_1 \prec a_2 \text{ and } a_3 \not\prec a_2, \text{ then } a_1 \prec a_3 \qquad (2.1)$$

for all $a_1, a_2, a_3 \in A$.

A rational preference relation is not only a partial ordering, but also satisfies the requirement that, if a_2 is judged better than a_1, and a_2 is *not* judged
better than a_3, then a_3 must be judged better than a_1. One explanation for
this property has been given. Another will become apparent when a new relation has been introduced.

The decision maker may well conclude that $a_1 \not\prec a_2$ ("a_2 is not better
than a_1") *and* $a_2 \not\prec a_1$ ("a_1 is not better than a_2"), in which case she will feel
unable to decide which of the alternatives is better or worse than the other.
Let us express these cases by a new relation on A, called the *indifference
relation*. Define the statement "a_1 is indifferent to a_2", denoted $a_1 \sim a_2$, as
follows:

$$a_1 \sim a_2 \text{ if } a_1 \not\prec a_2 \text{ and } a_2 \not\prec a_1 \qquad (2.2)$$

hold simultaneously. Which properties does the indifference relation have?

Lemma 2.1 *Let $\prec$ be a transitive relation on A. Then the indifference relation $\sim$ defined from $\prec$ by (2.2) is transitive if and only if $\prec$ satisfies (2.1).*

A proof of this lemma is in the appendix. It provides yet another justification for rational preferences. If a partial ordering satisfies (2.1), then its
induced indifference relation is transitive, and vice versa.

With the "better than" and the indifference relation, yet another relation
can be defined. For two alternatives $a_1, a_2 \in A$, the decision maker may feel
that "a_2 is better than a_1" *or* "a_2 is indifferent to a_1" (but she does not
commit to either of the two judgments)—in short that "a_2 is not worse than
a_1", denoted $a_1 \preceq a_2$. Define this "not worse than" relation by

$$a_1 \preceq a_2 \text{ if } a_1 \prec a_2 \text{ or } a_1 \sim a_2 \qquad (2.3)$$

for all $a_1, a_2 \in A$. The relation $\preceq$ is simply the union of $\prec$ and $\sim$ (when these
relations are viewed as subsets of $A \times A$).[1] A proof of the following result is
in the Appendix.

Proposition 2.1 *If $\prec$ is a rational preference relation on the set A of alternatives (Definition 2.2), then*
   *(a) the relation $\preceq$ defined by (2.3) is complete, reflexive, and transitive,
and vice versa;*
   (b) indifference, $\sim$ as defined by (2.2), is an equivalence relation.

---

[1]   Many texts start with the "not worse than" relation $\preceq$ and assume that it is transitive
and complete (and, therefore, reflexive). This approach is equivalent to the present.

Utility Functions   It is often convenient to describe preferences by means of a *utility function*. A utility function assigns a numerical value to each alternative, ranking the alternatives in accordance with preferences. It allows one to translate a decision problem into a numerical optimization problem. This is made precise in the following definition.

Definition 2.3 *A **utility function** is a real-valued function v on the set A of alternatives which **represents** the preference relation $\prec$ on A if*

$$a_1 \prec a_2 \text{ if and only if } v(a_1) < v(a_2) \tag{2.4}$$

for all $a_1, a_2 \in A$.

A utility function v representing a given preference relation $\prec$ is *not* unique. For, if v represents $\prec$ and $f : \mathbb{R} \to \mathbb{R}$ is any strictly increasing function, then the composition u of f and v, $u = f \circ v$ (viz., $u(a) = f(v(a))$, for all $a \in A$), also represents $\prec$. Only the ranking of alternatives matters, not the numerical values. Properties of utility functions that are preserved under any strictly increasing transformation are called *ordinal*. Any differences in the utility values between alternatives are *not* preserved under such transformations and are, therefore, not ordinal properties. (Those are rather called "cardinal".)

If $\prec$ is not a rational preference relation, then no such utility representation exists. Yet, a rational preference relation (Definition 2.2) and a *finite* set A of underlying alternatives are sufficient for the existence of a utility representation for preferences. Since most of the present theory will be concerned with a finite A, a rational preference relation is necessary *and* sufficient for this case. A proof of the following result is in the appendix.

Proposition 2.2 *(a) If a preference relation $\prec$ is representable by a utility function (Definition 2.3), then it is a rational preference relation (Definition 2.2).*

(b) If the set A of alternatives is finite and $\prec$ is a rational preference relation, then there exists a utility function representing the preferences $\prec$.

There are other sufficient conditions which guarantee the existence of a utility representation, apart from finiteness of A. Those are mainly continuity conditions on preferences with respect to the set A of alternatives, or a restriction on the number of equivalence classes generated by indifference. Some rational preference relations, however, cannot be represented by a utility function. A standard example is the lexicographic preference relation. Let $A = \mathbb{R}_+^2$ and $a_1 \prec a_2$ if "$a_{21} > a_{11}$, or $a_{21} = a_{11}$ and $a_{22} > a_{12}$", where $a_j = (a_{j1}, a_{j2}) \in A$, for $j = 1, 2$. This is a rational preference relation, but it fails to be continuous with respect to (the natural topology on) $A = \mathbb{R}_+^2$, and the set of equivalence classes generated by indifference is not countable. Hence, this rational preference relation is not representable by a utility function.

Summary

- A partial ordering on a set is an irreflexive and transitive relation; an equivalence relation is reflexive, symmetric, and transitive.

- A rational preference relation on a set of alternatives is a partial ordering for which the induced indifference relation is transitive.

- If a preference relation is representable by a utility function, then it is a rational preference relation.

- If the set of alternatives is finite, any rational preference relation can be represented by a utility function.

Appendix

Proof of Lemma 2.1   ("if") Suppose $\prec$ satisfies (2.1) and let $\sim$ be defined by (2.2). Let $a_1, a_2, a_3 \in A$ be such that $a_1 \sim a_2$ (i.e., $a_1 \not\prec a_2$ and $a_2 \not\prec a_1$) and $a_2 \sim a_3$ (i.e., $a_2 \not\prec a_3$ and $a_3 \not\prec a_2$). If $a_1 \prec a_3$, then $a_2 \not\prec a_3$ implies from (2.1) $a_1 \prec a_2$, in contradiction to $a_1 \not\prec a_2$. If $a_3 \prec a_1$, then $a_2 \not\prec a_1$ implies from (2.1) $a_3 \prec a_2$, in contradiction to $a_3 \not\prec a_2$. Hence, $a_1 \sim a_3$ and $\sim$ is transitive.

("only if") Assume $\sim$ is transitive. Suppose $a_1 \prec a_2$ and $a_3 \not\prec a_2$. Then either $a_2 \prec a_3$ or $a_2 \not\prec a_3$, because $\prec$ is a relation. In the first case $a_1 \prec a_3$ follows from transitivity of $\prec$ and verifies (2.1). In the second case $a_2 \sim a_3$. If $a_1 \not\prec a_3$ holds, then either $a_3 \not\prec a_1$ implies $a_1 \sim a_3$ and, by transitivity of indifference, $a_1 \sim a_2$, in contradiction to $a_1 \prec a_2$, or $a_3 \prec a_1$ implies, by transitivity of $\prec$, that $a_3 \prec a_2$, in contradiction to $a_2 \sim a_3$. Hence, $a_1 \prec a_3$ must hold, verifying (2.1). This completes the proof of Lemma 2.1.

Proof of Proposition 2.1   (a) Suppose $\prec$ is a rational preference relation (Definition 2.2). Then by Lemma 2.1 $\sim$ is transitive, and so is $\preceq$ as the union of $\prec$ and $\sim$. If there are $a_1, a_2 \in A$ such that neither $a_1 \preceq a_2$ nor $a_2 \preceq a_1$ holds, then $a_1 \not\prec a_2$ and $a_1 \not\sim a_2$ imply $a_2 \prec a_1$, and $a_2 \not\prec a_1$ and $a_2 \not\sim a_2$ imply $a_1 \prec a_2$, in contradiction to asymmetry. Hence, $\preceq$ must be complete. Completeness implies that every "diagonal" pair $(a, a) \in A \times A$ satisfies $a \preceq a$, so $\preceq$ is reflexive.

Conversely, assume that $\preceq$ is complete and transitive and define $\prec$ by $a_1 \prec a_2$ if $a_1 \preceq a_2$ and $a_1 \not\sim a_2$, where $a_1 \sim a_2$ if $a_1 \preceq a_2$ and $a_2 \preceq a_1$. By reflexivity of $\preceq$, every alternative $a \in A$ satisfies $a \sim a$, so $\prec$ is irreflexive. If $a_1 \prec a_2$ and $a_2 \prec a_3$, then by transitivity of $\preceq$ one has $a_1 \preceq a_3$. If also $a_3 \preceq a_1$ holds, then $a_3 \preceq a_1 \preceq a_2$ and $a_2 \preceq a_3$ imply $a_2 \sim a_3$, in contradiction to $a_2 \prec a_3$. Hence, $\prec$ is transitive. If $a_1 \sim a_2$ and $a_2 \sim a_3$, then $a_1 \preceq a_2 \preceq a_3$

and $a_3 \preceq a_2 \preceq a_1$ imply $a_1 \sim a_3$, so $\sim$ is transitive. Hence, by Lemma 2.1 the relation $\prec$ satisfies (2.1).

(b) If $\prec$ is a rational preference relation, by Lemma 2.1 indifference is transitive. By definition, (2.2), $\sim$ is symmetric. Since $\prec$ is irreflexive, every element $a \in A$ satisfies $a \not\prec a$, so $\sim$ is reflexive. This completes the proof of Proposition 2.1.

Proof of Proposition 2.2   (a) If $v : A \to \mathbb{R}$ represents $\prec$, then the fact that v is a function implies that $\prec$ is irreflexive. If $a_1 \prec a_2$ and $a_2 \prec a_3$, then $v(a_1) < v(a_2)$ and $v(a_2) < v(a_3)$ imply $v(a_1) < v(a_3)$, which implies $a_1 \prec a_3$, so $\prec$ is transitive. By Definition 2.3 and (2.2), $a_1 \not\prec a_2 \Leftrightarrow v(a_1) \geq v(a_2)$ and $a_2 \not\prec a_1 \Leftrightarrow v(a_2) \geq v(a_1)$, so $a_1 \sim a_2$ if and only if $v(a_1) = v(a_2)$. Hence, if $a_1 \sim a_2$ and $a_2 \sim a_3$, then $v(a_1) = v(a_2) = v(a_3)$ implies $a_1 \sim a_3$, so indifference is transitive. By Lemma 2.1, then, $\prec$ is a rational preference relation.

(b) If all alternatives $a \in A$ are indifferent, a constant utility function will do. Thus, assume that there are $a_1, a_2 \in A$ such that $a_1 \prec a_2$. Since A is finite, one can pick the largest subset A' of A such that no two elements in A' are indifferent. Let $K > 1$ be the (finite) number of elements in A' and number the elements of A' such that $a_1 \prec a_2 \prec ... \prec a_K$. Define $v : A' \to \mathbb{R}$ by $v(a_j) = j$. This function is strictly increasing in j and thus represents the preferences on A'. Extend v to A by assigning to every $a \in A$ the value $v(a) = v(a_j)$ for $a_j \in A'$ such that $a \sim a_j$. That this extended v represents preferences on *all* of A follows from the choice of A'. That (2.4) holds follows from Lemma 2.1 and Definition 2.2. This completes the proof of Proposition 2.2.

Exercises

Exercise 2.1 Show that any relation $\prec$ on A which is asymmetric and transitive is a (strict) partial ordering.

Exercise 2.2 A relation R on a set A is antisymmetric if $a_1 R a_2$ and $a_2 R a_1$ imply $a_1 = a_2$. Let $\prec$ be a relation on A and define the new relation $\preceq$ by $a_1 \preceq a_2$ if $a_1 \prec a_2$ or $a_1 = a_2$. Show that if $\prec$ is a (strict) partial ordering then $\preceq$ is reflexive, transitive, and antisymmetric.

Exercise 2.3 Suppose the function $v : \mathbb{R}^2_+ \to \mathbb{R}$ defined by $v(x) = x_1^{1/3} x_2^{2/3}$ represents a preference relation on the set $A = \mathbb{R}^2_+$ of two-dimensional consumption bundles $x = (x_1, x_2) \in A$. Verify that the preference relation represented by v is a rational preference relation.

Exercise 2.4 Verify that the lexicographic preference relation on $A = \mathbb{R}^2_+$ given by $a_1 \prec a_2$ if "$a_{21} > a_{11}$, or $a_{21} = a_{11}$ and $a_{22} > a_{12}$" where $a_j = (a_{j1}, a_{j2}) \in A$, for $j = 1, 2$, is a rational preference relation (Definition 2.2).

Exercise 2.5 Consider a "not worse than" relation $\preceq$ on a set A of alternatives which is complete and transitive. Define indifference by $a_1 \sim a_2$ if $a_1 \preceq a_2$ and $a_2 \preceq a_1$ and strict preference by $a_1 \prec a_2$ if $a_1 \preceq a_2$ and $a_1 \not\succeq a_2$, for all $a_1, a_2 \in A$. Show that, if $v(a_1) = v(a_2) \Rightarrow a_1 \sim a_2$ and $v(a_1) > v(a_2) \Rightarrow a_2 \prec a_1$, then v is a utility function representing $\preceq$.

Notes on the Literature: Extensive treatments of the preference approach to decision making are given by Richter [148] and Fishburn [59]. Violations of rational decision making in experiments with human subjects are reported by Kahneman and Tversky [93]. Fishburn ([59], Chapter 2) discusses the existence of utility representations when the number of equivalence classes generated by indifference is countable.

2.2   Expected Utility

Lotteries   Nothing has been said so far as to what the alternatives are over which preferences are defined. In particular, they may well be gambles or lottery tickets and, therefore, will carry risk; for example, they could be the various bets at a roulette table. Such risky or uncertain alternatives are of particular interest to game theory, because they provide one way to describe *decisions under uncertainty.*

Imagine that the decision maker faces such risky alternatives. Each risky alternative may result in one of a number of possible *consequences* or *outcomes,* but which one will occur is unknown at the time of the decision. Formally, let Z be the set of all possible consequences. Those could be wealth levels, consumption bundles, money prizes, or outcomes of a parlor game, like "win" or "lose". For the moment there is no need to be more specific and Z will be treated as an abstract set of consequences $z \in Z$.

One qualification about them is, however, important. Consequences are to be chosen such that how the decision maker feels about them does not depend on any action taken by her before the uncertainty about them resolves. Hence, preferences over risky alternatives should not be derived from ex ante actions. This can be accomplished by a suitable choice of what is regarded as a consequence.

Example 2.3 *Suppose you are invited to a dinner where either fish or meat will be served. And you wish to bring some wine. To formulate your decision problem about whether to bring red or white wine you should not think of the consequences as being "fish" and "meat", because, if you did, how you feel about fish may depend on which wine you brought. Rather, the appropriate consequences for the problem at hand are "meat and red wine", "meat and white wine", "fish and red wine", and "fish and white wine".*

Which risks are associated with a particular alternative will, for the moment, be assumed to be *objectively known.* Thus, given any alternative, the decision maker knows the random device which selects the consequence. For example, the alternatives may be monetary gambles on the spin of an unbiased roulette wheel. From unbiasedness the decision maker knows the objective probability of any number on the wheel. Betting on a (combination of) number(s) corresponds to choosing one from the menu of available *risky* alternatives. In fact, the term "risk" will be reserved for situations where the probabilities of the various consequences are objectively known.

Formally, we refer to a risky alternative with known probabilities as a *lottery.* Hence, lotteries consist of a set Z of underlying consequences together with a function assigning probabilities to (sets of) consequences.

Definition 2.4 *A **probability measure** on Z is a real-valued non-negative function p defined on subsets[2] of the set Z of consequences such that*

(a) $p(Z) = 1$, and

(b) if V_1 and V_2 are disjoint ($V_1 \cap V_2 = \emptyset$) and $V_1, V_2 \subseteq Z$, then $p(V_1 \cup V_2) = p(V_1) + p(V_2)$.[3]

*A probability measure is called **simple** if $p(V) = 1$ for some finite subset $V = \{z_1, ..., z_l\} \subseteq Z$.*

Property (a) is responsible for the prefix "probability" to the term "measure". Property (b) is the *additivity* property: the probability of the union of disjoint subsets of Z equals the sum of individual probabilities.

Example 2.4 *If there are only two consequences, like a tossed coin coming up heads or tails, $Z = \{heads, tails\}$, every probability measure corresponds to a single number between zero and one, say, $p(\{heads\})$, because $p(\{tails\}) = 1 - p(\{heads\})$. But things may be more complex.*

*Let the consequences be from the unit interval, $Z = [0, 1]$. The function that assigns to every half-open interval $(a, b] \subseteq [0, 1]$ the non-negative number $b - a$ is a probability measure, known as the **uniform** distribution.*

*If the consequences are non-negative integers, $Z = \{0, 1, ...\}$, and V is a subset of those, the function $p(V) = e^{-\mu} \sum_{z \in V} \mu^z / (z!)$ for some $\mu > 0$ is a probability measure, known as the **Poisson** distribution.*

*When there is some natural order on Z, like on the reals or integers, a probability measure is often given by what it assigns to intervals of the form $(-\infty, z]$. It is then referred to as a (cumulative) **distribution function**. For example, the uniform distribution on the unit interval is often given by assigning to every $z \in [0, 1]$ the probability of the associated interval, $p\left((-\infty, z]\right) = z$.*

The *support* of a probability measure p is the smallest (with respect to set inclusion) subset V of Z for which $p(V) = 1$. *Simple* probability measures are those with a finite support. Write $p(z) = p(\{z\})$ for the probability assigned by p to the singleton subset $\{z\}$ of Z . If p is simple, write supp(p) for the support, viz. the set of those consequences $z \in Z$ for which $p(z) > 0$. It can be shown (see e.g. [59], p. 106) that, if p is simple, then $p(z) = 0$ for all z which are not in the support of p, and

$$p(V) = \sum_{z \in V} p(z) \tag{2.5}$$

---

[2]   A probability measure may not be defined on *all* subsets of Z. If Z is a continuum, one picks an appropriate collection of subsets, known as a σ-algebra. This is a nonempty system of subsets of Z which contains Z and is closed under countable union and complementation.

[3]   If Z is a continuum, so p is defined on a σ-algebra, property (b) is strengthened by requiring that $p\left(\cup_{k=1}^{\infty} V_k\right) = \sum_{k=1}^{\infty} p(V_k)$ for any countable collection of disjoint subsets $V_k \subseteq Z$ which belong to the σ-algebra, for $k = 1, 2, ...$ In the finite case this "σ-additivity" boils down to (b).

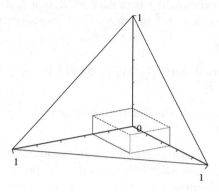

Figure 2.1: The 2-simplex in 3-space

for all subsets $V \subseteq Z$.

From now on the set Z of consequences will be fixed. Then a simple probability measure on Z is referred to as a *simple lottery*. Hence, a simple lottery consists of a finite number of consequences $\{z_1, ..., z_l\} \subseteq Z$, its support, and associated positive numbers $p_j = p(z_j) > 0$ which add up to 1, i.e., $\sum_{j=1}^{l} p_j = 1$. Denote by $\mathcal{L}$ the set of all simple lotteries.

Suppose for a moment that Z is a finite set—which will be indeed the relevant case. For concreteness, let $l > 1$ be the number of distinct consequences in Z. Then every lottery is simple. And it can be represented geometrically as a point in the $(l-1)$-dimensional simplex $\Delta^{l-1} = \left\{ p \in \mathbb{R}_+^l \middle| \sum_{j=1}^{l} p_j = 1 \right\}$ where $p_j = p(z_j)$, for all $j = 1, ..., l$.[4] Each vertex of the simplex stands for the degenerate lottery where one consequence has probability 1 and the others have probability 0. Each point in the (relative) interior[5] of the simplex represents a simple lottery where all consequences occur with positive probability.

The case $l = 3$ is easily illustrated. Figure 2.1 depicts the simplex Δ^2 in its ambient space $\mathbb{R}^3$ as the sloping triangle. The little box behind the simplex identifies the coordinates of a particular lottery $p = (p_1, p_2, p_3) \in \mathcal{L}$. Figure 2.2 depicts the simplex as a space of its own, as an equilateral triangle with altitude 1. Since in an equilateral triangle the sum of the perpendiculars from any point to the three sides equals its altitude, the lengths of the three dotted lines identify the coordinates of the lottery $p \in \mathcal{L}$. The length of the

---

[4] To distinguish by terminology, the representation of a simple probability measure by a non-negative vector (which adds up to 1) will be called a *probability distribution*.

[5] A relatively open subset of the simplex is the intersection of the simplex with an open set in the ambient space $\mathbb{R}^l$.

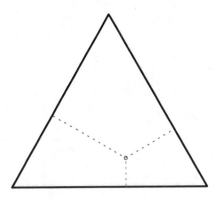

Figure 2.2: The 2-simplex as a space of its own

perpendicular from the side opposite vertex j equals p_j, for $j = 1, 2, 3$.

The consequences associated with a lottery $p \in \mathcal{L}$ need not necessarily be certain consequences $z \in Z$. They may themselves be lotteries. These more general cases are known as *compound* lotteries.

Definition 2.5 *Given two simple lotteries $p, q \in \mathcal{L}$ and a number $\alpha \in [0, 1]$, the **compound lottery** $(p, q; \alpha)$ is the risky alternative which yields the simple lottery p with probability α and the simple lottery q with probability $1 - \alpha$.*

The support of a compound lottery is the union of the supports of its constituent lotteries. If the latter are simple, the compound lottery is simple, because the union of two finite sets is finite.

For any compound lottery $(p, q; \alpha)$, there is a probability measure which assigns probability $\alpha p(V) + (1 - \alpha)q(V)$ to any subset V of Z and generates the same ultimate distribution over consequences. This probability measure is known as the *reduced lottery* associated with $(p, q; \alpha)$. It can be calculated by picking as its support the union $\text{supp}(p) \cup \text{supp}(q)$ and assigning $\alpha p(z) + (1 - \alpha)q(z) > 0$ to any $z \in \text{supp}(p) \cup \text{supp}(q)$.

Example 2.5 *Consider two urns, one with 8 black and 2 white balls, the other with 4 black and 6 white balls. A random draw from one of the urns constitutes a simple lottery with probability 1/5 for a white ball if the draw is from the first urn, and probability 3/5 for a white ball if it is from the second urn. If initially a (fair) coin is flipped to determine from which urn the draw comes, this constitutes a compound lottery with probability 1/2 for each of the urns. The associated reduced lottery has probability 2/5 for a white ball.*

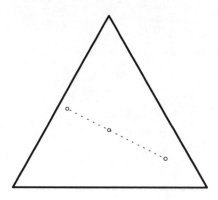

Figure 2.3: Compound lottery

If Z is finite and lotteries are depicted as points in the simplex Δ^{l-1}, the geometry of reduced lotteries is particularly simple. In Figure 2.3 the two extreme points on the dotted line represent the constituent lotteries of a compound lottery. The point on the dotted line between the two extremes represents the associated reduced lottery.

Hence, the reduced lottery associated with a compound lottery is simply a convex combination of the constituent simple lotteries. It can be calculated by vector addition:

$$
\alpha \begin{pmatrix} p_1 \\ \cdot \\ \cdot \\ \cdot \\ p_l \end{pmatrix} + (1-\alpha) \begin{pmatrix} q_1 \\ \cdot \\ \cdot \\ \cdot \\ q_l \end{pmatrix} = \begin{pmatrix} \alpha p_1 + (1-\alpha)q_1 \\ \cdot \\ \cdot \\ \cdot \\ \alpha p_l + (1-\alpha)q_l \end{pmatrix}
$$

for any two lotteries $p, q \in \mathcal{L}$. This linear structure is key for decisions under uncertainty and will be exploited below.

Preferences over Lotteries Having developed a way to represent risky alternatives, the next step is to introduce the decision maker's preferences over them. The preference relation to be introduced rests on a hypothesis about what the decision maker considers relevant. We will assume that the decision maker ultimately is interested only in *consequences* and their probabilities, but not in how these are brought about. This "consequentialist" premiss means that, for any risky alternative, only the *reduced* lottery over consequences is of relevance to the decision maker. Whether the probabilities of consequences arise from a simple lottery or from a more complex compound lottery has no significance.

In accordance with the consequentialist premiss, the set of alternatives over which preferences are defined is taken to be the set $\mathcal{L}$ of simple lotteries over consequences $z \in Z$. In the notation of the previous section this means $A = \mathcal{L}$. Hence, the decision maker's preferences are a relation $\prec$ on the set $\mathcal{L}$ with the following assumptions imposed.

Definition 2.6 *The preference relation $\prec$ on the set $\mathcal{L}$ of simple lotteries satisfies the **von Neumann-Morgenstern (v.N.-M.) axioms** if*

(a) *(rational preference) the relation $\prec$ is a rational preference relation on $\mathcal{L}$ (see Definition 2.2),*

(b) *(independence) if $p \prec q$, then $\alpha p + (1 - \alpha)r \prec \alpha q + (1 - \alpha)r$, for all $\alpha \in (0,1]$ and all $p, q, r \in \mathcal{L}$, and*

(c) *(continuity) if $p \prec q \prec r$, then there are $\alpha, \beta \in (0,1)$ such that $\alpha p + (1 - \alpha)r \prec q \prec \beta p + (1 - \beta)r$, for all $p, q, r \in \mathcal{L}$.*

Assumption (a) transfers the properties of rational preference relations studied in the previous section to preferences over risky alternatives. As previously, from the "better than" relation $\prec$ the associated indifference relation $\sim$ (see (2.2)) and the "not worse than" relation $\preceq$ (see (2.3)) can be defined and those will have the properties stated in Proposition 2.1.

It should be pointed out that, if anything, the rationality assumption (a) is stronger here than in a theory of decisions under certainty. The more complex the alternatives, the heavier the burden carried by the rationality postulates. In fact, its realism in the uncertainty context has been widely criticized.

Assumption (b) is often viewed as the heart of v.N.-M. decision theory. It is called an *independence* or "substitution" axiom. It says that, if we mix two lotteries with a third one, then the preference ordering of the two resulting mixtures does not depend on (is *independent* of) the particular third lottery used. This exploits, in a fundamental manner, the structure of uncertainty in the model.[6] The third lottery $r \in \mathcal{L}$ should be *irrelevant* to the decision maker's preference between the compound lotteries $\alpha p + (1 - \alpha)r$ and $\alpha q + (1 - \alpha)r$, because she does not end up with p or q together with r, but rather *instead* of it. She will face consequences governed by either p, or q, or r, where the "or" is exclusive. In essence, the independence axiom (b) says that the decision maker views compound lotteries as equivalent to (their) reduced lotteries, because she is only interested in final consequences.[7]

---

[6] Recall that consequences are chosen such that preferences do not depend on ex ante actions by the decision maker.

[7] Psychologically the two may not be identical, because you may well view the two-stage process of the compound lottery as more exciting.

Example 2.6 *Suppose the decision maker prefers the simple lottery q to the simple lottery p, i.e., $p \prec q$. Then $\frac{1}{2}p + \frac{1}{2}r$ can be thought of as the compound lottery arising from a coin toss which results in p if heads comes up and in r if tails does. Similarly, $\frac{1}{2}q + \frac{1}{2}r$ would be the coin toss where heads results in q and tails in r. Conditional on heads, lottery $\frac{1}{2}q + \frac{1}{2}r$ gives q and is, therefore, better than $\frac{1}{2}p + \frac{1}{2}r$ which gives p conditional on heads. But conditional on tails the two give identical results. The independence axiom (b) requires the conclusion that $\frac{1}{2}q + \frac{1}{2}r$ is better than $\frac{1}{2}p + \frac{1}{2}r$.*

The third assumption, (c), is a *continuity* condition, sometimes called an "Archimedean axiom" (for obscure reasons). It says that small changes in probabilities do not change the nature of the ordering between two lotteries. If a "beautiful and uneventful car trip" is preferred to "staying home", then a mixture between a "beautiful and uneventful car trip" with a sufficiently small but positive probability of "death by car accident" is still better than "staying home". Continuity rules out that the decision maker has lexicographic ("safety first") preferences for alternatives with a zero probability of some bad consequence.

The Expected Utility Theorem The v.N.-M. axioms are intimately linked to the representability of preferences on simple lotteries by a utility function which has the form of a mathematical expectation. Such a utility function, called a *von Neumann-Morgenstern (v.N.-M.) expected utility function,* has the convenient property of being linear in the probabilities. It is given by a real-valued function U on the set $\mathcal{L}$ of simple lotteries defined by

$$U(p) = \sum_{z \in \mathrm{supp}(p)} p(z)u(z) \qquad (2.6)$$

for some real-valued function u on the set Z of consequences. The latter is sometimes called the *Bernoulli utility function* associated with the v.N.-M. utility function U. Hence, the v.N.-M. utility function U resembles a mathematical expectation of the Bernoulli utility function u with respect to the probabilities of the lottery p. This completes the preparations for the central theorem on decisions under uncertainty.

Theorem 2.1 *(Expected utility theorem) There exists a real-valued function u on the set Z of consequences which satisfies*

$$p \prec q \text{ if and only if} \sum_{z \in \mathrm{supp}(p)} p(z)u(z) < \sum_{z \in \mathrm{supp}(q)} q(z)u(z) \qquad (2.7)$$

for all simple lotteries $p, q \in \mathcal{L}$ if and only if the relation $\prec$ on $\mathcal{L}$ satisfies the v.N.-M. axioms (a)-(c) from Definition 2.6. Moreover, the function u is unique up to a positive linear transformation; i.e., some (other) real-valued *function v on Z satisfies (2.7) for all simple lotteries if and only if there are numbers $a > 0$ and b such that $v(z) = au(z) + b$ for all $z \in Z$.*

The somewhat lengthy proof of this theorem is in the appendix to this section. At this point it may be helpful to attempt an intuitive understanding of why it is true.

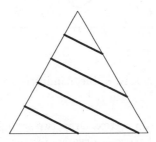

Figure 2.4: Linear parallel indifference curves

Suppose Z has only three elements, so $\mathcal{L}$ can conveniently be depicted as the two-dimensional simplex. An *indifference curve* through a point $p \in \Delta^2$ is the set of lotteries that are indifferent to p. If there are no indifferences among (the degenerate lotteries which result with probability one in one of) the consequences, the indifference curve through any point will indeed be a curve. But under the v.N.-M. axioms it will, moreover, be a *straight line,* and indifference "lines" through two different points (lotteries that are not indifferent) will be *parallel.* Figure 2.4 depicts four of those indifference curves, with the "better" direction towards the north-east. Linear and parallel indifference curves are at the heart of the expected utility theorem.

Figure 2.5 shows an indifference curve (the thick parabolic) that has curvature. Therefore, the straight (broken) line which connects the (rightward) point q *on* the indifference curve with the (leftward) point p *off* the indifference curve has a second intersection (apart from q) with the indifference curve. This second intersection corresponds to a compound lottery $(p, q; \alpha)$ which, by construction is indifferent to q; i.e., there is some $\alpha \in (0,1)$ such that $\alpha p + (1 - \alpha)q \sim q$. But then, from $q \sim \alpha q + (1 - \alpha)q$ and transitivity of indifference, it follows that the independence axiom implies $p \sim q$, in contradiction to the choice of p off the indifference curve through q.[8] Hence, indifference curves must be straight lines.

Figure 2.6 illustrates why indifference curves must be parallel. The thick line is again an indifference curve. Pick a lottery $r \in \mathcal{L}$, say, the degenerate lot-

[8] To see this, observe that the implication in the independence axiom can equivantly be rewritten as $\alpha p + (1 - \alpha)r \not\prec \alpha q + (1 - \alpha)r \Rightarrow p \not\prec q$ (see Exercise 2.9).

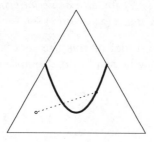

Figure 2.5: Non-linear indifference curve

tery corresponding to the top vertex. Draw two straight (broken) lines through r which intersect the indifference curve at different points. Then extend the two straight lines beyond their intersections with the indifference curve by doubling the distance (of the intersection with the indifference curve) to r. This gives two new lotteries, say, p and q. By construction, for $\alpha = 1/2$ we have $\alpha p + (1 - \alpha)r \sim \alpha q + (1 - \alpha)r$. Then the independence axiom implies $p \sim q$. But the latter simply says that p and q are on the same indifference curve, which, therefore, must be parallel to the original one by construction.

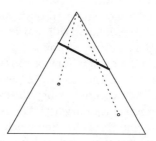

Figure 2.6: Construction of parallel indifference curves

The expected utility property (2.7) is a *cardinal* property. While in general (i.e. without independence) any strictly increasing transformation of a utility function which represents preferences is also a representation of preferences, the second part of the theorem says that this is not the case here. Only *positive linear* (or "affine") transformations are allowed. As a consequence, differences of utilities have meaning (which, in general, they do not). For example, if there are four consequences, the statement "the difference in utility between consequences z_1 and z_2 is greater than the difference between consequences z_3 and z_4", $u(z_1) - u(z_2) > u(z_3) - u(z_4)$, is equivalent to

$$\frac{1}{2}u(z_1) + \frac{1}{2}u(z_4) > \frac{1}{2}u(z_2) + \frac{1}{2}u(z_3).$$

This is the statement that lottery $q = (1/2, 0, 0, 1/2)$ is preferred to lottery $p = (0, 1/2, 1/2, 0)$ and this ranking is preserved by all positive linear transformations.

The expected utility theorem underlies most of the economics of information and uncertainty and, of course, game theory. As a descriptive theory, however, it is not without difficulties.

Example 2.7 *(The Allais paradox) Suppose there are three money prizes, $z_1 = 2,500,000$ dollars, $z_2 = 500,000$ dollars, and $z_3 = 0$ dollars. A human subject is asked to take the following two decisions. First, choose between the two simple lotteries $p = (p(z_1), p(z_2), p(z_3)) = (0, 1, 0)$ and $q = (q(z_1), q(z_2), q(z_3)) = (.10, .89, .01)$. Second, choose between lotteries $p' = (0, .11, .89)$ and $q' = (.10, 0, .90)$. It is common for individuals to express the preference $q \prec p$ and $p' \prec q'$.*

The first choice means that the subject prefers $500,000$ dollars for sure to a lottery with expected value $695,000$ dollars but a tiny risk of getting nothing. The second choice means that the subject prefers a $1/10$ chance of getting $2,500,000$ dollars to a $11/100$ chance of getting $500,000$ dollars. But this is inconsistent with expected utility. The choice $q \prec p$ implies in terms of expected utility that

$$u(z_2) > \frac{1}{10}u(z_1) + \frac{89}{100}u(z_2) + \frac{1}{100}u(z_3).$$

Adding $\frac{89}{100}[u(z_3) - u(z_2)]$ to both sides of the inequality yields

$$\frac{11}{100}u(z_2) + \frac{89}{100}u(z_3) > \frac{1}{10}u(z_1) + \frac{9}{10}u(z_3),$$

which are precisely the probabilities of the lotteries p' and q', demonstrating that expected utility implies $q' \prec p'$.

Despite such empirical doubts about the validity of the v.N.-M. axioms, the use of expected utility is pervasive in economics. One reason is certainly its simplicity. A second, at least in game theory, is that the theory deals with rational decision makers whose capabilities extend far beyond those of human subjects. And those will foresee the following possibility.

If the decision maker's preferences are such that there are three lotteries with $p \prec q$ and $r \prec q$, but in violation of the independence axiom $q \prec \alpha p + (1 - \alpha)r$ for some $\alpha \in (0, 1)$, then the decision maker can be exploited.

If she initially holds lottery q, then she would be willing to pay a small fee to trade q for a compound lottery yielding p with probability α and r

with $1 - \alpha$. But as soon as the first stage of the compound lottery has been realized, she holds either p or r and would be willing to pay again to trade this for lottery q. At this point she would be back to her original position at a loss.[9]

If a rational decision maker cannot be exploited, preferences like those just described can be ruled out. Since the focus of game theory is strategic interaction, and not inconsistencies in choice behavior, the v.N.-M. axioms seem an acceptable simplification.

Nevertheless, their realism remains debatable, and decision theories that use weaker axioms have been an active area of research. This has led to a variety of generalizations of expected utility, sometimes called non-expected utility models, which use less stringent assumptions. Yet, these more general models do not necessarily beat expected utility when fitted to experimental data (see Hey and Orme [86]).

Summary

- Risky alternatives are modeled as lotteries. A lottery is an objectively known probability measure on a fixed set of consequences.

- A lottery is simple if it has finite support. A compound lottery has other lotteries as its prizes. It is simple if its constituent lotteries are simple.

- The von Neumann-Morgenstern (v.N.-M.) axioms require the preferences over simple lotteries to be a rational preference relation which satisfies the independence axiom and a continuity condition.

- The expected utility theorem states that the v.N.-M. axioms are equivalent to the representability of preferences by a utility function which has the (expected utility) form of a mathematical expectation.

- The expected utility property is a cardinal property.

Appendix

The proof of the expected utility theorem works via a series of auxiliary results, leading up to a statement which is slightly stronger than the expected utility theorem, called a "mixture set theorem". That the latter is indeed

---

[9]   This is really only an argument for the convexity of worse-sets, i.e., for $\alpha p + (1-\alpha)r \prec q$ whenever $p \prec q$ and $r \prec q$. This property is implied by the v.N.-M. axioms, but is weaker. Similar arguments for the full independence axiom are possible, but more involved (see Green [71]).

slightly stronger than the expected utility theorem is because the arguments to follow rely only on a few properties of convex combinations of lotteries and not on those being defined by vector addition (for details see Fishburn [59], p. 111).

Lemma 2.2 *If the preference relation $\prec$ on the set $\mathcal{L}$ of simple lotteries satisfies the v.N.-M. axioms (Definition 2.6), then for all $p, q \in \mathcal{L}$ with $p \prec q$:*

(a) $\beta p + (1 - \beta)q \prec \alpha p + (1 - \alpha)q$ *if and only if* $0 \leq \alpha < \beta \leq 1$;

(b) *if* $p \preceq r \preceq q$, *then there is a unique* $\alpha \in [0, 1]$ *such that* $\alpha p + (1 - \alpha)q \sim r$.

Proof　First observe that by the independence axiom (Definition 2.6(b)) with $r = p$ one has that $p \prec q$ implies $p = \alpha p + (1 - \alpha)p \prec \alpha q + (1 - \alpha)p$, for all $\alpha \in (0, 1)$. Likewise, by independence with $r = q$ one has $(1 - \alpha)p + [1 - (1 - \alpha)]q = \alpha q + (1 - \alpha)p \prec (1 - \alpha)q + \alpha q = q$, for all $\alpha \in (0, 1)$. Therefore, $p \prec q$ implies $p \prec \alpha p + (1 - \alpha)q \prec q$, for all $\alpha \in (0, 1)$.

(a) Now suppose $p \prec q$ and $0 \leq \alpha < \beta \leq 1$. Write the lottery $\beta p + (1 - \beta)q$ as

$$\beta p + (1 - \beta)q = \frac{\beta - \alpha}{1 - \alpha}p + \frac{1 - \beta}{1 - \alpha}\left[\alpha p + (1 - \alpha)q\right].$$

By the initial step, $p \prec \alpha p + (1 - \alpha)q$. Hence, applying the initial step once again,

$$\frac{\beta - \alpha}{1 - \alpha}p + \frac{1 - \beta}{1 - \alpha}\left[\alpha p + (1 - \alpha)q\right] \prec \alpha p + (1 - \alpha)q$$

or $\beta p + (1 - \beta)q \prec \alpha p + (1 - \alpha)q$, as desired.

Conversely, suppose $0 \leq \beta \leq \alpha \leq 1$. If $\alpha = \beta$, then $\beta p + (1 - \beta)q \sim \alpha p + (1 - \alpha)q$, because the two are identical. So, assume $\beta < \alpha$. Applying the same argument as in the previous paragraph with the roles of α and β reversed yields $\alpha p + (1 - \alpha)q \prec \beta p + (1 - \beta)q$. This verifies statement (a).

(b) Suppose first that $p \sim r$. Then $r \sim 1 \cdot p + 0 \cdot q \prec \alpha p + (1 - \alpha)q$ for all $\alpha \in [0, 1)$, by (a). By transitivity of $\prec$ this $\alpha = 1$ is the unique α for which $r \sim \alpha p + (1 - \alpha)q$ holds. A symmetric argument shows that $\alpha = 0$ if $r \sim q$. If $p \prec r \prec q$ and there is *no* $\alpha \in (0, 1)$ with $r \sim \alpha p + (1 - \alpha)q$, then by (a) and continuity (Definition 2.6(c)) there exists $\beta \in (0, 1)$ such that

either　$r \prec \alpha' p + (1 - \alpha')q$, for all $0 \leq \alpha' \leq \beta$, and
　　　　$\alpha' p + (1 - \alpha')q \prec r$, for all $\beta < \alpha' \leq 1$,

or　$r \prec \alpha' p + (1 - \alpha')q$, for all $0 \leq \alpha' < \beta$, and
　　$\alpha' p + (1 - \alpha')q \prec r$, for all $\beta \leq \alpha' \leq 1$.

Consider, for example, the latter case (the former case is similar). By hypothesis, $\beta p + (1 - \beta)q \prec r \prec q$. Then by continuity (Definition 2.6(c)) there exists $\alpha \in (0, 1)$ such that

$$\alpha\left[\beta p + (1 - \beta)q\right] + (1 - \alpha)q \prec r \Leftrightarrow \alpha\beta p + (1 - \alpha\beta)q \prec r.$$

But since $\alpha\beta < \beta$, this contradicts $r \prec \alpha'p + (1 - \alpha')q$, for all $\alpha' < \beta$. Hence, $r \sim \alpha p + (1 - \alpha)q$ for some $\alpha \in (0,1)$. If $r \sim \alpha_i p + (1 - \alpha_i)q$ for $i = 1,2$, then transitivity of indifference implies $\alpha_1 p + (1 - \alpha_1)q \sim \alpha_2 p + (1 - \alpha_2)q$, which can be true only if $\alpha_1 = \alpha_2$, by (a). This completes the proof of Lemma 2.2.

One more step is required. It states that two indifferent lotteries which are mixed with the same third lottery remain indifferent.

Lemma 2.3 *If the preference relation $\prec$ on the set $\mathcal{L}$ of simple lotteries satisfies the v.N.-M. axioms (Definition 2.6) and $p \sim q$, then $\alpha p + (1 - \alpha)r \sim \alpha q + (1 - \alpha)r$ for all $r \in \mathcal{L}$ and all $\alpha \in [0, 1]$.*

Proof For $\alpha = 0$ or $\alpha = 1$, the statement follows directly from $p \sim q$. Thus, assume $\alpha \in (0,1)$. Consider first the case $r = q$. If $\alpha p + (1 - \alpha)q \prec q$, then $p \sim q$ implies $\alpha p + (1 - \alpha)q \prec p$.
But then, by independence (Definition 2.6(b)),

$$
\begin{aligned}
\alpha\left[\alpha p + (1 - \alpha)q\right] + (1 - \alpha)\left[\alpha p + (1 - \alpha)q\right] &= \alpha p + (1 - \alpha)q \\
\prec \alpha p + (1 - \alpha)\left[\alpha p + (1 - \alpha)q\right] &= \alpha(2 - \alpha)p + (1 - \alpha)^2 q, \text{ and} \\
(1 - \alpha)\left[\alpha p + (1 - \alpha)q\right] + \alpha p &= \alpha(2 - \alpha)p + (1 - \alpha)^2 q \\
\prec (1 - \alpha)q + \alpha p &= \alpha p + (1 - \alpha)q
\end{aligned}
$$

yields a contradiction. A similar argument applies if $\alpha p + (1 - \alpha)q \prec p$. Thus, by (2.2) and transitivity, $\alpha p + (1 - \alpha)q \sim p \sim q$.

Next consider the case $r \sim p$. Then by the above and transitivity, $\alpha p + (1 - \alpha)r \sim p \sim q \sim \alpha q + (1 - \alpha)r$. Finally, consider the case $r \prec p$. (The case $p \prec r$ is analogous.) Then by independence (Definition 2.6(b)) $r \prec \alpha p + (1 - \alpha)r$. If $\alpha p + (1 - \alpha)r \prec \alpha q + (1 - \alpha)r$ holds, then by Lemma 2.2(b) there is a unique $\beta \in (0,1)$ such that $(1 - \beta)r + \beta\left[\alpha q + (1 - \alpha)r\right] \sim \alpha p + (1 - \alpha)r$ or, equivalently, $(1 - \alpha\beta)r + \alpha\beta q \sim \alpha p + (1 - \alpha)r$. Since $p \sim q$ implies $r \prec q$, independence implies $(1-\beta)r+\beta q \prec q \sim p$. Therefore, again by independence, $\alpha\left[\beta q + (1 - \beta)r\right]+(1-\alpha)r \prec \alpha p+(1-\alpha)r$ or $(1-\alpha\beta)r+\alpha\beta q \prec \alpha p+(1-\alpha)r$, in contradiction to the previous finding. Hence, $\alpha p+(1-\alpha)r \not\prec \alpha q+(1-\alpha)r$ must hold. A similar argument establishes $\alpha q + (1 - \alpha)r \not\prec \alpha p + (1 - \alpha)r$, thus demonstrating the statement of Lemma 2.3.

This completes the preparations and allows us to state the main theorem. Its proof is given under the assumption that there exists a best and a worst element in the set of lotteries. This saves on notation and entails no true loss of generality. (For how to extend the proof if the assumption does not hold, see Fishburn [59], p. 114.)

Theorem 2.2 *The preference relation $\prec$ on the set $\mathcal{L}$ of simple lotteries satisfies the v.N.-M. axioms (Definition 2.6) if and only if there exists a function*

$U : \mathcal{L} \to \mathbb{R}$ *which represents preferences (Definition 2.3) on lotteries and is linear, i.e.*

$$U\left(\alpha p + (1 - \alpha)q\right) = \alpha U(p) + (1 - \alpha)U(q), \qquad (2.8)$$

for all $\alpha \in [0, 1]$ and all $p, q \in \mathcal{L}$. Moreover, some other function $V : \mathcal{L} \to \mathbb{R}$ represents preferences and is linear, (2.8), if and only if there are numbers $a > 0$ and b such that $V(p) = aU(p) + b$, for all $p \in \mathcal{L}$.

Proof (a) Let p^* resp. q^* be a worst resp. best element in $\mathcal{L}$, i.e., $p^* \preceq p \preceq q^*$ for all $p \in \mathcal{L}$. Assume that $\prec$ satisfies the v.N.-M. axioms. Define $U : \mathcal{L} \to [0, 1]$ by $U(p^*) = 0$, $U(q^*) = 1$ and

$$p \sim [1 - U(p)]\, p^* + U(p)q^*, \text{ for all } p \in \mathcal{L},$$

where $U(p)$ exists and is unique by Lemma 2.2(b). Clearly, if $p \sim p^*$ ($p \sim q^*$), then $U(p) = 0$ ($U(p) = 1$). If $p \prec q$, then by (the "only if" part of) Lemma 2.2(a) and transitivity, $U(p) < U(q)$. Conversely, if $U(p) < U(q)$, then by the construction of U, transitivity, and (the "if" part of) Lemma 2.2(a), $p \prec q$. Hence, U represents preferences.

By construction of U one has, for $p, q \in \mathcal{L}$ and $\alpha \in [0, 1]$,

$$\alpha p + (1 - \alpha)q \sim [1 - U\left(\alpha p + (1 - \alpha)q\right)]\, p^* + U\left(\alpha p + (1 - \alpha)q\right) q^*. \qquad (2.9)$$

Applying Lemma 2.3 twice yields

$$\alpha p + (1 - \alpha)q \sim \alpha\left[(1 - U(p))\, p^* + U(p)q^*\right] + (1 - \alpha)\left[(1 - U(q))\, p^* + U(q)q^*\right]$$

which implies

$$\alpha p + (1 - \alpha)q \sim [1 - \alpha U(p) - (1 - \alpha)U(q)]\, p^* + [\alpha U(p) + (1 - \alpha)U(q)]\, q^*.$$

Comparing the last expression with (2.9) and invoking Lemma 2.2(b) yields

$$U\left(\alpha p + (1 - \alpha)q\right) = \alpha U(p) + (1 - \alpha)U(q)$$

for all $\alpha \in [0, 1]$ and all $p, q \in \mathcal{L}$, viz. (2.8). Hence, if $\prec$ satisfies the v.N.-M. axioms, then U is a *linear* representation of preferences.

To see the converse, assume that U represents preferences and satisfies (2.8). Then by Proposition 2.2 $\prec$ is a rational preference relation. If $p \prec q$, then by (2.4) $U(p) < U(q)$ implies $\alpha U(p) + (1 - \alpha)U(r) < \alpha U(q) + (1 - \alpha)U(r)$ for all $\alpha \in (0, 1]$ and all $r \in \mathcal{L}$, which implies independence (Definition 2.6(b)). Likewise, $U(p) < U(q) < U(r)$ implies that there are $\alpha, \beta \in (0, 1)$ such that

$$\alpha U(p) + (1 - \alpha)U(r) < U(q) < \beta U(p) + (1 - \beta)U(r)$$

which establishes continuity (Definition 2.6(c)).

(b) Suppose U is a linear representation of preferences and $V(p) = aU(p) + b$, for all $p \in \mathcal{L}$, for some $a > 0$. Then it is easily verified that V is a linear representation of preferences.

To go the other way, suppose V and U are linear representations of preferences. If U is constant on L, then so is V, and they differ only by an additive constant. Otherwise, i.e. if $p^* \prec q^*$, define for any $p \in \mathcal{L}$

$$f_1(p) = \frac{U(p) - U(p^*)}{U(q^*) - U(p^*)} \text{ and } f_2(p) = \frac{V(p) - V(p^*)}{V(q^*) - V(p^*)}.$$

Since f_1 and f_2 are positive linear transformations of U and V, both satisfy (2.8) and represent preferences. Moreover, $f_i(p^*) = 0$ and $f_i(q^*) = 1$ for $i = 1, 2$. If $p \sim p^*$ or $p \sim q^*$, then $f_1(p) = f_2(p)$. If $p^* \prec p \prec q^*$, then $f_1(p) = f_2(p)$ by $p \sim f_i(p)q^* + (1 - f_i(p))\, p^*$ and Lemma 2.2(b). Hence, $f_1 \equiv f_2$. Then by construction of the f_i's

$$V(p) = \frac{V(q^*) - V(p^*)}{U(q^*) - U(p^*)} U(p) + V(p^*) - U(p^*) \frac{V(q^*) - V(p^*)}{U(q^*) - U(p^*)}$$

establishes that V is a positive linear transformation of U. This completes the proof of Theorem 2.2.

The expected utility theorem follows from this by setting $u(z) = U(p_z)$ where $\mathrm{supp}(p_z) = \{z\}$, for all $z \in Z$. If $\mathrm{supp}(p) = \{z_1, ..., z_l\}$, then repeated applications of (2.8) with $p_z \in \mathcal{L}$ such that $p_z(z) = 1$ give

$$U(p) = U\left(\sum_{z \in \mathrm{supp}(p)} p(z)p_z \right) = \sum_{z \in \mathrm{supp}(p)} p(z)U(p_z) = \sum_{z \in \mathrm{supp}(p)} p(z)u(z)$$

such that (2.7) holds and the expected utility theorem is verified.

Exercises

Exercise 2.6 Let p be a simple probability measure on Z. Show that the smallest (with respect to set inclusion) subset V of Z for which $p(V) = 1$ coincides with the largest (again with respect to set inclusion) subset V' of Z for which $p(z) > 0$ for all $z \in V'$.

Exercise 2.7 Show that Definition 2.4(b) implies for a probability measure p on Z that if $V_1, V_2 \subseteq Z$, then $p(V_1 \cup V_2) = p(V_1) + p(V_2) - p(V_1 \cap V_2)$.

Exercise 2.8 Show that if preferences $\prec$ over simple lotteries L with three consequences are representable by a utility function (Definition 2.3) which has the expected utility form (2.6), then indifference curves are parallel straight lines. (Assume there are no ties between consequences.)

Exercise 2.9 Assume that the preference relation $\prec$ on the set L of simple lotteries satisfies the independence axiom (Definition 2.6(b)). Show that if there is some $\alpha \in (0,1]$ such that $\alpha p + (1-\alpha)r \sim \alpha q + (1-\alpha)r$, then $p \sim q$, for any $p, q, r \in L$. (Hint: Recall (2.2).)

Exercise 2.10 (The St. Petersburg paradox) Consider the following lottery. A fair coin will be tossed repeatedly until heads comes up. If this happens in the t-th toss, the lottery yields a money prize of 2^t dollars. The probability of this outcome is $1/2^t$. Discuss how much an expected utility maximizer may be willing to pay for participation in this lottery. (Hint: This problem can be used to show that the Bernoulli utility function of an expected utility maximizer will be bounded.)

Notes on the Literature: The expected utility theorem and the independence axiom are originally due to von Neumann and Morgenstern ([134], Chapter 3). Later modifications are by Savage [159] and by Luce and Raiffa ([105], Chapter 2), among others. The idea of an expected utility representation, however, dates back to Bernoulli [19]. The proof of the expected utility theorem given below, which rests on a mixture set theorem, follows Fishburn ([59], Chapter 8). The Allais paradox (Example 2.7) originates in Allais [3]. A compact survey on theories of decisions under uncertainty and their problems is by Machina [106]. That game theory can accommodate decision theories that are more general than expected utility is exemplified by Dow and Werlang [51] and Ritzberger [150], among others. Generalizations of expected utility have been investigated empirically with experimental data by Hey and Orme [86], among others.

2.3   Risk Attitude

The proof of the expected utility theorem suggests that the Bernoulli utility function u is, in a sense, redundant. But in many economic settings the Bernoulli utility function carries important information about the decision maker's risk attitude.

Risk Aversion   For this section it will be assumed that consequences $z \in Z$ are wealth levels or money prizes, so Z can be identified with the real line $\mathbb{R}$. The Bernoulli utility function u is then a function from the real line to itself. The v.N.-M. axioms (Definition 2.6) alone do not place any restrictions on u. For applications, however, the analytical power of the expected utility formulation hinges on specifying the Bernoulli utility function u in such a way that it captures interesting attributes of choice behavior. With consequences being wealth levels, plausible postulates are that u is (strictly) *increasing* and *continuous*. These two assumptions will be maintained throughout this section.

Beyond that, the Bernoulli utility function can capture the decision maker's *risk attitude*. A property commonly assumed in economic models is that agents wish to avoid risks, even at a cost. We start with a definition of risk aversion which does not presume an expected utility formulation.

Definition 2.7 *A decision maker is **risk averse** if, for any nondegenerate lottery $p \in \mathcal{L}$, the degenerate lottery which yields the expected value (or "mean")*

$$E_p z = \sum_{z \in \mathrm{supp}(p)} p(z)z \tag{2.10}$$

*with certainty is always preferred to p. She is **risk neutral** if she is always indifferent between these two lotteries. She is **risk loving** if she always prefers p to the lottery that yields the expected value $E_p z$ for sure.*

If preferences admit an expected utility representation (2.6) with Bernoulli utility function u, the definition of risk aversion implies that the decision maker is risk averse if and only if

$$\sum_{z \in \mathrm{supp}(p)} p(z)u(z) < u \left(\sum_{z \in \mathrm{supp}(p)} p(z)z \right) = u\left(E_p z\right) \tag{2.11}$$

for all nondegenerate lotteries $p \in \mathcal{L}$. This inequality, also called *Jensen's inequality*, is the defining property of a (strictly) *concave* function. This makes

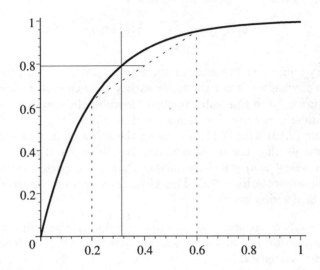

Figure 2.7: Concavity represents risk aversion

sense. Strict concavity means that the marginal utility (viz. the slope of u) of wealth is declining as wealth increases. Hence, at any level of wealth z, the utility gain from one extra dollar is smaller than the (absolute value of the) utility loss from one dollar less. It follows that the risk of gaining or losing one dollar with even probability is not worth taking.

Figure 2.7 illustrates. The horizontal depicts wealth levels z and the vertical the values of the Bernoulli utility function u. The thick concave curve shows (the graph of) u and the two broken verticals indicate two wealth levels, z_1 and z_2, to which the lottery p assigns probability $1/2$ each. The expected value $E_p z$ of this lottery is, therefore, in the middle between z_1 and z_2 (at 0.4). The thin solid horizontal line indicates the utility from this lottery (approximately 0.8).

The definition of risk aversion, however, refers to the *expected value* (or "mean") of money that a given lottery yields and not to its *expected utility* (except when the decision maker is risk neutral). Yet, there is also a sure amount of money that is precisely as good for the decision maker as the lottery.

This is known as the *certainty equivalent*. It is the amount of money, denoted $z(p)$, for which the decision maker is indifferent between the gamble $p \in \mathcal{L}$ and the certain amount $z(p)$, i.e. $p \sim p_{z(p)}$, where $p_{z(p)}$ denotes the degenerate lottery with $p_{z(p)}(z(p)) = 1$. In expected utility terms, the certainty equivalent $z(p)$ is the solution to the equation

$$u(z(p)) = \sum_{z \in \text{supp}(p)} p(z)\,u(z). \qquad (2.12)$$

The certainty equivalent depends on the lottery $p \in \mathcal{L}$ *and* on preferences (irrespective of whether or not the latter satisfy the expected utility hypothesis). In Figure 2.7 the thin solid vertical identifies the certainty equivalent $z(p)$. This, under concavity of u, is smaller than $E_p z$.

Comparing (2.12) with (2.11) and using the assumption that u is strictly increasing reveals that the decision maker is risk averse if and only if the certainty equivalent $z(p)$ is strictly smaller than the expected value $E_p z$ for all nondegenerate lotteries $p \in \mathcal{L}$. This gives three ways to describe the risk attitude of the decision maker.

Proposition 2.3 *Suppose the decision maker is an expected utility maximizer with Bernoulli utility function u on wealth levels z. Then the following three statements are equivalent:*

(a) The decision maker is risk averse (risk neutral) [risk loving];

(b) the Bernoulli utility function u is strictly concave (linear) [strictly convex];

(c) $z(p) < E_p z$ ($z(p) = E_p z$) [$z(p) > E_p z$] for all nondegenerate lotteries $p \in \mathcal{L}$.

Risk aversion is a common assumption in economics, because if it did not generally hold, economic agents would not buy insurance. The use of the risk aversion concept is illustrated in the following example which constitutes a basic insight from the economics of insurance.

Example 2.8 *Suppose a risk averse individual holds initial wealth w, but runs the risk of a loss of d dollars, $0 < d < w$, with probability $\varepsilon \in (0,1)$. It is possible, however, for the individual to buy insurance. One unit of insurance costs c dollars and pays off one dollar if the damage occurs (nothing otherwise). If $x \geq 0$ units of insurance are bought, the individual's wealth will be $w - cx$ if no damage occurs and $w - cx - d + x$ if damage occurs. The expected value of wealth with x units of insurance is $E_p w = w - \varepsilon d + x(\varepsilon - c)$.*

If the price c is actuarially fair, in the sense of being equal to the expected cost of insurance, $c = \varepsilon$, expected wealth becomes $E_p w = w - \varepsilon d$, irrespective of the level x of insurance. Since, under $c = \varepsilon$, the decision maker can always get this expected wealth level by setting $x = d$, the definition of risk aversion implies that $x = d$ is the optimal level of insurance. Hence, if insurance is actuarially fair, a risk averse decision maker insures completely.

Expected utility is a cardinal property. Risk aversion, as a property of an expected utility function, therefore, can be quantified. Proposition 2.3 suggests that a quantitative measure of the degree of risk aversion will have to do with the curvature of the Bernoulli utility function.

Given a twice continuously differentiable Bernoulli utility function u, the *Arrow-Pratt coefficient of absolute risk aversion* at $z \in Z = \mathbb{R}$ is defined as (minus) the ratio of the second to the first derivative of u at z, i.e. by $-u''(z)/u'(z)$. The *coefficient of relative risk aversion* at z is defined by $-zu''(z)/u'(z)$. Both these coefficients measure the curvature of u in such a way that they are invariant with respect to positive linear transformations (as allowed under expected utility).

These measures of risk aversion are, of course, tailored to expected utility. More general decision theories under uncertainty require different measures of risk aversion.

Stochastic Dominance   With the assumption that consequences are money prizes, $Z = \mathbb{R}_+$, the probability distributions on the real line can be (partially) ordered so as to reflect the preferences of *all* expected utility maximizers.

Let, for the moment, $F(z)$ and $G(z)$ denote two (cumulative) distribution functions on the non-negative reals which take the value 0 at $z = 0$ and the value 1 for all z greater or equal to some (common) upper bound, are right-continuous, and nondecreasing.[10] Then expected utility can be written as the integral (from zero to infinity) of the Bernoulli utility function with respect to these distribution functions and the expected value is the integral of the random payment with respect to the distribution function.

Definition 2.8 *The distribution function F **first-order stochastically dominates** the distribution G if, for every nondecreasing function $u : \mathbb{R}_+ \to \mathbb{R}$, we have*

$$\int_0^\infty u(z)\, dF(z) \geq \int_0^\infty u(z)\, dG(z). \tag{2.13}$$

*The distribution F **second-order stochastically dominates** the distribution G if $E_F\, z = E_G\, z$ and inequality (2.13) holds for every nondecreasing concave function $u : \mathbb{R}_+ \to \mathbb{R}$.*[11]

Hence, if F first-order stochastically dominates G, then *no* expected utility maximizer will prefer G to F. In this case the mean of F cannot be smaller than the mean of G, because otherwise a risk neutral decision maker would prefer G. The second-order criterion, therefore, looks at distributions with the same mean and risk aversion. If F and G have the same mean and F second-order stochastically dominates G, then *no risk averse* expected utility maximizer will prefer G to F. The following proposition is proved in the appendix.

---

[10]   For a simple probability measure the cumulative distribution function will be a step-function with (upward) "steps" at any point where probability mass sits.

[11]   If F and G have the same mean, first-order stochastic dominance implies second-order stochastic dominance, but not vice versa.

Proposition 2.4 *(a) The distribution function F first-order stochastically dominates the distribution G if and only if $F(z) \leq G(z)$, for all $z \geq 0$.*

(b) The distribution function F second-order stochastically dominates the distribution G if and only if they have the same mean and the Lorenz curve of G never lies below the Lorenz curve of F, i.e.,

$$\int_0^z F(x)\ dx \leq \int_0^z G(x)\ dx,\ for\ all\ z \geq 0. \tag{2.14}$$

Stochastic dominance can, therefore, be defined purely in terms of the distribution functions, without having to specify utilities explicitly.

Summary

- If consequences are monetary payoffs, risk aversion means that for any nondegenerate lottery the decision maker definitely prefers the expected value to the lottery itself.

- The certainty equivalent of a lottery is the sure money prize which is indifferent to the lottery. Risk aversion means that the expected value of any nondegenerate lottery exceeds its certainty equivalent.

- In terms of the Bernoulli utility function (on monetary payoffs), risk aversion is equivalent to (strict) concavity.

- Stochastic dominance orders distribution functions in accordance with any v.N.-M. preferences.

Appendix

Proof of Proposition 2.4   (a) "if": Assume that u is continuously differentiable.[12] If $F(z) \leq G(z)$ for all $z \geq 0$, then, integrating by parts,

$$
\begin{aligned}
\int_0^\infty u(z)\ dF(z) - \int_0^\infty u(z)\ dG(z) &= u(z)\left[F(z) - G(z)\right]|_0^\infty \\
&\quad - \int_0^\infty u'(z)\left[F(z) - G(z)\right]\ dz \\
&= \int_0^\infty u'(z)\left[G(z) - F(z)\right]\ dz \geq 0.
\end{aligned}
$$

---

[12]   Though it is not proved here, this assumption does not entail any loss of generality if u is continuous, as always assumed in this section.

"only if": If there is $\bar{z} \geq 0$ such that $F(\bar{z}) > G(\bar{z})$, then $\bar{z} > 0$, and with $u(z) = 1$, for all $z > \bar{z}$, and $u(z) = 0$, for all $z \leq \bar{z}$, we obtain

$$\int_0^\infty u(z)\, dF(z) - \int_0^\infty u(z)\, dG(z) = G(\bar{z}) - F(\bar{z}) < 0.$$

(b) "if": Assume that u is twice continuously differentiable.[13] So, concavity is equivalent to $u''(z) \leq 0$, for all $z \geq 0$. If (2.14) holds, then, integrating by parts twice,

$$\begin{aligned}
\int_0^\infty u(z)\, dF(z) - \int_0^\infty u(z)\, dG(z) &= u(z)\left[F(z) - G(z)\right]\big|_0^\infty \\
&\quad - \int_0^\infty u'(z)\left[F(z) - G(z)\right] dz \\
&= \int_0^\infty u'(z)\left[G(z) - F(z)\right] dz \\
&= u'(z)\int_0^z \left[G(x) - F(x)\right] dx \big|_0^\infty \\
&\quad - \int_0^\infty u''(z)\int_0^z \left[G(x) - F(x)\right] dx\, dz \\
&= \int_0^\infty u''(z)\int_0^z \left[F(x) - G(x)\right] dx\, dz \geq 0,
\end{aligned}$$

because of the equality of means, i.e.

$$\begin{aligned}
0 = \int_0^\infty z\, dF(z) - \int_0^\infty z\, dG(z) &= z\left[F(z) - G(z)\right]\big|_0^\infty \\
&\quad - \int_0^\infty \left[F(z) - G(z)\right] dz \\
&= \int_0^\infty \left[G(z) - F(z)\right] dz.
\end{aligned}$$

"only if": If there is $\bar{z} \geq 0$ such that

$$\int_0^{\bar{z}} F(z)\, dz > \int_0^{\bar{z}} G(z)\, dz,$$

then $\bar{z} > 0$. Considering the concave utility function given by $u(z) = \bar{z}$, for all

---

[13] The previous footnote applies. Note, however, that concavity is sufficient to guarantee continuity.

$z > \overline{z}$, and $u(z) = z$, for all $z \leq \overline{z}$, we obtain

$$
\begin{aligned}
\int_0^\infty u(z)\, dF(z) - \int_0^\infty u(z)\, dG(z) &= \int_0^{\overline{z}} z\, dF(z) - \int_0^{\overline{z}} z\, dG(z) \\
&\quad + \overline{z}\left[G\left(\overline{z}\right) - F\left(\overline{z}\right)\right] \\
&= z\left[F(z) - G(z)\right]\big|_0^{\overline{z}} \\
&\quad - \int_0^{\overline{z}} \left[F(z) - G(z)\right]\, dz \\
&\quad + \overline{z}\left[G(\overline{z}) - F(\overline{z})\right] \\
&= \int_0^{\overline{z}} \left[G\left(z\right) - F\left(z\right)\right]\, dz < 0.
\end{aligned}
$$

This completes the proof of Proposition 2.4.

Exercises

Exercise 2.11 Reconsider the insurance problem (Example 2.8), but assume that insurance is not actuarially fair. Show that if $c > \varepsilon$ a risk averse expected utility maximizer will not fully insure; i.e., she will accept some risk. (Hint: Assume that the Bernoulli utility function is continuously differentiable and consider first order conditions for a utility maximum.)

Exercise 2.12 Let u be a real-valued increasing function on the real line. Show that if u is concave, then it must be continuous. (Hint: Use a graphical argument.)

Exercise 2.13 Calculate the coefficients of absolute and relative risk aversion for the two (Bernoulli) utility functions $u_1(z) = -e^{-\alpha z}$ and $u_2(z) = z^\alpha$ for some $\alpha \in (0, 1)$.

Exercise 2.14 Let u_1 be a Bernoulli utility function on wealth levels z and $g : R \to R$ a strictly increasing and strictly concave function. Define a new Bernoulli utility function u_2 by $u_2(z) = g(u_1(z))$ for all $z \in Z = R$. Show that u_2 exhibits "more" risk aversion than u_1 in the sense of the coefficient of absolute risk aversion.

Exercise 2.15 Suppose an expected utility maximizer can invest her wealth $w > 0$ into an asset that always yields a (per unit) return of 1 and another asset the (per unit) return of which is given by a simple lottery $p \in L$ with $E_p z > 1$. If x and y denote the amounts of wealth invested in the safe and risky assets respectively, the budget constraint is $x + y = w$. Show that if the decision maker is risk averse she will still invest in the risky asset, i.e., $y > 0$ at the optimum. (Hint: Assume that the Bernoulli utility function is continuously differentiable and consider first order conditions for a utility maximum.)

Notes on the Literature: The classical references on risk aversion are Arrow [6] and Pratt [143]. A comprehensive account of theories of risk attitude is given by Kreps [103]. Notions of risk aversion in the context of generalizations of expected utility are studied by Quiggin [144], Chew, Karni, and Safra [31], Yaari [184], and Wakker [178], among others. Stochastic dominance criteria have been introduced to the field of economics by Rothschild and Stiglitz [154].

2.4   States

Up to now the choice among risky alternatives was based solely on the consequences (and their probabilities), like the monetary payoffs from the previous
section. In essence, the underlying cause of the consequence was of no im-
portance. If the cause is one's state of health, this assumption is unlikely to
be fulfilled. The probabilities of a lottery over consequences are then not the
appropriate objects of choice.

States   Here we turn to the possibility that the decision maker cares not
only about consequences, but also about the underlying events, or *states of
the world*, that cause them. This is of particular interest to game theory. In a
game of chess, for instance, the players may be concerned not only about the
terminal arrangement of pieces on the board and who wins, but also about
how elegantly the game was played.

Moreover, there are situations in which something that may be formalized
as a lottery is known to be generated by some underlying cause. In particular,
in game theory lotteries are often generated by the opponents' strategies which
makes the formulation with states attractive for the current purpose.

Example 2.9 *Suppose a girl wishes to bring a boyfriend to a party and can
ask one or both of two, call them Adam and Bob, to accompany her. Suppose
also that she can bring only one and that she does not know whether Adam
or Bob will agree to join her; she has, however, the discretion to ask either
Adam, or Bob, or both, and to select one if both agree to join (if asked). This
can be viewed as a choice from lotteries over consequences $Z = \{bring\ Adam,
bring\ Bob,\ attend\ alone\}$. If she decides to ask only Adam (or only Bob), then
the probability of "bring Bob" (or "bring Adam") is zero. If she decides to ask
both, the probability of who she will bring depends on the probabilities of the
boys' answers and on which one she will select if both agree to join.*

*But the probabilities of the boys' answers are not necessarily generated
from a coin flip, though they might be. A more fundamental cause of the
consequences is the boys' decisions on how to respond to the girl's invitation.
These can be expressed as* states *of the world. In this small world the possible
states are given by the set $\Omega = \{only\ Adam\ agrees,\ only\ Bob\ agrees,\ both
agree,\ neither\ agrees\}$. Assuming that the girl will always bring a boy if at
least one agrees, she has at her disposal five* acts *or* strategies: *namely, "ask
only Adam", "ask only Bob", "ask both and choose Adam if both agree", "ask
both and choose Bob if both agree", and "ask neither". Given a probabilistic
belief concerning the states, each of these acts translates into a lottery over
consequences. For example, if she "asks both and chooses Adam if both agree"*

the probability of bringing Adam is the sum of the probabilities of the states "only Adam agrees" and "both agree", while the probability of bringing Bob is the probability of the state "only Bob agrees". If she "asks neither", the probability of "bring Adam" (or "bring Bob") is zero.

The formulation of uncertainty that emerges from such considerations has three constituent parts. The first is a set Z of *consequences* with which we are already familiar. The second is a set Ω of *states* ω of the world. Each such state is "... a description of the world, leaving no relevant aspect undescribed" (Savage [159], p. 9). The third part is a probability measure π on the set Ω of states, expressing the decision maker's *beliefs* about which state obtains.

The delicate part is the specification of states. In particular, the state space needs to be sufficiently rich for what is being modeled. States are "all inclusive" descriptions of the world; i.e., they incorporate *all* decision-relevant factors about which the decision maker may be uncertain. This assumes that it is possible to enumerate all possible contingencies in the first place—certainly not a mild assumption, but one that is consistent with the noncooperative approach as the theory of games with complete rules. States need to satisfy the following three conditions:

- they must be *mutually exclusive*; i.e., at most one state $\omega \in \Omega$ can obtain;

- they must be *collectively exhaustive*; i.e., the decision maker cannot conceive of none of the states obtaining;

- the state that obtains must *not depend* on the decision maker's choice.[14]

Random Variables With this concept of states, an uncertain alternative can be described as a function on the set Ω of states. This is also known as a *random variable*. Moreover, the next definition introduces the notion of an *event*. An event is simply a set of states, i.e. a subset of Ω. For instance, in the previous example "precisely one boy agrees" is an event, namely the event {only Adam agrees, only Bob agrees}.

Definition 2.9 *A **random variable** is a function f on the set Ω of states. An **event** is a subset of the set Ω of states.*

This is, admittedly, an obscure definition, because it leaves unspecified the range (the set from which the values of f come) of the random variable. This is because there is some freedom here. Depending on what is being modeled, random variables may, for instance, be *real-valued*, so the range of f is the

---

[14] This is reminiscent of the earlier condition that preferences over consequences do not depend on ex ante actions of the decision maker.

real line $\mathbb{R}$. A typical example would be the wealth levels from the previous section.

A random variable may also assign to each state a *consequence* $z \in Z$.[15] The most prominent example of such random variables are the acts or strategies which the decision maker may have at her disposal. In the previous example, each of the girl's five acts can be thought of as a function from states to consequences. If, for instance, she "asks both and chooses Adam if both agree", the states "only Adam agrees" and "both agree" will both map into the consequence "bring Adam". Under the same act, the state "only Bob agrees" maps into the consequence "bring Bob". And "neither agrees" results in "attend alone".

Every random variable f that maps into consequences gives rise to a lottery over consequences where the probability of a particular consequence z is the probability of the event (subset of states) that the random variable takes the value z; i.e., it is given by the probability $\pi\left(f^{-1}(z)\right)$ of the preimage $f^{-1}(z) = \{\omega \in \Omega \,|\, f(\omega) = z\} \subseteq \Omega$. But there is a loss of information in going from the random variable representation to the lottery representation. In the latter there is no record of which states give rise to a particular consequence, and only the aggregate probability of every consequence is retained.

Finally, a random variable may map into some *signal* space, where "signals" or "reports" are representations of information that the decision maker may receive. Let R denote the set of all possible signals or reports $r \in R$. Observing a particular realization $r \in R$ of such a random variable $f : \Omega \to R$ conveys information to the decision maker about which states can obtain.

To see this, first observe that the collection of events $H_f = \left\{f^{-1}(r)\right\}_{r \in R}$ *partitions* the state space Ω, because f is a function. (Formally, $f^{-1}(r_1) \cap f^{-1}(r_2) = \emptyset$ if $r_1 \neq r_2$ and $\Omega \subseteq \cup_{r \in R} f^{-1}(r)$.) Hence, if the decision maker observes $\bar{r} \in R$ as the realization of the random variable f, then she knows that the true state $\bar{\omega}$ must be an element of the event $f^{-1}(\bar{r}) = \{\omega \in \Omega \,|\, f(\omega) = \bar{r}\}$, i.e., $\bar{\omega} \in f^{-1}(\bar{r}) \subseteq \Omega$. In other words, upon observing $\bar{r} = f(\omega)$ the decision maker learns that only states in the preimage $f^{-1}(\bar{r})$ are still *possible* and all states in the complement $\Omega \setminus f^{-1}(\bar{r})$ are deemed impossible.

The same applies for *two* random variables, f and g, mapping into a signal space R. The collection of all events, of the form

$$f^{-1}(r_1) \cap g^{-1}(r_2) = \{\omega \in \Omega \,|\, f(\omega) = r_1 \text{ and } g(\omega) = r_2\}$$

for $r_1, r_2 \in R$, partitions the state space. Observing the realizations r_1 and r_2 of f and g amounts to the information that only states in $f^{-1}(r_1) \cap g^{-1}(r_2)$ are possible.

However, if it so happens that for every r_1 there is some r_2 such that $f^{-1}(r_1) \subseteq g^{-1}(r_2)$ then, of course, observing realizations of g does not add

---

[15]   More generally, a random variable may assign to each state a (simple) lottery over consequences. This amounts to a kind of compound lottery where first a state realizes and then a roulette wheel is spun (see Anscombe and Aumann [4]).

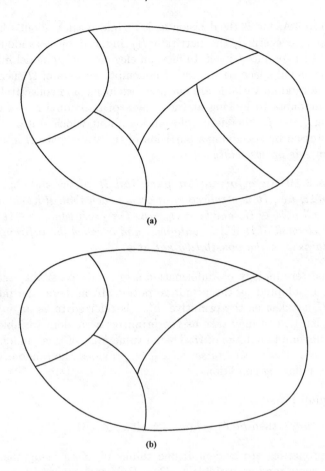

(a)

(b)

Figure 2.8: (a) Finer and (b) coarser partitions

anything to the information contained in realizations of f. In such a case we say that g is *measurable* with respect to the partition H_f induced by f. Measurability (of g with respect to H_f) means that, if f is constant on some set, then so is g, and therefore g cannot contain more information than f. Rather, the values of g can perfectly be predicted if a realization of f has been observed.

Figure 2.8 illustrates. The oval area depicts the state space Ω. Part (a) shows the partition H_f induced by the random variable f. Part (b) shows the partition H_g induced by g. Clearly, H_f is a *finer* partition than H_g in the sense that H_f partitions further the lower right element of H_g. Therefore, g is measurable with respect to H_f. More intuitively, H_f contains any information that H_g contains, and sometimes more.

Information   As far as the decision maker's information about which states may obtain is concerned, the partition H_f induced by a random variable $f : \Omega \to R$ is as good as f itself. In fact, an elegant way to model information is to let the signal space be the set of nonempty subsets of Ω such that the values of the random variable are mutually exclusive and collectively exhaustive. This amounts to looking at the images of individual states under the composition $f^{-1} \circ f$. Therefore, information about which states may obtain can be expressed by specifying a partition of the state space Ω into so-called *information sets* or *possibility sets* h.

Definition 2.10 *An* **information partition** *H of the state space Ω is a partition of Ω; i.e., H is a collection of events such that, if $h \cap h' \neq \emptyset$, then $h = h'$ for all $h, h' \in H$, and $\Omega = \cup_{h \in H} h$. For every state $\omega \in \Omega$ the event $h(\omega)$ is the element of H which contains ω and is called the* **information set** *which contains ω or the* **possibility set** *at ω.*

The dual terminology of "information sets" and "possibility sets" comes from what is adopted as the primitive notion. If, as here, the information partition H is taken as the primitive, $h(\omega)$ is referred to as an *information set*. Equivalently, one may take as the primitive a random variable h which maps into nonempty subsets of Ω. Then a value $h(\omega)$ of this random variable is referred to as the *possibility set* at $\omega \in \Omega$, provided the random variable h satisfies the following conditions:

(h.1) $\omega \in h(\omega)$ for all $\omega \in \Omega$, and

(h.2) if $\omega' \in h(\omega)$, then $h(\omega') = h(\omega)$, for all $\omega, \omega' \in \Omega$.

These properties are evident if one thinks of H as being the partition induced by some random variable $f : \Omega \to R$. If one thinks of h as a random variable mapping into nonempty subsets of Ω, describing what is regarded as possible, then properties (h.1) and (h.2) deserve explanation.
Property (h.1) simply says that the state which obtains is always regarded as possible. Property (h.2) says, first, that if something is possible, $\omega' \in h(\omega)$, then anything that would be regarded as possible in that state ω' must also be regarded as possible in the current state ω, i.e., $h(\omega') \subseteq h(\omega)$. Second, (h.2) says that anything that is regarded as possible in the current state ω must also be regarded as possible in a possible state ω', i.e., $h(\omega) \subseteq h(\omega')$.
Condition (h.2) is more difficult than (h.1), because of its second part. It turns out that this second part, $h(\omega) \subseteq h(\omega')$, requires that, if the decision maker does *not* know that she does *not* know something, then she knows it. This is often called the "axiom of wisdom" and is highly controversial. On the other hand, this second part of (h.2) is responsible for possibility sets to partition Ω and, hence, for the equivalence of the approaches via information partitions and via the "possibility operator" h. And, indeed, to fulfill (h.2) in

concrete cases is often simply a matter of specifying a sufficiently rich state space.

An argument for the full (h.2) is as follows. Suppose the decision maker finds that only states in $h(\omega)$ are possible. If $\omega' \in h(\omega)$ and $h(\omega') \neq h(\omega)$, then she can deduce by introspection that it is impossible that ω' obtains. For, if ω' were to obtain, then her view (at ω') of what is possible would differ from the view that she actually holds (at ω). But then there is a basis on which the decision maker can distinguish between ω and ω', implying that ω' is not possible. Thus, $h(\omega') \neq h(\omega)$ implies that $\omega' \notin h(\omega)$ or, equivalently, $\omega' \in h(\omega)$ implies that $h(\omega') = h(\omega)$.

Knowledge What is regarded possible, of course, is not the same as what is "known" to the decision maker upon observing the realization of a random variable.

Example 2.10 *Suppose the girl in Example 2.9 is about to call Adam. Hearing his answer amounts to the observation of a random variable that takes the values "Yes" if one of the states "only Adam agrees" and "both agree" obtains, and the value "No" if the states "only Bob agrees" or "neither agrees" obtain. These two sets of states define the girl's information partition H. If Adam's answer is "Yes", the girl regards only states in the set {only Adam agrees, both agree} possible. But then she knows the event "at least one boy will agree", which amounts to {only Adam agrees, only Bob agrees, both agree}. This is because every possible state is an element of this event. She does not, however, know the event "precisely one boy agrees", because the possible state "both agree" is inconsistent with only one boy agreeing.*

Knowledge, therefore, is derived from what is regarded as possible. Formally, an event $E \subseteq \Omega$ is *known at the state* $\omega \in \Omega$ if $h(\omega) \subseteq E$. In words, an event is *known* at some state $\omega \in \Omega$ if everything possible implies it. Hence, what is known to the decision maker depends on the state. At first this may seem hard to reconcile with the intuitive meaning of "knowledge". But from a probabilistic viewpoint it becomes more transparent.

Upon observing the realization of a random variable, the decision maker will revise or "update" her probability assignment π over states. Call π the *prior* beliefs and the revised probability assignment $\tilde{\pi}$ the *posterior* beliefs. That the decision maker regards states outside of $h(\omega)$ as impossible means, in probabilistic terms, that the complement of $h(\omega)$ gets assigned posterior probability zero, $\tilde{\pi}(\Omega \setminus h(\omega)) = 0$. Therefore, $\tilde{\pi}(h(\omega)) = 1$ (see Definition 2.4). That she regards all states in $h(\omega)$ as possible means, in probabilistic terms, that $\tilde{\pi}(\omega') > 0$ for all $\omega' \in h(\omega)$. That an event is "known", in probabilistic terms, means that it is believed with probability 1. So, what are the events that get assigned posterior probability 1? They are precisely those that contain the set $h(\omega)$ where the posterior beliefs must concentrate all the probability mass. If some part of $h(\omega)$ were left out, the probabilities would not add up to 1.

Bayes' Rule   But, how precisely does the decision maker update her beliefs about which state obtains? Given her prior beliefs π, upon observing the realization of a random variable h, she will have to calculate for any event $E \subseteq \Omega$ the *conditional probability of E given $h(\omega)$*. Since $h(\omega)$ depends on the state, this conditional probability assignment is itself a random variable. In the abstract the conditional probability of an event $E \subseteq \Omega$, given some other event $E' \subseteq \Omega$, denoted $\pi(E|E')$, is given by

$$\pi(E|E') = \frac{\pi(E \cap E')}{\pi(E')} \text{ if } \pi(E') > 0, \tag{2.15}$$

and is arbitrary otherwise. Hence, if state $\omega \in \Omega$ obtains, the decision maker's *posterior beliefs at ω* are given by $\tilde{\pi}(E) = \pi(E|h(\omega))$ for all events $E \subseteq \Omega$.

Ultimately, the decision maker will care about random variables mapping into consequences $z \in Z$, like acts or strategies. Therefore, we now turn to how the observation of a random event improves the information about a random variable mapping into consequences.

Let $f : \Omega \to Z$ be such a random variable and $H_f = \{f^{-1}(z)\}_{z \in Z}$ the associated partition of the state space. Suppose the decision maker observes the realization of a random variable h (a "possibility operator" satisfying (h.1) and (h.2) as above) with associated information partition H. For simplicity, assume that Z contains only finitely many elements. Then, because H_f is a partition of Ω, for each $h \in H$ the collection $\{f^{-1}(z) \cap h\}_{z \in Z}$ partitions the set h. Therefore,

$$\pi(h) = \sum_{z \in Z} \pi\left(h \cap f^{-1}(z)\right)$$

holds for all $h \in H$. Combining this with

$$\pi\left(h \cap f^{-1}(z)\right) = \pi\left(f^{-1}(z)\right) \pi\left(h|f^{-1}(z)\right)$$

for all $h \in H$ and $z \in Z$ from (2.15) yields

$$\pi\left(f^{-1}(z)|h\right) = \frac{\pi\left(h|f^{-1}(z)\right) \pi\left(f^{-1}(z)\right)}{\sum_{z' \in Z} \pi\left(h|f^{-1}(z')\right) \pi\left(f^{-1}(z')\right)} \tag{2.16}$$

whenever the denominator is positive, for all $z \in Z$ and all $h \in H$. (If the denominator is zero, $\pi\left(f^{-1}(z)|h\right)$ is arbitrary.)

Formula (2.16) is known as *Bayes' rule* and is a key ingredient of decision theory. It gives the conditional probability of a consequence $z \in Z$ given the information $h \in H$ which the decision maker receives and, therefore, the posterior beliefs about the lottery over consequences at information set $h \in H$.

Bayes' rule is, of course, not confined to random variable mapping into consequences. It generally asserts that the *conditional probability* of an event E_k, out of mutually exclusive events $E_1, ..., E_K$, *given* another event h, denoted $\text{Prob}(E_k | h)$, is the *joint* probability of E_k and h, $\text{Prob}(E_k \cap h) = \text{Prob}(h | E_k) \text{Prob}(E_k)$, divided by the *marginal* probability of h, denoted $\text{Prob}(h) = \sum_{k=1}^{K} \text{Prob}(E_k \cap h)$, i.e., $\text{Prob}(E_k | h) = \text{Prob}(E_k \cap h) / \text{Prob}(h)$.

The following example illustrates an application of Bayes' rule.

Example 2.11 *(Herding) A number of people (at least three, say) arrive in a town, one after the other. They all wish to have a good dinner. There are two restaurants in town, call them A and B. Which of the two offers the better value is unknown to the prospective customers. But each of them has some private information about the restaurants' quality. Assume that these private signals are, conditional upon the truth, independently and identically distributed. Specifically, a private signal takes the value a (resp. b) with probability $1 - \delta$ if A (resp. B) is the better restaurant, where $1/2 > \delta > 0$. On top of that, each of the arriving people can observe which restaurant the person arriving immediately before her chooses (but only this). All share the common (prior) belief that, in the absence of private signals, the probability that A is the better restaurant is p, where $1 - \delta > p > 1/2$.*

Bayes' rule implies that, upon observing $k \geq 0$ signals a and $m \geq 0$ signals b, the conditional (posterior) probability of A being the better restaurant is

$$\pi_{km} = \frac{p(1 - \delta)^k \delta^m}{p(1 - \delta)^k \delta^m + (1 - p)(1 - \delta)^m \delta^k}$$

for all $k, m = 0, 1, ...$ Hence, given these signals, it is preferable to choose restaurant A if and only if

$$\frac{p}{1 - p} > \left(\frac{\delta}{1 - \delta}\right)^{k - m}$$

which holds if and only if $k \geq m$, because $1 - \delta > p$. In other words, a rational decision maker who has access to all private signals will choose A if and only if the number of a's is at least as large as the number of b's.

Now consider the person arriving first. Her only information is her private signal, so she will choose A if and only if her signal is a. (This is because $1 - \delta > p > \delta$.) Hence her decision perfectly reveals her signal. The person arriving second, therefore, effectively observes two private signals, her own and her predecessor's. Consequently, the second person will choose A if the first did (irrespective of her own signal), or if the first chose B but her own signal is a. In other words, the second person will choose B if and only if both she and her predecessor observed a b-signal.

If the third person sees the second choosing B, she, therefore, knows that (at least) two out of three available signals are b's. Hence, upon observing the

*second choosing B, the third will also choose B, whatever her private signal
is. If she sees the second choosing A, however, her own signal may make a
difference. By Bayes' rule the conditional probability that A is better, given
that the second person has chosen A and the private signal of the third person
is a, is given by*

$$\pi(a) = \frac{p\left(1 - \delta\right)^2 \left(1 + \delta\right)}{p\left(1 - \delta\right)^2 \left(1 + \delta\right) + \left(1 - p\right)\delta^2\left(2 - \delta\right)} \, ,$$

*while the posterior, given that the second has chosen A and the third's signal
is b, is*

$$\pi(b) = \frac{p(1 + \delta)}{p(1 + \delta) + (1 - p)\left(2 - \delta\right)} \, .$$

*One can check that, if the third person's signal is a (and the second has chosen
A), she too will choose A. (This is because $1/2 > \delta$.) If, moreover, parameters
are such that $3p > 2 - \delta$, then the third person will choose A (if the second
did), even if her private signal is b. In this case the third person does precisely
the same as the second person, irrespective of her signal.*

*But then the fourth person arriving cannot infer the signal of her imme-
diate predecessor. The only information conveyed to the fourth through the
action of the third is how the second behaved. Hence, the informational position of the person arriving fourth is precisely the same as for the person
arriving third. Thus, the fourth will behave like the third; i.e., she will imitate her predecessor (if the above parameter restriction holds). A fortiori, this
holds for all later arrivals. From the third arrival onwards, therefore, we will
see "herding", i.e. people imitating their predecessors and ignoring their private information. Note that this example is purely decision theoretic; i.e., all
potential strategic interaction has been assumed away.*

Summary

- States of the world are "all inclusive" descriptions of the underlying
  causes for the uncertainty that the decision maker faces. They are mutually exclusive, collectively exhaustive, and the act which the decision
  maker chooses has no effect on the state that obtains.

- A random variable is a function on the set of all states. An event is a
  subset of the set of all states.

- A random variable induces a partition of the set of all states into information sets.

- The observation of a realization of a random variable can be used to update information and beliefs. The former is accomplished by information
  sets, the latter by Bayes' rule.

Exercises

Exercise 2.16 Each of three ladies in a railway carriage may have a dirty face. The faces of the other ladies are visible to each lady, but not her own face. Formalize this by (eight) states of the world and give the information partition for each lady based on her observing the faces of other two. Each lady blushes if and only if she knows her own face to be dirty. Will any one of them blush without further information at any of the states?

Exercise 2.17 Reconsider the three ladies from Exercise 2.16 and suppose that a clergyman enters the carriage and remarks that there is at least one lady in the carriage with a dirty face. Formalize the clergyman's remark as a random variable which improves upon the ladies' information. Are there now states in which some of the ladies will blush? (Hint: The information contained in the observation of the other ladies' faces and the clergyman's announcement is formalized by taking intersections of the two partitions induced by the two random variables.)

Exercise 2.18 There are two jobs for which Ms Miller can apply. She assigns prior probability 0.6 to being accepted for the first job and prior probability 0.5 to being accepted for the second job. If she were accepted for the first job, her posterior probability for the second job would be 0.7. What are the appropriate states and what are their prior probabilities? What is Ms Miller's prior probability of landing precisely one job?

Exercise 2.19 Suppose a fraction $\varepsilon \in (0,1)$ of the population carries the gene κ, so in the absence of any other information you assign probability ε to your carrying κ (your prior belief). An imperfect test on the presence of κ becomes available. The test is positive in a fraction $1 - \delta \in (0,1)$ of the subjects who carry κ and in a fraction $\nu \in (0,1)$ of subjects who do not carry κ. You test positive. What (posterior) probability should you assign to your carrying κ according to Bayes' rule? What value does your posterior belief take when $\varepsilon = 0.001$, $\delta = 0.01$, and the test is negative for 99% of subjects who do not carry κ?

Exercise 2.20 Let V_1 and V_2 be two, not necessarily disjoint, events, $h \in H$ an information set, and π a belief over states. Show that $\pi (V_1 \cup V_2 |h) + \pi (V_1 \cap V_2 |h) = \pi (V_1 |h) + \pi (V_2 |h)$. (Hint: Use Definition 2.4.)

Exercise 2.21 Let the signal space R be large enough so that there exists a random variable $f : \Omega \to R$ which is one-to-one; i.e., $f(\omega) = f(\omega')$ implies $\omega = \omega'$. Show that every other random variable $g : \Omega \to R$ is measurable with respect to the partition H_f induced by f.

Exercise 2.22 Often individuals choose in more than one stage; e.g., you first choose a restaurant and then a dish from the menu. The first stage amounts to choosing between sets of alternatives. This requires an extension of preferences on alternatives to a preference relation on sets of alternatives. Let $z \in Z$ be alternatives (meals) and $x \in X$ nonempty sets of alternatives (restaurants). From the preference relation $\preceq$ on Z, define a preference relation $\overset{o}{\preceq}$ on X by

$$x' \overset{o}{\preceq} x \text{ if for all } z' \in x' \text{ there is } z \in x \text{ such that } z' \preceq z. \qquad (2.17)$$

The new preference relation then satisfies

$$\text{if } x' \overset{o}{\preceq} x \text{ then } x \overset{o}{\sim} x \cup x' \qquad (2.18)$$

where $\overset{o}{\sim}$ is the induced indifference relation on X. Show that a binary relation $\overset{o}{\preceq}$ on X is complete and transitive and satisfies (2.18) if and only if there is some complete and transitive $\preceq$ on Z such that (2.17) holds.

Notes on the Literature: The concepts relating to states of the world originate in the work of Ramsey [145] and Savage [159]. Fishburn ([59], pp. 163) suggests an interesting reinterpretation of states as functions from acts to consequences. An excellent discussion of knowledge and, in particular, the interpersonal concept of *common knowledge* is given by Binmore ([22], Chapter 4; Exercise 2.16 is based on it). Anscombe and Aumann [4] is the key reference for expected utility with subjective probabilities when acts map states into lotteries over consequences. Example 2.11 is based on Banerjee [11] (see also Bikhchandani, Hirshleifer, and Welch [21]). Exercise 2.22 is from Kreps [100].

2.5 State Dependent Utility

Finally, we turn to representations of preferences when states of the world matter. This provides an interesting example of the power of the expected utility theorem. But, more importantly, it is of interest to game theory, because such preferences come up in games of "incomplete information" (see Section 3.5). This section aims at deriving a state dependent expected utility representation of preferences which later serves to model uncertainty about the game itself.

Horse Lotteries More precisely, preferences will be considered which are defined on the, rather general, space of random variables mapping into simple lotteries over consequences. Such random variables are sometimes called "horse lotteries" (see [4]) because of the following (pseudo-operational) interpretation.

   If such a random variable is selected and state ω (the outcome of a "horse race") obtains, then the value of this random variable (a simple lottery over consequences) at ω determines the resultant gamble over consequences $z \in Z$. This is like a compound lottery that assigns lotteries over consequences as prizes contingent on the realization of the state.

   Of course, these random variables encompass the "acts" chosen by the decision maker from the previous subsection, which also amount to (possibly degenerate) lotteries over consequences in each state. But the formulation is more general.

Example 2.12 *Imagine that there are objective randomizing devices—fair coins to flip, unbiased roulette wheels, or fair dice—which can be used to construct any objective probability distribution over consequences. "Objective" here means that all decision makers will agree on the probabilities. For example, to create a lottery which gives a 10 dollar monetary win with probability 4/5 and a zero win with probability 1/5, we might roll a fair die, paying 10 dollars if it comes up with one through four spots up, paying zero if it comes up with five spots up, and rolling again (until it lands with at most five spots up) if it comes up with six spots up. Next, imagine that there takes place a horse race between two horses. It may result in either one of the horses winning or the race being a dead heat. This gives three states of the world.*

   You are now offered bets on "objective" random devices, depending on the states of the world. For example, such a betting ticket may give you −2 dollars for sure (i.e., you lose 2 dollars for sure) if the first horse wins the race. If the second horse wins, you face a coin flip which gives you 3 dollars if the coin comes up heads and 1 dollar if it comes up tails. If the race is a dead heat, a symmetrical die will be rolled which gives you 6 dollars if it comes up with one or two spots and −4 dollars (a loss of 4 dollars) if it comes up with three through six spots.

On the other hand, to simplify things, it will be assumed that there are only *finitely* many states $\omega \in \Omega$. There are extensions of the theory to infinite state spaces (see [59], Chapter 14), but for the present purposes of game theory the relevant case is a finite state space.

Formally, let $\mathcal{F}$ be the set of all random variables $f : \Omega \to \mathcal{L}$ mapping into (simple) lotteries over consequences $z \in Z$. The value of such a random variable $f \in \mathcal{F}$ at a particular state $\omega \in \Omega$ is a simple lottery $f(\omega)$ over consequences and is denoted by the simple probability measure $p_f (\,.\,|\omega) \in \mathcal{L}$ on consequences. In somewhat confusing notation, this says that $f(\omega)(z) = p_f (z|\omega)$ for all $z \in Z$, all $\omega \in \Omega$, and all $f \in \mathcal{F}$. Recall that $f(\omega) \in \mathcal{L}$ means that $\operatorname{supp}(f(\omega))$ is finite.

Convex combinations of random variables are defined in the natural way. For any two $f, g \in \mathcal{F}$ and $\alpha \in [0, 1]$, the convex combination of f and g is the random variable which gives the lottery $\alpha f(\omega) + (1 - \alpha)g(\omega)$ in state ω, for all $\omega \in \Omega$. Using the previous notation, and abbreviating $\alpha f + (1 - \alpha)g = h$, this says

$$p_h (z|\omega) = \alpha p_f (z|\omega) + (1 - \alpha)p_g (z|\omega), \text{ for all } z \in Z \text{ and all } \omega \in \Omega.$$

Clearly, if $f, g \in \mathcal{F}$, then their convex combination $\alpha f + (1 - \alpha)g$ is in $\mathcal{F}$. It follows that $\mathcal{F}$ is a mixture set in the sense of Theorem 2.2.

While under some $f \in \mathcal{F}$ every state results in objective (extraneous) probabilities for consequences, the likelihood that the decision maker assigns to the various states can be viewed as a property of her preferences. This leads to an approach known as "subjective probability" theory (see [4]). To disentangle these subjective probabilities from the (Bernoulli) utility function, however, requires an extra axiom (known as "state uniformity") which is of limited help for the goals of game theory. Therefore, this presentation will be confined to the case where state probabilities $\pi(\omega) > 0$ are given for all $\omega \in \Omega$. Those will be referred to as the decision maker's *beliefs* about which state obtains.

State Dependent Utility The primitive datum of the theory is now a preference relation $\prec$ on the set $\mathcal{F}$ of "horse lotteries". And onto these preferences on the enlarged domain $\mathcal{F}$ the von Neumann-Morgenstern axioms are imposed. This is to say that Definition 2.6 with $\mathcal{L}$ replaced by $\mathcal{F}$ (and $p, q, r \in \mathcal{L}$ replaced by $f, g, h \in \mathcal{F}$) is assumed to hold, so that $\prec$ is a rational preference relation which satisfies "independence" and "continuity". The arguments in favor of the independence and the continuity axioms are the same as before. But, of course, if they were hard to buy if imposed on preferences on

$\mathcal{L}$, they are even harder to believe in if imposed on preferences on $\mathcal{F}$, because the latter represents even more complex alternatives.

Yet, if the v.N.-M. axioms are imposed on preferences $\prec$ on "horse lotteries" in $\mathcal{F}$, the analytical gain is considerable.

Theorem 2.3 *Suppose that Ω is finite and the preference relation $\prec$ on $\mathcal{F}$ is a rational preference relation satisfying the independence and continuity axioms (Definition 2.6). Then there is a real-valued function u on $Z \times \Omega$ such that*

$$f \prec g \text{ if and only if}$$
$$\sum_{\omega \in \Omega} \pi(\omega) \sum_{z \in \text{supp}(f(\omega))} p_f(z \,|\, \omega) \, u(z, \omega) \qquad (2.19)$$
$$< \sum_{\omega \in \Omega} \pi(\omega) \sum_{z \in \text{supp}(g(\omega))} p_g(z \,|\, \omega) \, u(z, \omega)$$

for all $f, g \in \mathcal{F}$. Moreover, u is unique up to positive linear transformations.

In the appendix a proof of this theorem is given which is based on Theorem 2.2. In this sense, Theorem 2.3 is almost a corollary to the expected utility theorem. And, indeed, comparing (2.19) and (2.7) reveals that the former is like a two-stage version of the latter.

To understand what Theorem 2.3 says, compare it with the original expected utility theorem. If only consequences mattered and preferences on simple lotteries satisfied the v.N.-M. axioms, then the expected utility theorem would lead to a state-independent (or "state-uniform") expected utility representation, where the Bernoulli utility function did not depend on the state. In the present formulation, it does depend on the state.

In this sense, (2.19) is a generalization of (2.7). Of course, instead of viewing u as a function on $Z \times \Omega$, you may think of a different (Bernoulli) utility function u_ω for each state (recall that Ω is finite), to express the idea that underlying causes for uncertainty do matter.

Example 2.13 *Recall from Example 2.8 that a risk averse expected utility maximizer will always fully insure if actuarially fair insurance is available. To formulate the insurance problem with states, only two states are required. In the first state, ω_1, no damage occurs; in the second state, ω_2, the decision maker suffers a loss of $d \in (0, w)$.*

In the absence of insurance, the individual's wealth is a (real-valued) random variable f with $f(\omega_1) = w$ and $f(\omega_2) = w - d$. An insurance contract is also a real-valued random variable g with $g(\omega_1) = y_1$ and $g(\omega_2) = y_2$ specifying the net change in wealth in the two states (the insurance payoff minus any premium paid). If the individual buys x units of the insurance contract g, her final wealth position is a random variable h with $h(\omega_1) = w + xy_1$ and

$h(\omega_2) = w - d + xy_2$. *The insurance policy is* actuarially fair *if its expected payoff is zero, i.e. if* $(1 - \varepsilon)y_1 + \varepsilon y_2 = 0$ *where* $\varepsilon = \pi(\omega_2) = 1 - \pi(\omega_1)$.

Assume that the decision maker's preferences are representable by a state dependent Bernoulli utility function $u : Z \times \Omega \to \mathbb{R}$ *which is differentiable in its first argument. The "horse lotteries" which are to be considered concern the amount* x *of insurance and are all degenerate (in the sense that the only uncertainty arises from which state will obtain). Denoting* $u_i(.) = u(., \omega_i)$, *for* $i = 1, 2$, *the first-order condition for the optimal amount* x^* *of insurance is*

$$(1 - \varepsilon)y_1 u_1'(w + x^* y_1) + \varepsilon y_2 u_2'(w - d + x^* y_2) \leq 0$$

with equality if $x^* > 0$. *If insurance is actuarially fair and the individual does insure,* $x^* > 0$, *this reduces to* $u_1'(w + x^* y_1) = u_2'(w - d + x^* y_2)$. *For the regular risk averse expected utility maximizer, for whom* $u_1 = u_2$ *holds, this implies that she fully insures;* $x^*(y_2 - y_1) = d$. *But with state dependent utility, this is no longer so.*

If the partial derivative u_1' *is relatively higher than* u_2' *and* u *is concave in wealth, expressing risk aversion, this creates a desire to have a greater wealth in state* ω_1, *implying that the individual will* not *fully insure even if insurance is actuarially fair. This may be due, say, to the fact that the damage is a health problem which forbids the individual to consume her wealth.*

This example illustrates that a state dependent expected utility representation may often be more "realistic" than the regular expected utility function. On the other hand, there are difficulties with (2.19) as well.

Example 2.14 *(The Ellsberg paradox) There are two urns, each with 100 balls. The first urn contains 49 white balls and 51 black balls. The second urn contains an unspecified assortment of balls. A ball has been randomly picked from each urn, but their colors have not been disclosed. A decision maker will be offered prizes which depend on the colors of the two balls. States of the world correspond to the color of the ball from the second urn.*

Now consider two decision problems. In both the decision maker must choose between the ball drawn from the first and the ball drawn from the second urn. After her choice, the color will be disclosed. In the first situation a money prize of 100 dollars is won if the chosen ball is black. In the second situation the same 100 dollars is won if the ball is white.

In experiments with human subjects, most people choose the ball from the first urn in the first situation. If this says something about the beliefs about which state obtains, it should mean that people assign at least *probability .49 to the ball from the second urn being white, because*

$$.51 \geq \pi \, (black) \quad \text{if and only if} \quad \pi \, (white) \geq .49.$$

Hence, most people should choose the ball from the second urn in the second situation. However, it turns out that this does not happen overwhelmingly in experiments. Subjects understand that by choosing the ball from the first urn in the second situation they have only a .49 chance of winning. But this chance is "safe" and well understood. The uncertainties incurred by choosing the ball from the second urn are much less clear.

Knight [96] suggested distinguishing between *risk* and *uncertainty* according to whether or not the probabilities are objectively given. The Ellsberg paradox suggests that there may be something to this distinction. This is an active area of research.

Summary

- "Horse lotteries" are random variables mapping states into simple lotteries over consequences with extraneous probabilities.

- If the von Neumann-Morgenstern axioms are imposed onto preferences on "horse lotteries", this results in a state dependent expected utility representation.

- A state dependent expected utility function has the Bernoulli utility function depending on the state of the world, but it is also linear in probabilities.

Appendix

Proof of Theorem 2.3 By Theorem 2.2 there is a real-valued function U on $\mathcal{F}$ such that $f \prec g \Leftrightarrow U(f) < U(g)$ and $U(\alpha f + (1-\alpha)g) = \alpha U(f) + (1-\alpha)U(g)$, for all $\alpha \in [0,1]$, which is unique up to positive linear transformations.

Fix some $f \in \mathcal{F}$ and define for every $g \in \mathcal{F}$ and $\omega \in \Omega$ the new random variable $g_\omega \in \mathcal{F}$ by $g_\omega(\omega) = g(\omega)$ and $g_\omega(\omega') = f(\omega')$ for all $\omega' \in \Omega \setminus \{\omega\}$. Let m denote the number of states, $|\Omega| = m$. Then $\frac{1}{m}g + \left(1 - \frac{1}{m}\right)f = \frac{1}{m}\sum_{\omega \in \Omega} g_\omega$, because, for any $\omega' \in \Omega$,

$$\frac{1}{m}\sum_{\omega \in \Omega} g_\omega(\omega') = \frac{1}{m}g_{\omega'}(\omega') + \frac{1}{m}\sum_{\omega \in \Omega \setminus \{\omega'\}} g_\omega(\omega')$$

$$= \frac{1}{m}g(\omega') + \left(1 - \frac{1}{m}\right)f(\omega').$$

This together with the linearity of U yields

$$\frac{1}{m}U(g) + \left(1 - \frac{1}{m}\right)U(f) = \frac{1}{m}\sum_{\omega \in \Omega} U(g_\omega). \tag{2.20}$$

Define now for each $\omega \in \Omega$ the function $U_\omega : \mathcal{L} \to \mathbb{R}$ by $\pi(\omega)U_\omega(g(\omega)) = U(g_\omega) - (1 - \pi(\omega))U(f)$. Summing over ω and using (2.20) yields

$$\sum_{\omega \in \Omega} \pi(\omega)U_\omega(g(\omega)) = \sum_{\omega \in \Omega} U(g_\omega) - (m-1)U(f) = U(g). \qquad (2.21)$$

From (2.21) it follows that, for $g, h \in \mathcal{F}$ and $\alpha \in [0, 1]$,

$$\begin{aligned}
U\left(\alpha g_\omega + (1-\alpha)h_\omega\right) = {} & \pi(\omega)U_\omega\left(\alpha g(\omega) + (1-\alpha)h(\omega)\right) \qquad (2.22) \\
& + \sum_{\omega' \in \Omega \setminus \{\omega\}} \pi(\omega')U_{\omega'}\left(f(\omega')\right)
\end{aligned}$$

for all $\omega \in \Omega$. In addition, by linearity and (2.21),

$$\begin{aligned}
U\left(\alpha g_\omega + (1-\alpha)h_\omega\right) = {} & \alpha U(g_\omega) + (1-\alpha)U(h_\omega) \qquad (2.23) \\
= {} & \pi(\omega)\left[\alpha U_\omega\left(g(\omega)\right) + (1-\alpha)U_\omega\left(h(\omega)\right)\right] \\
& + \sum_{\omega' \in \Omega \setminus \{\omega\}} \pi(\omega')U_{\omega'}\left(f(\omega')\right).
\end{aligned}$$

Comparing (2.22) and (2.23) yields the linearity of U_ω by

$$U_\omega\left(\alpha g(\omega) + (1-\alpha)h(\omega)\right) = \alpha U_\omega\left(g(\omega)\right) + (1-\alpha)U_\omega\left(h(\omega)\right) \qquad (2.24)$$

for all $\alpha \in [0, 1]$ and all $\omega \in \Omega$.

It follows that there exists a real-valued function u on $Z \times \Omega$ defined by the condition

$$\text{if } \operatorname{supp}\left(g(\omega)\right) = \{z\}, \text{ then } u(z, \omega) = U_\omega\left(g(\omega)\right),$$

for all $z \in Z$ and all $\omega \in \Omega$. Then (2.24) and the construction of u yield that, for any $\omega \in \Omega$ and $g \in \mathcal{F}$, the function $U_\omega\left(g(\omega)\right)$ is simply the expectation of u with respect to $g(\omega)$. Writing $g(\omega) = p_g\left(\,.\,|\omega\right)$ for all $\omega \in \Omega$ and $g \in \mathcal{F}$, this is to say that

$$\begin{aligned}
U_\omega\left(g(\omega)\right) = {} & \sum_{z \in \operatorname{supp}(g(\omega))} p_g\left(z \,|\, \omega\right) u(z, \omega) \text{ and} \\
U(g) = {} & \sum_{\omega \in \Omega} \pi(\omega)U_\omega\left(g(\omega)\right)
\end{aligned}$$

which is the expected utility representation (2.19) required by the theorem.

Uniqueness up to positive linear transformations follows from the fact that U is unique up to positive linear transformations. This completes the proof of Theorem 2.3.

Exercises

Exercise 2.23 Reconsider the experiment in Example 2.14. Suppose the decision maker has a Bernoulli utility function u on money with $u(0) = 0$ and $u(100) = 1$. Her acts or strategies are to choose a ball from the first urn (f) or a ball from the second urn (g). Her belief that the ball from the second urn is white is a number $\pi \in [0, 1]$. But now assume that she has not a single belief, but a set of beliefs given by a subset Π of $[0, 1]$. For each of the two choice situations, now define her utility functions over acts f and g as follows.

- For the first situation (where a black ball pays 100 dollars), $U_1(f) = .51$ and $U_1(g) = \min_{\pi \in \Pi}(1 - \pi)$.

- For the second situation (where a white ball pays 100 dollars), $U_2(f) = .49$ and $U_2(g) = \min_{\pi \in \Pi} \pi$.

This is to say that her utility derived from choosing the ball from the first urn (f) is the expected utility of 100 dollars calculated with the extraneous probabilities. On the other hand, her utility from choosing the ball from the second urn (g) is the expected utility of 100 dollars with the most pessimistic belief in Π. Show that, if $\Pi = \{\pi\}$ for some $\pi \in [0, 1]$, then U_1 and U_2 are derived from a v.N.-M. utility function and $U_2(f) > U_2(g)$ if and only if $U_1(f) < U_1(g)$. Moreover, find a set $\Pi \subseteq [0, 1]$ such that $U_2(f) > U_2(g)$ and $U_1(f) > U_1(g)$.

Exercise 2.24 Reconsider Example 2.13. Suppose that $u_1(.) = \beta u_2(.)$ for some $\beta \in (0, 1)$. Will the decision maker with this state dependent expected utility function over- or underinsure if insurance is actuarially fair?

Exercise 2.25 (The winner's curse) A supplier is willing to sell an indivisible object if she is offered a price p at least as large as her reservation value v (which she knows). To a potential buyer the object is worth $3v/2$, but she does not know v. The buyer only knows that v is uniformly distributed on the unit interval $[0, 1]$. If the buyer gets the object by bidding a sufficiently high price $p \in [0, 1]$, her payoff is $(3v/2) - p$. If her bid falls short of v, her payoff is zero. How much should the buyer bid?

Exercise 2.26 Under the conditions of Theorem 2.3, show that $u(z, \omega)$ is constant in $z \in Z$ if and only if $f \sim g$ whenever $f(\omega') = g(\omega')$, for all $\omega' \in \Omega \setminus \{\omega\}$, and all $f, g \in F$. (The latter condition is sometimes referred to as "state ω being null".)

Exercise 2.27 Consider random variables $f : \Omega \to Z$ mapping into consequences. If f is such a random variable and $\pi = (\pi(\omega))_{\omega \in \Omega}$ is (the vector of) the decision maker's beliefs (where Ω is finite), then

$$p_f(z) = \sum_{\omega \in f^{-1}(z)} \pi(\omega)$$

is the (aggregate) probability of consequence $z \in Z$ induced by f (where the preimage $f^{-1}(z)$ is the set of all ω such that $f(\omega) = z$). Suppose the decision maker's preferences are represented by a state dependent expected utility function (2.19). Calculate an expression for the expected utility function which represents utility as an expectation of an appropriate Bernoulli utility function with respect to p_f. (Hint: Not only the probabilities of consequences may depend on f, but also the Bernoulli utility function.)

Notes on the Literature: State dependent expected utility representations originate in Savage's work [159]. The term "horse lotteries" was coined by Anscombe and Aumann [4]. The proof of Theorem 2.3 below follows Fishburn ([59], Section 13.1). Example 2.14 is due to Ellsberg [54]. Theories that account for the Knightian [96] distinction between risk and uncertainty are presented by Gilboa [65], Gilboa and Schmeidler [67] (on which Exercise 2.23 is based), Schmeidler [161], Dow and Werlang [50], and Kelsey [94], among others. Exercise 2.25 is based on Bazerman and Samuelson [15].

Part II

Game Representations

3

Games in Extensive Form

This chapter deals with how to represent the *rules* of a game. The most detailed representation is known as an extensive form. It constitutes a way of verifying that the rules of a game are complete, and therefore is an indispensable part of the theory. Effectively, a game has complete rules if and only if it can be represented in extensive form.

The formalism that comes with extensive forms may seem like a burden. But it does provide a transparent model for game representations. It should also be noted that the treatment in this chapter is not quite standard—if there is a standard treatment of extensive forms. It generalizes more conventional definitions and allows for more compactness. Moreover, the present treatment allows more flexibility in representing simultaneous decisions.

3.1 Perfect Information

Trees Since non-cooperative game theory studies games with complete rules, all possible histories or *plays* can be enumerated—a play being a complete history of what happens in the game from the start to the end. Moreover, all events that happen during a play can be enumerated. Therefore, it is possible to specify until which event certain plays are indistinguishable. This gives rise to the concept of a *tree*.

Most of the time trees are introduced as connected graphs without loops and a distinguished node. Since this is not quite appropriate for decision theory, an alternative (and more general) definition is given here.

First, some terminology is developed. Let W be a (nonempty) set and N a collection of nonempty subsets of W. The collection N is naturally ordered by set inclusion.[1] The subcollection $M \subseteq N$ of (sub)sets (of W) in N is called a

---

[1] For two sets x and y write $x \subseteq y$ if y contains x, i.e. if all elements of x also belong to y. Write $x \subset y$ if y properly contains x, i.e. if $x \subseteq y$ but not $y \subseteq x$.

chain if, for all $x, y \in M$, either $x \subset y$ or $y \subseteq x$ holds, i.e. if the ordering by set inclusion on the subset M of N is complete (see Section 2.1). A chain $M \subseteq N$ in N has an *infimum* (resp. *supremum*) if there is $\underline{z} \in N$ (resp. $\overline{z} \in N$) such that $x \supseteq \underline{z}$ (resp. $x \subseteq \overline{z}$) for all $x \in M$, and $x \supseteq y \in N$ (resp. $x \subseteq y \in N$) for all $x \in M$ implies $\underline{z} \supseteq y$ (resp. $\overline{z} \subseteq y$). A chain M in N has a *minimum* (resp. *maximum*) if it has an infimum (resp. supremum) that belongs to M.

With these preparations, the formal definition is as follows.[2]

Definition 3.1 *A **tree** $\mathcal{T}$ is a pair $\mathcal{T} = (W, N)$, where W is a set of **plays** and N is a collection of nonempty subsets of W, called **nodes**, such that $\{w\} \in N$ for all $w \in W$,*

(T.1) the collection $M \subseteq N$ is a chain if and only if there is $w \in W$ such that $w \in x$ for all $x \in M$, and

(T.2) every chain in $X \equiv N \setminus \{\{w\}\}_{w \in W}$ has a maximum and either also a minimum or an infimum in $E \equiv \{\{w\}\}_{w \in W}$.

*A tree (W, N) is **rooted** if $W \in N$. The set $W \in N$ is then called the **root**. Nodes in $E \equiv \{\{w\}\}_{w \in W}$ are called **terminal** and nodes in $X \equiv N \setminus E$ are called **moves**.*

This definition is, admittedly, somewhat difficult, because it accounts for infinite cases. But next result (below) shows that in the finite case the definition of a tree becomes very simple: it is characterized by the property that, if two nodes of a tree intersect, one must contain the other.

The second part of the result below states that from any (finite) node it is possible to uniquely reconstruct how that node has been reached—by iterating an *(immediate) predecessor function*. Such a predecessor function assigns to each (finite) node (except the root) its smallest proper superset in the set of nodes. Iterating this predecessor function sufficiently often from a node $x \in N$ results in a superset of x which contains any other node that properly contains x. (A proof is given in the appendix to this section.)

Proposition 3.1 *Let (W, N) be a (nonempty) set W together with a collection N of nonempty subsets of W such that $\{w\} \in N$ for all $w \in W$.*

(a) If W is a finite set, then (W, N) is a tree if and only if it satisfies

$$\text{if } x \cap y \neq \varnothing, \text{ then either } x \subset y \text{ or } y \subseteq x, \tag{3.1}$$

for all $x, y \in N$; i.e., whenever two nodes intersect, one contains the other. Moreover, every tree (finite or not) satisfies (3.1) for all $x, y \in N$.

---

[2] I am grateful to Carlos Alos-Ferrer for active cooperation on the following definition and the arguments in the appendix to this section.

*(b) If (W, N) is a rooted tree, then there exists an **immediate predecessor** function $P : F(N) \to X$ which is onto[3], satisfies $P(W) = W$, and*

$$x \subset P(x) \text{ and if } x \subset y \in N \text{ then } P(x) \subseteq y \subseteq \cup_{t=1}^{\infty} P^t(x) \qquad (3.2)$$

for all $x \in F(N) \setminus \{W\}$, where $P^t(x) = P\left(P^{t-1}(x)\right)$ for all $t = 1, 2, \ldots$ and $P^0(x) = x$ for all $x \in F(N)$, and where

$$F(N) = \left\{ x \in N \,\middle|\, x \subset \bigcap_{x \subset y \in N} y \right\} \qquad (3.3)$$

*is the set of **finite elements** of N.*

Condition (3.1) is the key property of a tree. Every tree satisfies it, even if W is not finite. And if W is finite, then (3.1) together with the presence of all singletons is sufficient for (W, N) to be a tree.

As Proposition 3.1(b) shows, the more involved conditions (T.1) and (T.2) of Definition 3.1 are required only to deal with infinite cases. For those, they ensure the existence of an immediate predecessor $P(x)$ for every node $x \in F(N)$. Since P is a function, it identifies *uniquely* how a given (finite) node has been reached.

In a rooted tree the elements $w \in W$ are called *plays*, because they can be identified with maximal chains in N. (A chain $M \subseteq N$ is *maximal* if there is no $x \in N \setminus M$ such that $M \cup \{x\}$ is a chain.) Sets $x \in N$ of plays are called *nodes*, the universal element $W \in N$ is called the *root*, singleton sets of plays $(x = \{w\} \in N$, for some $w \in W)$ are called *terminal* nodes, and all nodes that contain more than one play, i.e. nodes in $X \equiv N \setminus \{\{w\}\}_{w \in W}$, are called *moves*.

In the appendix to this section it is shown that every move $x \in X$ is a finite element $x \in F(N)$ of the tree; i.e., $X \subseteq F(N)$. In particular, if W is finite, then $F(N) = N$. Moreover, every move $x \in X$ properly contains some *finite* successor $y \in F(N)$.

Moves represent events up to which the plays contained in them are indistinguishable, because they evolve identically up to that event. The root, where all plays are still possible, represents the start of the game. Terminal nodes are events where the game ends, because a single play has materialized. The plays contained in a node will occasionally be referred to as the plays *passing through* the node.

The nodes in a rooted tree are partially ordered by set inclusion. The value $P(x) \in X$ of the function P at $x \in F(N)$ (the smallest proper superset of x) is called the *immediate predecessor* of node $x \in N$. The function P operates "backwards", identifying for each (finite) node a unique other node which

---

[3]   A function is onto (surjective) if every point in its range is the image of some point in the domain.

immediately precedes the first, viz. the "latest" event after which the (event represented by the) node will become distinguishable.

Nodes in the preimage $P^{-1}(x) = \{y \in F(N) \,|\, P(y) = x\}$ are called *immediate successors* of node x. Say that node $x \in N$ *comes after* (resp. *before*) node $y \in N$ if $x \subset y$ (resp. $y \subset x$); i.e., if x comes after y, then x represents more precise ("finer") information than y, because x is (properly) contained in y.

The function P and its inverse, the correspondence P^{-1}, are easily extended to the power set[4] of $F(N)$ by taking unions, i.e.

$$P(M) = \{x \in N \,|\, \exists y \in M : x = P(y)\} \text{ and}$$
$$P^{-1}(M) = \{x \in N \,|\, P(x) \in M\}, \qquad (3.4)$$

for all subsets $M \subseteq F(N)$.[5] In words, $P(M)$ is the set of nodes that are immediate predecessors of *some* node in M and $P^{-1}(M)$ is the set of nodes that are immediate successors of *some* node in M.

Graphically, the tree amounts to a "flow chart" representation of an outside observer's view of the rules of the game. Nodes are drawn as points on a plane and the immediate predecessor function P is represented by drawing a line (an "edge") from each node to its immediate predecessor. If the tree is rooted, every node will be connected to the root by a unique sequence of edges. Plays are one-to-one to terminal nodes and will show up as sequences of connected edges from the root to a terminal node. Such a graph cannot have any cycles (sequences of edges that connect a node to itself), because of (3.1), and every node will be connected by an edge to some other node. This depicts how the various plays may evolve. The following example illustrates the graphical device.

Example 3.1 *A simple example is the class of m-Marienbad games (see Section 1.1). These are games with two players where m^2 matches are placed on the table in $m \geq 1$ rows, one match in the first row, three matches in the second row, five matches in the third row, ..., and $2m - 1$ matches in the last row. The resulting pattern looks like a triangle. The player who starts the game, say, player 1, picks as many matches as she wishes (but at least one) from one, but only one, of the rows. Then the other player, player 2, picks as many matches as she wishes (at least one) from only one of the remaining rows. Then it is again player 1's turn, and the game continues as before with the remaining matches. This goes on until one of the players picks the last match and has lost. Figure 3.1 illustrates the cases for $m = 2$ and $m = 4$.*

For simplicity, consider the 2-Marienbad game. Its tree has 19 nodes, 7 of which are moves and 12 of which are terminal, so there are precisely 12 possible plays. The root, which corresponds to player 1 being confronted with

---

[4]   The power set 2^X of a set X is the set of all the subsets of X.
[5]   Make sure you understand that sets of nodes are sets of sets of plays.

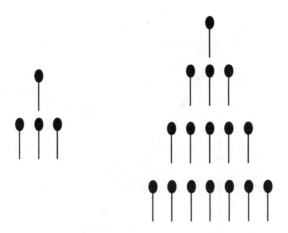

Figure 3.1: Marienbad games

the initial pattern of matches, has four immediate successors, one of which is terminal (where player 2 has lost, because she is confronted with a single match in the first row). The three moves immediately succeeding the root correspond to player 2 being confronted with $(0,3)$, $(1,2)$, and $(1,1)$, where (m_1, m_2) denotes $m_1 \in \{0,1\}$ matches in the first row and $m_2 \in \{0,1,2,3\}$ matches in the second row. The first of those moves has three immediate successors, two of which are terminal. The second has also three immediate successors, but only one of those is terminal. The third comes immediately before two terminal nodes. Consult Figure 3.2 and see how the pattern $(0,2)$ is represented by two different moves depending on whether it is reached from $(0,3)$ or from $(1,2)$. How many (terminal) nodes correspond to the pattern $(0,1)$?

As the example illustrates, moves represent "partial histories", i.e. sets of plays that have (up to the move) evolved identically. Moves do *not* represent the current position of players (like the arrangement of matches on the table in the example), if those have been brought about by different histories. In the tree representing a chess game, two different sequences of moves that lead to the same arrangement of stones on the board are represented by different nodes—despite the fact that only the current arrangement of stones matters for the players. This is necessary to express contingent rules that condition on the history of, say, an arrangement of stones on the board. In chess a triple repetition of the same arrangement of stones may lead to a draw. Because a node records one-to-one its history (see Proposition 3.1(b)), it also records whether the same arrangement of stones has already preceded it twice, making it a terminal node.

Likewise, immediate successors of different moves may well represent the

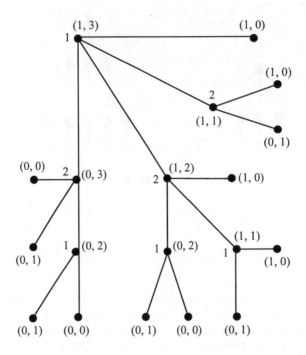

Figure 3.2: Tree for 2-Marienbad game

same "action", despite being different nodes. For example, in the 2-Marienbad game of Figure 3.2, player 2 may opt for the action "take away one match in the second row" *at each* of her moves. Still this action is represented by three different nodes, depending on the histories after which this action is taken.

The reason for modeling nodes as sets of plays is decision theoretic. Plays can be perceived as "states" and nodes as (elementary) "events" (see Exercise 3.1 below). From this point of view, the succession of nodes reached as the game unfolds corresponds to progressive narrowing down of the selection among plays. The root corresponds to an event where all plays are still possible. Other moves correspond to an event where only plays passing through this move are possible and where a decision is due that will further narrow down the possible plays.

In most applications the set W of plays will be finite. This implies that there are only finitely many nodes as well. The definition, however, also allows infinite trees. And indeed, some applications do require trees where some plays "never end".

Example 3.2 *Suppose two persons bargain over the distribution of a perfectly divisible pie of original size 1 in the following orderly way. First player 1 proposes a share $a_1 \in [0,1]$ of the pie for herself. Player 2, upon hearing the proposal, may either accept it, in which case it is implemented and the game terminates, or reject it. In the latter case the pie depreciates by a factor $\delta \in (0,1)$ and player 2 makes a counteroffer by proposing a share $a_2 \in [0,1]$ of the remaining pie (which is now of size δ) for herself. To that player 1 can react by accepting it or not. Acceptance implements the proposal, otherwise the pie depreciates again (to size δ^2) and player 1 makes a counteroffer. This procedure is repeated until either an acceptance implements the currently valid proposal or the pie has disappeared.*

If players always reject their opponents' proposals, this game may continue forever, because by (geometric) depreciation it takes infinitely long for the pie to disappear. Hence, this game tree not only has a continuum of immediate successors for all of its moves, but it also allows for plays of infinite length. The set of plays here consists of all sequences (of variable length) of (rejected) offers, i.e., $W = \cup_{t=1}^{\infty} W_t$ where

$$W_t = \left\{ (a_\tau)_{\tau=1}^{t} \, | \, a_\tau \in [0, \delta^{\tau-1}], \text{ for all } \tau = 1, ..., t \right\},$$

for all $t = 1, 2, ...$, and the set of nodes consists of the set of all plays (the root), all moves following a rejected offer, and all (singleton sets of) plays that end with an acceptance; i.e., N consists of W together with

$$\left((\{w \in \cup_{\tau=t}^{\infty} W_\tau \, | \, a_\tau = \bar{a}_\tau, \, \forall \tau = 1, ..., t\})_{(\bar{a}_1, ..., \bar{a}_t) \in \times_{\tau=1}^{t} [0, \delta^{\tau-1}]}, (\{w\})_{w \in W_t} \right)_{t=1}^{\infty}$$

where the first group of nodes are moves where offers are about to be made (after a rejection) and the second group consists of terminal nodes.

The last example shows that it may not always be possible to physically draw the trees of certain games. They can nevertheless be defined via subtrees. In a rooted tree one can always pick a move $x \in X$ and treat it as the root of a tree consisting of the plays passing through x and all nodes (of the original tree) that are subsets thereof. This new object is also a (rooted) tree, a *subtree*. In a rooted tree every move is the root of a subtree.

To define a complicated game, one often resorts to defining subtrees together with an instruction on how to "glue" these. If in a rooted tree T_1 a terminal node x is replaced by another rooted tree T_2 (with the root of T_2 "glued" to the terminal node x of T_1), the result is a larger rooted tree T_3. In T_3 the "glued" tree T_2 shows up as a subtree. This is what phrases like "This procedure is repeated until ..." formally amount to (see Exercise 3.2 below).

Perfect Information Games   A rooted tree is almost sufficient to describe a simple class of games where players, whenever they have to take a decision, know the entire history; that is to say, the players have precisely the same information as an outside observer, who can see all events. Such games are known as games with *perfect information*. All that is missing is a specification of which player gets to decide and which decisions are exogenous or random ("chance moves").

The first of these is specified by enumerating all moves where a particular player i will be called upon to choose, for all $i = 1, ..., n$. This is on the understanding that, if $x \in X$ is assigned to player i, then player i will choose among the immediate successors $y \in P^{-1}(x)$ of move x. The interpretation is that, by picking precisely one of the immediate successors of x, say, $y \in P^{-1}(x)$, player i chooses among all plays that pass through x precisely those that also pass through y. That player i's choices at $x \in X$ can be modeled as immediate successors of x is due to the perfect information assumption, that i knows precisely the history that leads from the root to x. It will be seen that this model does not suffice when more complicated information structures are accounted for.

Not all moves are necessarily assigned to some personal player. In chess the two players may first flip a coin on who will play the white stones. How the coin comes up decides between the set of plays where player 1 has the first move and those where 2 has the first move. Such moves are known as *moves of nature* or *chance moves*. They are treated like any other move, except that they are assigned to an artificial extra player, usually "called" $i = 0$. Since chance has no preferences, these moves have to come with probability distributions that specify which immediate successor of a chance move will occur with which probability. Formally, then, we have the following.

Definition 3.2 *An n-player (n $\geq$ 1) **extensive form** with **perfect information** is a triple $F = (T, \mathcal{X}, p)$ where*

- $T = (W, N)$ *is a **rooted tree**,*

- $\mathcal{X} = (X_0, X_1, ..., X_n)$ *is a partition[6] of the set X of moves, known as the **assignment of decision points**, and*

- $p : P^{-1}(X_0) \to \mathbb{R}_{++}$ *is a function assigning probabilities to **chance moves** such that[7]*

$$\sum_{y \in P^{-1}(x)} p(y) = 1, \text{ for all } x \in X_0.$$

[6]   A partition $\mathcal{X}$ of X is a collection of subsets of X such that (a) any two distinct subsets in $\mathcal{X}$ are disjoint, $X_i \cap X_j = \emptyset$, and (b) the union of all sets in $\mathcal{X}$ covers X, i.e., $X \subseteq \cup_{i=0}^n X_i$ (the latter inclusion is, in fact, an equality).
[7]   When $P^{-1}(X_o)$ is infinite, the function p has to assign probability measures with support in $P^{-1}(x)$ to all chance moves $x \in X_o$. For the definition of a probability measure, see Section 2.2.

The interpretation is as follows. The game starts at the root. If $W \in X_i$ for some personal player i (≥ 1), then i is called upon to choose between the immediate successors of the root—meaning that from all plays she can choose those that pass through one particular immediate successor of the root. The successor of the root chosen by i is the new "state" of the game. If $W \in X_0$ is a chance move, then the game continues randomly to an immediate successor of the root according to the probability distribution specified by p. Proceeding inductively, if the playing of the game has reached a move $x \in X$ and $x \in X_i$ for some personal player i, then i chooses a node in $P^{-1}(x)$ which becomes the new "state" of the game. If $x \in X_0$ a random device governed by p picks an immediate successor of x. This continues until finally a single play has materialized.

For example, in the 2-Marienbad game (Example 3.1) the root and every move that has two predecessors is assigned to player 1 by the assignment of decision points $\mathcal{X}$. Player 2 owns all other moves, all of which have a single predecessor, the root. (In Figure 3.2 nodes are labeled with the identity of the player to whom they are assigned by $\mathcal{X}$ and with the pattern of matches remaining on the table. The pattern of matches is, however, irrelevant information as far as the representation of the game is concerned, because the tree keeps track of everything relevant.) In the bargaining game (Example 3.2) player 1 owns all moves after an even number of proposals (zero counts as even) and player 2 is assigned all moves after an odd number of (rejected) proposals. In none of these examples does chance get to move (X_0 is empty).

The function p represents either the objective probabilities for outcomes of random devices, like coins, dices, or roulette wheels, or the personal players' beliefs about decisions that are not under the control of any (personal) player in the game. It takes only positive values, on the understanding that zero probability events will not be represented in the tree, or that zero probability subtrees will be pruned.

There are two important qualifications in Definition 3.2. First, it defines extensive forms with *perfect information*. This means that when a player is called upon to choose she knows the move, the complete history of it, and which plays are still possible. In particular, she knows what all her opponents, who have chosen before her, have done. Second, there is no move that is assigned to two (or more) players at the same time. Formally this is expressed by the partition property where $X_i \cap X_j = \varnothing$, for all $j \neq i$. So, *simultaneous moves* are *excluded* by the perfect information assumption.[8]

Given that it is known how the game can possibly unfold and who controls which moves, all that remains is to specify preferences for players. Since players

---

[8]   A few remarks on terminology. First, "perfect information" is to be carefully distinguished from "complete information". (The latter means that the rules of the game and the payoff functions are common knowledge.) Second, what is here called an extensive form is sometimes referred to as an "extensive game form". The middle word has been dropped here for brevity.

may not only be interested in and motivated by final outcomes, like "win" and "lose", but they may react to what happens in the process of playing, the appropriate domain for preferences are *plays* $w \in W$. Of course, by the definition of a tree, this is the same as preferences over terminal nodes.

But, since chance may be present in the extensive form, players need to have preferences not only on plays, but also on lotteries over plays. More precisely, players are required to have preferences on lotteries over plays that conform to the von Neumann-Morgenstern axioms. Then, according to the expected utility theorem, preferences on plays are represented by a Bernoulli utility function and preferences on lotteries over plays are represented by the expected value of the Bernoulli utility function with respect to the probabilities of plays induced by chance moves (see Section 2.2). In this sense von Neumann-Morgenstern expected utility theory is implicit in the following, where only the Bernoulli utility function is mentioned.

Definition 3.3 *An n-player **extensive form game** with **perfect information** is a pair $G = (F, v)$ where F is an n-player extensive form with perfect information and $v = (v_1, ..., v_n) : W \to \mathbb{R}^n$ is a payoff function mapping plays $w \in W$ into player i's payoff $v_i(w) \in \mathbb{R}$, for all personal players $i = 1, ..., n$.*

Graphically, the values of the payoff function $v(w)$ are represented by an n-vector of payoff numbers at the terminal node which (uniquely) represents the play $w \in W$. If plays can be of infinite length, this, of course, cannot always be done.

Example 3.3 *Consider again the bargaining game (Example 3.2). To any play that ends with an acceptance after t proposals, $t = 1, 2, ...,$ one associates the payoff vector $\left(\delta^{t-1}a_1, \delta^{t-1}(1 - a_1)\right)$ if t is odd and the vector $\left(\delta^{t-1}(1 - a_2), \delta^{t-1}a_2\right)$ if t is even where a_i is the last offer made, $i = 1, 2,$ and the first entry is player 1's payoff and the second is player 2's payoff. To the infinitely long plays that result from continued rejections the payoff vector $(0, 0)$ is associated.*

Due to perfect information, the tree structure transfers, for this class of games, to the consecutive decisions by players. Not surprisingly, therefore, there is an amusing way to find a Nash equilibrium in such games by working backwards from single person decision problems in simple subtrees to the full game. This result, while often incorrectly attributed to a 1913 paper by Zermelo [185] on chess, is known as *Kuhn's algorithm*. It is a prime example of a more general idea known as *backwards induction*.

Kuhn's Algorithm More formally, for a finite extensive form game with perfect information let $Y = P^{-1}(X) = \cup_{x \in X} P^{-1}(x) = N \setminus \{W\}$ be the set of immediate successors of moves (all nodes except the root). To describe a Nash equilibrium, define a *recommendation* as a function $r : \cup_{i=1}^{n} X_i \to Y$ such that $P(r(x)) = x$, for all $x \in \cup_{i=1}^{n} X_i$. A recommendation is a proposal, say, by an umpire, to all personal players on which choices to make at their decision points.[9] It maps every move assigned to a personal player into precisely one (because it is a function) immediate successor (because $P(r(x)) = x$) of the move. Any personal player may follow this recommendation when making her moves, or she may deviate in the sense of picking some $y \in P^{-1}(x)$ with $y \neq r(x)$ at some move x which is assigned to her. A proof of the following theorem, based on breaking the game down to single-player optimization problems, is given in the appendix to this section.

Theorem 3.1 *(Kuhn's algorithm) For every finite n-player extensive form game with perfect information (and without simultaneous moves), there is a recommendation such that no player has an incentive to deviate from it, if she expects no opponent to deviate; i.e., every such game has a Nash equilibrium.*

Kuhn's algorithm applies to finite games only. But sometimes a similar argument applies also to infinite extensive form games, in particular when the tree consists of infinitely many copies of the same underlying tree which are "glued" in such a way that the underlying tree is repeated over and over again.

Example 3.4 *(resumed) Consider again Example 3.2. Suppose the recommendation r already assigns choices for all subtrees that start with a decision point of player 1 after t moves, where t is an even integer. Let $u_i^t(r)$ be player i's payoff obtained in such subtrees under r for $i = 1, 2$. Consider any offer $a \in [0, 1]$ made by player 2 after $t - 1$ moves. If $1 - a < \delta u_1^t(r)$ it maximizes 1's payoff to reject such an offer, and if $1 - a \geq \delta u_1^t(r)$ it maximizes 1's payoff to accept it. Since $u_1^t(r) + u_2^t(r) \leq 1$ and $0 < \delta < 1$, it is payoff maximizing for 2 to offer $a = 1 - \delta u_1^t(r)$ which is larger than $\delta u_2^t(r)$. Therefore, after an offer $a' \in [0, 1]$ by 1 after $t - 2$ moves, it is payoff maximizing for 2 to reject any $a' > 1 - \delta [1 - \delta u_1^t(r)]$ and to accept any $a' \leq 1 - \delta [1 - \delta u_1^t(r)]$. It follows that $u_1^{t-2}(r) = 1 - \delta + \delta^2 u_1^t(r)$.*

A (stationary) solution for the last equation is $u_1^t(r) = 1/(1 + \delta)$ and $u_2^t(r) = 1 - u_1^t(r) = \delta/(1 + \delta)$, by a symmetric argument for player 2, for all t. This entails all players always claiming shares $1/(1 + \delta)$ of the remaining pie for themselves, always accepting offers that give them at least a share of $\delta/(1 + \delta)$ of the remaining pie, and always rejecting offers that give them smaller shares than $\delta/(1 + \delta)$ of the remaining pie. With this explicit specification of the recommendation r, it is not difficult to verify that no player has an incentive to deviate from r if she expects the opponent not to deviate. Note that if δ approaches 1, so the pie almost does not depreciate, the shares of both

---

[9] It will be seen below that such a recommendation is like a combination of (pure) strategies, one for each player.

players approach 1/2. *Similarly, if the pie depreciates very rapidly,* $\delta \to 0$, *player* 1 *takes it all.*

Kuhn's algorithm and its variant in the last example single out one particular Nash equilibrium. But there may be many more such equilibria, because what a player can do in an extensive form game may be quite complex. It is important to realize that what players can do depends exclusively on the tree and the assignment of decision points. In particular, there is *no* room for extra restrictions on what players can do. For example, in the bargaining game one *cannot* in general require that a player always claim the same share for herself when its her turn to propose, simply because the game she is confronted with always "looks" the same. This would mean that the player is forbidden to condition on the history, which is what the perfect information assumption explicitly allows her to do.

Summary

- The rules of any non-cooperative game can be represented by a rooted tree. In a rooted tree every node represents one-to-one the history by which it is reached and the plays that remain possible at this node.

- An extensive form (game) with perfect information (and without simultaneous moves) consists of a rooted tree, an assignment of (all) decision points to personal players or chance, and probability distributions for chance moves (and a payoff function for all personal players).

- For an extensive form game with perfect information, Kuhn's algorithm detects a recommendation which is such that no player wishes to deviate from it if she expects no other player to deviate.

Appendix

In this appendix Proposition 3.1 and Theorem 3.1 are proved. We begin with Proposition 3.1. The first lemma reveals that condition (T.1) of Definition 3.1 is a combination of (3.1) with a boundedness condition.

Lemma 3.1 *Let* (W, N) *be a set* W *together with a collection* N *of nonempty subsets of* W. *Then* (W, N) *satisfies condition (T.1) of Definition 3.1 if and only if it satisfies (3.1) for all* $x, y \in N$ *and if, for every chain* $M \subseteq N$, *there is* $w \in W$ *such that* $w \in x$ *for all* $x \in M$ *(i.e., every chain in* N *has a lower bound in* W).

Proof ("if") Suppose (3.1) holds for all $x, y \in N$ and every chain M in N has a lower bound in W. Then, if $M \subseteq N$ is a chain, there is $w \in W$ such that $w \in x$ for all $x \in M$. Conversely, if $w \in x$ for all $x \in M$ for some arbitrary collection $M \subseteq N$, then, if $x, y \in M$, the fact that $w \in x \cap y$ implies from (3.1) that either $x \subset y$ or $y \subseteq x$ holds; i.e., M is a chain. This verifies (T.1).

("only if") Suppose (W, N) satisfies (T.1). Then for any chain $M \subseteq N$ its "only if" part implies that there is $w \in W$ such that $w \in x$ for all $x \in M$. Furthermore, if $x, y \in N$ are such that $x \cap y \neq \emptyset$, then that there is $w \in x \cap y$ implies from the "if" part of (T.1) that $\{x, y\} \subseteq N$ is a chain; i.e., either $x \subset y$ or $y \subseteq x$, verifying (3.1). This completes the proof of Lemma 3.1.

Proof of Proposition 3.1(a) ("if") If (W, N) satisfies (3.1) for all $x, y \in N$ and $M \subseteq N$ is a chain, then $\cap_{x \in M} x \neq \emptyset$, because N (and, thus, M) is finite whenever W is. Hence, there exists $w \in W$ such that $w \in x$ for all $x \in M$. Conversely, if $w \in x$ for all $x \in M$ for some arbitrary collection $M \subseteq N$, then, for all $x, y \in M$, that $w \in x \cap y$ implies from (3.1) either $x \subset y$ or $y \subseteq x$. Thus, M is a chain. This verifies condition (T.1) of Definition 3.1. Condition (T.2), that every chain in X has a maximum and a minimum, follows directly from finiteness.

("only if") If (W, N) is a tree (not necessarily finite), then that (3.1) holds for all $x, y \in N$ follows directly from Lemma 3.1.

To prove part (b) of Proposition 3.1, three more auxiliary results are helpful.

Lemma 3.2 *If (W, N) is a tree, then $X \subseteq F(N)$.*

Proof Let $z \in N \backslash F(N)$ be an infinite node, i.e. $z = \cap_{z \subset y \in N} y$ (see (3.3)). By condition (T.1) of Definition 3.1 the set $M_z = \{x \in N \,|\, z \subset x\}$ is a chain which, by condition (T.2), has an infimum x. Since $z \in N \setminus F(N)$ and x is an infimum (greatest lower bound), $z \subseteq x$. If $z \neq x$, then $z \subset x \subseteq \cap_{z \subset y \in N} y$ is a contradiction. Therefore, $z = x$ is the infimum of the chain M_z. If $z \in X$, then by condition (T.2) of Definition 3.1 z is a minimum, so $z \in M_z$ contradicts the construction of M_z. Therefore, $z \in E = \{\{w\}\}_{w \in W}$.

This shows that $N \setminus F(N) \subseteq E$ or, equivalently, $X = N \setminus E \subseteq F(N)$, and completes the proof of Lemma 3.2.

Lemma 3.3 *If (W, N) is a tree and $z \in N$ is the infimum of a chain M in N, then $z = \cap_{x \in M} x$.*

Proof Since an infimum is a lower bound, $z \subseteq \cap_{x \in M} x$. Consider any $w \in \cap_{x \in M} x$. Then $\{w\} \subseteq x$ for all $x \in M$. Since an infimum is the *greatest* lower bound, this implies $\{w\} \subseteq z$; i.e., $w \in z$. Therefore, $z = \cap_{x \in M} x$ and Lemma 3.3 is proved.

Lemma 3.4 *If (W, N) is a tree and $x \in X = N \setminus \{\{w\}\}_{w \in W}$, then there is $y \in F(N)$ such that $y \subset x$.*

Proof Let $x \in X = N \setminus \{\{w\}\}_{w \in W}$ and suppose to the contrary that all $y \in N$ which satisfy $y \subset x$ are infinite elements; i.e., $y = \cap_{y \subset z \in N} z$. Then by Lemma 3.2 all $y \in N$ that satisfy $y \subset x$ are terminal; i.e., $y = \{w\}$ for some $w \in W$.

But, on the other hand, $x \subseteq \cap_{\{w\} \subset y \in N} y$ for all $w \in x$. For, if $w \in x$ and $\{w\} \subset y \in N$, then by Lemma 3.1 either $x \subseteq y$ or $y \subset x$. But, because $\{w\} \subset y$, the latter implies $y \in X \subseteq F(N)$ (by Lemma 3.2), in contradiction to the hypothesis. Hence, $w \in x$ and $\{w\} \subset y \in N$ imply $x \subseteq y$, so that $x \subseteq \cap_{\{w\} \subset y \in N} y$ for all $w \in x$.

But by $x \in X$ then $\{w\} \subset x \subseteq \cap_{\{w\} \subset y \in N} y$ implies $\{w\} \in F(N)$ for all $w \in x$, a contradiction. This completes the proof of Lemma 3.4.

Proof of Proposition 3.1(b) For $W \in F(N)$, set $P(W) = W$. If $x \in F(N) \setminus \{W\}$, let $M_x = \{y \in N \mid x \subset y\}$. By condition (T.1) of Definition 3.1 M_x is a chain which, by condition (T.2), has an infimum z, so that $x \subseteq z$. If $z \in E = \{\{w\}\}_{w \in W}$, then $x = z$. But then, by Lemma 3.3, $x = z = \cap_{y \in M_x} y$ yields a contradiction to $x \in F(N)$. Therefore, $z \in X$ is a minimum of M_x and $x \subset z$. Define $P(x) = z$; i.e., $P(x)$ is the minimum of the chain M_x. This verifies the first part of (3.2) by construction.

Suppose now that $x \subset y \in N$. Then $y \in M_x$ and, because $z = P(x)$ is the minimum of this chain, $z \subseteq y$. This verifies the first half of the second part of (3.2).

Next, consider the chain $\{P^t(x)\}_{t=0}^{\infty}$. By condition (T.2) of Definition 3.1 this chain has a maximum, say, $P^k(x)$. Hence, $P^k(x) = P^{k+1}(x)$. But, unless $P^k(x) = W$, this contradicts $P^k(x) \subset P\left(P^k(x)\right) = P^{k+1}(x)$, which was established before. Thus, $\cup_{t=0}^{\infty} P^t(x) \supseteq P^k(x) = W \supseteq y$ verifies the second half of the second part of (3.2).

It remains to show that P is onto. If $x \in X$, then by Lemma 3.4 there is $y \in F(N)$ such that $y \subset x$. Consider the chain $\{P^t(y)\}_{t=1}^{\infty}$. Since $x \subseteq \cup_{t=1}^{\infty} P^t(y)$ by the above, there exists $k \geq 1$ such that $P^{k-1}(y) \subset x \subseteq P^k(y)$. But, since $P^{k-1}(y) \subset x$ implies $P^k(y) = P\left(P^{k-1}(y)\right) \subseteq x$, it follows that $x = P^k(y) = P\left(P^{k-1}(y)\right)$. Since $x \in X$ was arbitrary, P is onto. This completes the proof of Proposition 3.1.

Next, we turn to the proof of Kuhn's algorithm.

Proof of Theorem 3.1 Fix a *finite* perfect information extensive form game $G = (F, v)$. Consider any move $x \in X$ such that x is assigned to a personal player, say, player i, and all its (immediate) successors are either themselves terminal or chance moves leading to terminal nodes. Identify each immediate successor $y \in P^{-1}(x)$ of x with the lottery over plays induced by the chance move that comes after y or with the play that ends at y. By the expected utility theorem, there exists an immediate successor y of x which maximizes player i's expected utility over those lotteries and plays. Eliminate now all plays in x except one which maximizes player i's expected utility (over plays that pass through x) and make the remaining immediate successor of x the value of $r(x)$. Do this for all such moves $x \in \cup_{i=1}^{n} X_i$. This gives rise to a new game G', which differs from G by the elimination of some plays, and pins down the values of the recommendation r for those moves where subtrees have been pruned. The utility function for the new game G' is the restriction of v to the remaining plays.

Apply the same procedure to G'. This will pin down more values of the recommendation r and produce yet another game, G'', to which the same procedure gets applied once again. Do this until r specifies an instruction for *all* $x \in \cup_{i=1}^{n} X_i$ in the original tree. This will be the case after finitely many steps, because the tree is finite.

By construction, if any player expects all her opponents to choose according to the recommendation r in G, then she has no incentive to deviate from s. This is so, because at any decision point r maximizes the expected payoff of the player who owns the move, under the assumption that r is followed at all other moves of personal players. This completes the proof of Kuhn's algorithm.

Exercises

Exercise 3.1 Let $(\hat{N}, \hat{P})$ be a pair consisting of a finite set $\hat{N}$ (of "nodes") and $\hat{P} : \hat{N} \to \hat{N}$ be a function (the "predecessor function") which satisfies that there exists $\hat{x}_0 \in \hat{N}$ (the "root") such that

$$\text{for all } \hat{x} \in \hat{N} \text{ there is } k = 0, 1, \ldots \text{such that } \hat{P}^h(\hat{x}) = \hat{x}_0,$$

for all $h = k, k+1, \ldots$, where $\hat{P}^k(\hat{x}) = \hat{P}\left(\hat{P}^{k-1}(\hat{x})\right)$, for all $k = 1, 2, \ldots$, and $\hat{P}^0(\hat{x}) = \hat{x}$, for all $\hat{x} \in \hat{N}$. Define plays $\hat{w} \in \hat{W}$ as sequences $\{\hat{x}_t\}_{t=0}^{\tau}$ such that $\hat{x}_t = \hat{P}(\hat{x}_{t+1})$, for all $t = 0, 1, \ldots, \tau$, and $\hat{P}^{-1}(\hat{x}_\tau) = \emptyset$. For any $\hat{x} \in \hat{N}$, let $W(\hat{x}) = \left\{\hat{w} \in \hat{W} \,|\, \hat{x} \in \hat{w}\right\}$ be the plays passing through $\hat{x}$. Show that the pair $\left(\hat{W}, (W(\hat{x}))_{\hat{x} \in \hat{N}}\right)$ is a rooted tree (as in Definition 3.1). Conversely, given a finite rooted tree $T = (W, N)$, define the predecessor function P as in (3.2) and show that the pair (N, P) satisfies the above conditions.

Exercise 3.2 In Example 3.2 define the tree via subtrees; i.e., define the tree for what happens in one "period" and "glue" these trees appropriately.

Exercise 3.3 Let $W = \{w_1, w_2, w_3\}$ and

$$N = \{W, \{w_1, w_2\}, \{w_2, w_3\}, \{w_1\}, \{w_2\}, \{w_3\}\}.$$

Draw the graph corresponding to the pair $T = (W, N)$ and verify that this is not a rooted tree. (Why?)

Exercise 3.4 Let $W = [0, 1]^2$ be the unit square and

$$N = \left\{ [0, 1]^2, (\{x_1\} \times [0, 1])_{x_1 \in [0,1]}, (\{w\})_{w \in [0,1]^2} \right\}.$$

Is the corresponding pair $T = (W, N)$ a rooted tree?

Exercise 3.5 Let $W = [0, 1]$ and N be the following collection of subsets of the unit interval:

$$N = \left\{ [0, 1], ([0, x])_{x \in [0,1]}, (\{w\})_{w \in [0,1]} \right\}.$$

Is the pair (W, N) a rooted tree?

Exercise 3.6 Let $W = [0, 1]$ and N be the following collection of subsets thereof:

$$N = \left\{ [0, 1], \left[0, \frac{3-h}{3}\right], \left[\frac{h}{3}, 1\right] \left(\left[\frac{1}{3} + \frac{1}{k}, \frac{2}{3} - \frac{1}{k}\right]\right)_{k=6}^{\infty}, (\{w\})_{w \in [0,1]} \right\}.$$

Is the pair (W, N) a rooted tree for $h = 1$ or $h = 2$?

Exercise 3.7 Let $W = [0, 1]$ and $N = \{(\{w\})_{w \in W}, (x_t)_{t=1}^{\infty}\}$ where

$$x_t = [0, 1/(t + 1)] \cup [t/(t + 1), 1]$$

for all $t = 1, 2, \ldots$ Is the pair (W, N) a tree?

Exercise 3.8 The following grim spectacle is staged in ancient Rome. A Christian is put into an arena and a queue of n hungry lions is lined up in a narrow corridor leading into the arena. Then the door is opened and the first of the lions is released into the arena. Once in the arena the lion can decide whether to eat the Christian or hold out. The lion, of course, prefers to eat, but also faces a problem. A lion after having eaten becomes tired and slow. So, if he goes for the Christian, the next lion in the row will be released into the arena and will most likely eat the first lion (unless the second lion decides to hold out). On the other hand, if the first lion holds out, he will remain hungry, but cannot be eaten by another lion. The procedure of releasing a new lion into the arena whenever his predecessor has eaten is repeated for all lions in the queue. Use Kuhn's algorithm to find out whether the Christian will survive or not.

Exercise 3.9 (Centipede game) Let $W = \{w_1, \ldots, w_K\}$ and

$$N = \left\{ (\{w_k, \ldots w_K\})_{k=1}^{K}, (\{w_k\})_{k=1}^{K} \right\}$$

for some integer $K > 1$. Verify that the pair $T = (W, N)$ is a rooted tree. What are the moves? Show that moves are completely ordered (by set inclusion) and number them accordingly $x_0, x_1, x_2, ...$, where $x_0 = W$ is the root. Assign all even-numbered moves to player 1 (zero counts as even) and all odd-numbered moves to player 2. If the game terminates (immediately) after an even-numbered move at $\{w_k\}$ for $k < K$, players obtain payoffs $((k + 1)/2, (k - 1)/2)$ where the first entry is player 1's and the second 2's payoff. If the game terminates (immediately) after an odd-numbered move at $\{w_k\}$ for $k < K$, players obtain payoffs $((k - 2)/2, (k + 2)/2)$. If the game terminates at $\{w_K\}$ (immediately) after an even-numbered move, the players obtain payoffs $((K - 2)/2, (K + 2)/2)$, and if it terminates at $\{w_K\}$ (immediately) after an odd-numbered move, players obtain $((K + 1)/2, (K - 1)/2)$. Draw the extensive form and apply Kuhn's algorithm to it. Are you surprised?

Exercise 3.10 Two people have to share 50 dollars which come in five 10-dollar bills. They do not have change, so what each player gets must be a multiple of 10 dollars. The procedure for sharing is as follows. Player 1 makes a proposal (the number of 10-dollar bills she will receive). Hearing it, player 2 decides whether to accept the proposal, in which case it is implemented, or to reject it. In the latter case it again becomes player 1's turn, and she now has to choose between "giving in", which yields both players zero payoff, and seeking "revenge", which yields player 1 zero payoff but costs player 2 one dollar. Sketch the extensive form and find self-enforcing recommendations by applying Kuhn's algorithm. How many such recommendations can you find?

Notes on the Literature: The technique to represent games by a tree is due to von Neumann and Morgenstern ([134] , Chapters 8-10), who draw on a contribution by von Neumann [133]. Example 3.2 is based on Rubinstein [155]. For variations see Shaked and Sutton [166]; Binmore, Rubinstein, and Wolinsky [24]; Binmore and Dasgupta [23]; Rubinstein and Wolinsky [156]; and Sutton [173], among others. Kuhn's algorithm is described in Theorem 3 and Corollary 1 of Kuhn ([104]; see also Dalkey [35]). This paper is also the source for the more popular definition of extensive forms which generalizes that of von Neumann and Morgenstern. Exercise 3.9 (the Centipede game) is based on Rosenthal [153].

3.2   Imperfect Information

Extensive Forms   While in a perfect information game players at every move know the full history of that move, in many applications one may want to model players deciding under less information. The way to model this is by making it impossible for players to distinguish between several moves and, thereby, between several histories. This, of course, has to be done in such a way that the player cannot infer from her available choices what the histories are.

Example 3.5 *Consider a Stackelberg leader-follower game. In its simplest version this is a duopoly game between two firms, one of which, the "leader" $i = 1$, first chooses an output level which the "follower" $i = 2$ can observe. Then the follower chooses its output level, and finally market forces bring about a market-clearing price at which both firms sell their outputs. (Both firms are aware of how the market price depends on aggregate supply.) Two complications can arise in this setting, both of which can be described in a generalized version. Let there now be n (> 2) firms, where firm $i = 1$ is the leader and firms $i = 2, ..., n$ are the followers. All firms serve the same market and seek to maximize profits, given their cost functions. The leader chooses its output level first. Then followers hear about the leader's choice, but now possibly with some noise. Finally, all followers simultaneously choose their output levels (and market forces determine the price).*

There are two sources of uncertainty here. First, followers do not precisely know whether what they heard about the leader's choice is indeed the truth (with positive probability it may not be). Second, when followers choose their output levels they do this in ignorance about what other followers will do. The consequences of these two informational restrictions are, however, the same. A follower i that chooses an output level q_{it} upon hearing the noisy signal t about the leader's choice will have to choose the output level q_{it} irrespective of whether the signal t was preceded by the leader's choice $q_1 = t$ (the signal corresponds to the truth) or by the leader's choice $q_1 \neq t$ (the signal was wrong). And the firm will have to choose q_{it} irrespective of what other followers might already secretly have chosen.

Such a situation cannot be represented by simply assigning decision points (in a tree) to players. Effectively, players need to be constrained to picking the same actions after certain *classes* of histories (rather than after a single history). But then the definition of choices cannot be kept implicit, as it is under perfect information. Recall that under perfect information a "choice" of a player was simply an immediate successor of one of this player's decision points.

Under *imperfect information* we need to turn "choices" into *sets* of nodes, expressing the idea that, if a player chooses the immediate successor y after her decision point x, then the rules of the game force her to choose a particular immediate successor y' (in the same "choice") after some other decision point x' that is assigned to her. In the noisy Stackelberg game followers can condition their quantity choices only on the signals about the leader's output, but not on the leader's true choice. Hence, after *all* of the leader's quantity choices, which are by coincidence followed by the *same* signal, followers will have to choose the *same* output levels. This corresponds to a "hardwiring" of choices, the formalization of which requires machinery beyond perfect information games.

The following definition is not quite standard, but is simpler and more general than the popular definitions.

Definition 3.4 *An **extensive form** with n players is a triple* $F = (\mathcal{T}, C, p)$ *where*

- $\mathcal{T} = (W, N)$ *is a **rooted tree**,*

- $C = (C^1, ..., C^n)$, *the **set of choices**, consists of n partitions C^i of the set $Y \equiv F(N) \setminus \{W\}$ which satisfy the condition that, if two **choices** $c, c' \in C^i$ of the same player i have one common immediate predecessor, then they have all predecessors in common; i.e.,*

$$\text{if } P(c) \cap P(c') \neq \emptyset, \text{ then } P(c) = P(c'), \text{ for all } c, c' \in C^i, \qquad (3.5)$$

for all players $i = 1, ..., n,$[10] and

- *p is a positive real-valued function defined on Y which assigns **probability distributions** over the set of immediate successors of a move selected by the combination of one (available) choice for each of the personal players, such that p satisfies*

$$\text{if } x \in \cap_{i=1}^{n} P(c_i), \text{ then } \sum_{y \in [\cap_{i=1}^{n} c_i] \cap P^{-1}(x)} p(y) = 1, \qquad (3.6)$$

for all $(c_1, ..., c_n) \in C^1 \times ... \times C^n.$[11]

This definition is considerably more involved than Definition 3.2 and deserves explanation. The first object, $\mathcal{T}$, is, however, easy. It is the same as under perfect information, a rooted tree.

---

[10] Recall that for a set c of nodes $P(c) = \cup_{x \in c} P(x)$ is the set of all immediate predecessors of nodes in c.

[11] Again, if $[\cap_{i=1}^{n} c_i] \cap P^{-1}(x)$ contains infinitely many nodes, p assigns a probability measure supported on this set.

The second object expresses the hardwiring of choices. The set C of choices is an n-tuple of partitions C^i of the set Y (all finite nodes except the root), for all players $i = 1, ..., n$. An element c of a partition C^i for some $i = 1, ..., n$ is called a *choice* of player i. It models how a player can select a continuation of a history when she is at an immediate predecessor of a choice. Formally, say that choice $c \in C^i$ is *available at move $x \in X$* (to player i) if x is an immediate predecessor of c, i.e. if $x \in P(c)$. Picking a choice c which is available at x then means that from all plays in x those are selected that are also in c. That a player "chooses", therefore, means that she narrows down the range of possible plays.

That C^i is a partition expresses that picking different choices implies picking different successors (and, hence, disjoint sets of plays), because partitions consist of non-overlapping subsets of Y. These partitions have to satisfy property (3.5), which allows us to interpret the hardwiring of choices as a consequence of informational restrictions. Condition (3.5) guarantees that if the two choices $c, c' \in C^i$, for some i, are both available at some common move $x \in P(c) \cap P(c')$, then these two choices are available at *identical* moves. This ensures that the player cannot distinguish between decision points (moves) on the basis of which choices are available.

A further restriction on choices is in fact buried in condition (3.6) on the probability assignment for chance moves. In principle, it is possible that $x \in \cap_{i=1}^n P(c_i)$, but the intersection $P^{-1}(x) \cap [\cap_{i=1}^n c_i]$ is empty, for some profile of choices $(c_i)_{i=1}^n \in \times_{i=1}^n C^i$ for the different players. For example, when $P(x_j) = W$ for $j = 1, 2$ and $c_i = \{x_i\}$ for $i = 1, 2$, then $P(c_1) = P(c_2) = \{W\}$, but $c_1 \cap c_2$ is empty. Since the sum over the empty set is zero, condition (3.6) implicitly says that if $x \in \cap_{i=1}^n P(c_i)$, for some profile of choices $(c_i)_{i=1}^n \in \times_{i=1}^n C^i$, then $P^{-1}(x) \cap [\cap_{i=1}^n c_i]$ always contains at least one node.

But condition (3.6) says more. It says that if at some move $x \in X$ all players picking a particular one of their available choices (resulting in a profile $(c_i)_{i=1}^n \in \times_{i=1}^n C^i$) is not sufficient to single out one particular immediate successor of x, then chance will assign a probability distribution over those immediate successors of x that are consistent with the players' choice profile. In this way, if the playing of the game is at move $x \in X$ and all players have made their choices, then either there is only a single successor consistent with those choices to which the playing will proceed (which then gets assigned probability 1), or chance will spin a roulette wheel to decide on which successor will be chosen.

The interpretation of an extensive form is as follows. The game starts at the root where all players $i = 1, ..., n$ are told which choices they have available (viz. $W \in P(c_i)$) and are asked to choose among them. The choices selected determine a nonempty set $P^{-1}(W) \cap [\cap_{i=1}^n c_i]$ of immediate successors of the root. If this set contains only a single node, this single successor is the new "state" of the game with probability 1. If it contains several nodes, a random device picks one of its elements according to the probability distribution assigned by p. In any case the playing will proceed to a successor of the root. Proceeding inductively, if the playing has reached a move $x \in X$, players are told which choices they have available at x and are asked to choose among those. They are *not*, however, told that they get to choose at x, but only which choices are available to them at x (viz. $x \in P(c_i)$). Therefore, if some player has the same choices available at some other move $x' \in X$, she cannot distinguish whether she is at x or at x'. The choices selected at x determine a nonempty set $P^{-1}(x) \cap [\cap_{i=1}^{n} c_i]$ of immediate successors of x. If this set is a singleton set, its single element is reached with probability 1. If it contains several nodes, a random device picks one of its elements according to p.

The definition of choices, however, contains a redundant part. Obviously, every choice c is a subset of the set of immediate successors of its immediate predecessors; i.e., $c \subseteq P^{-1}(P(c))$, for all $c \in C^i$. But this inclusion may not be proper, in which case player i effectively has "no choice", because for every move $x \in P(c)$ all immediate successors will be in the same choice c; i.e., $P^{-1}(x) \subseteq c$. Such cases are allowed for by the definition, but they are uninteresting. Hence, call a choice $c \in C^i$ for some $i = 1, ..., n$ *trivial*, if $P^{-1}(P(c)) \subseteq c$. (The latter inclusion is, in fact, an equality.) Otherwise call it *nontrivial* and let C_i denote the set of nontrivial choices of player i.

It may seem a bit abstract to think of choices as sets of nodes rather than as certain actions that come from the underlying situation being modeled. But the underlying situation is where the tree comes from, so it effectively dictates which sets of nodes have to be collected in a choice. An example will clarify.

Example 3.6 *Consider the noisy generalized Stackelberg oligopoly from Example 3.5 with three firms, one leader and two followers. For simplicity, suppose that each of the three firms $i = 1, 2, 3$ can choose only between two different output levels, $0 < q_{iL} < q_{iH}$. Suppose further that if the leader chooses $q_1 = q_{1L}$, followers observe a common signal $t = H$ with probability $\varepsilon > 0$ and a common signal $t = L$ with probability $1 - \varepsilon > 0$. If the leader chooses $q_1 = q_{1H}$, then followers will observe the common signal H with probability 1. Thus, from seeing L followers can deduce that the leader has chosen $q_1 = q_{1L}$. But upon observing H, followers do not know what the leader has chosen. Their output levels must, therefore, be the same after the leader firm has chosen $q_1 = q_{1H}$ and after it has chosen $q_1 = q_{1L}$ and chance has sent the signal H.*

Figure 3.3 depicts the tree for this extensive form. It has 16 nodes where nodes $x_0, x_1, ..., x_4$ are moves (with $x_0 = W$ the root) and all other nodes are terminal. Letting $x_j = \{w_{j-4}\}$ denote terminal nodes, for $j = 5, ..., 16$, the moves are

$$x_0 = \quad W, \ x_1 = \{w_5, ..., w_{12}\}, \ x_2 = \{w_9, ..., w_{12}\},$$
$$x_3 = \quad \{w_1, ..., w_4\}, \ and \ x_4 = \{w_5, ..., w_8\}.$$

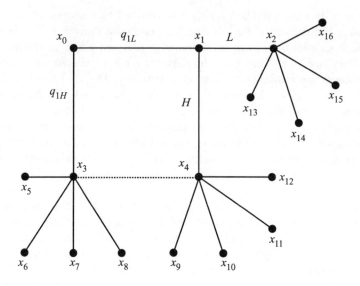

Figure 3.3: Noisy Stackelberg game

The probability assignment p is given by $p(x_2) = 1 - \varepsilon$, $p(x_4) = \varepsilon$, and $p(x) = 1$, for all $x \in \{x_1, x_3, x_5, ..., x_{16}\}$. What does C look like? For the leader this is simple: $C^1 = (\{x_1\}, \{x_3\}, \{x_2, x_4, x_5, ..., x_{16}\})$. The leader's first (resp. second) choice corresponds to $q_1 = q_{1L}$ (resp. $q_1 = q_{1H}$); its last choice is trivial. For followers it is a little more subtle. They cannot distinguish between x_3 and x_4 (the first representing the leader's choice $q_1 = q_{1H}$ and the second the signal H after $q_1 = q_{1L}$) as indicated by the broken line. After these two moves, therefore, their choices need to be hardwired; e.g.,

$$C^2 = (\{x_1, ..., x_4\}, \{x_5, x_6, x_9, x_{10}\}, \{x_7, x_8, x_{11}, x_{12}\}, \{x_{13}, x_{14}\}, \{x_{15}, x_{16}\})$$

where the first choice is trivial, the second choice corresponds to $q_2 = q_{2H}$ after seeing signal H, the third corresponds to $q_2 = q_{2L}$ after signal H, the fourth to $q_2 = q_{2H}$ after signal L, and the last to $q_2 = q_{2L}$ after signal L. Given that p assigns probability 1 to all nodes x_j with $j \geq 5$, conditions (3.5) and (3.6) now pin down firm 3's choice set to

$$C^3 = (\{x_1, ..., x_4\}, \{x_5, x_7, x_9, x_{11}\}, \{x_6, x_8, x_{10}, x_{12}\}, \{x_{13}, x_{15}\}, \{x_{14}, x_{16}\})$$

with a similar interpretation as for firm 2. The payoffs now have to be assigned to plays (which are one-to-one represented by terminal nodes) so as to reflect this interpretation.

In this example the moves x_3, x_4, and x_2 are controlled by two players, $i = 2$ and $i = 3$. This means that they decide *simultaneously*. This simultaneity could also be represented by, say, splitting x_2 into three nodes, x_2, x_2', and x_2'', where $x_2' \cup x_2'' = x_2$ such that $P(x_{13}) = P(x_{14}) = x_2'$ and $P(x_{15}) = P(x_{16}) = x_2''$. Then the decision between x_2' and x_2'' could represent firm 2's output decision, and the decision between $\{x_{13}, x_{15}\}$ and $\{x_{14}, x_{16}\}$ would represent firm 3's output decision. The same could be done for x_3 and x_4. (Draw the tree!)

The latter is, in fact, the conventional way to represent simultaneous decisions in extensive forms. This has the advantage that at every move one personal player at most has nontrivial choices. But it has the disadvantage that it cannot represent simultaneous decisions by the same move. Definition 3.4 does not take a position on this issue. It allows both ways.

That representing simultaneous decisions by an immediate succession of separate sets of moves (between which the decisive player cannot distinguish) is as good as assigning moves to several players is due to von Neumann-Morgenstern decision theory. By the independence axiom a compound lottery, represented by a succession of "information sets", is equivalent (from the players' point of view) to the reduced lottery, represented by a single "information set". Without an independence axiom, it may make a difference whether a representation like that in the left part of Figure 3.4 (below) or like that in the right part of the figure is adopted.

Conventional extensive forms also consider only nontrivial choices $c \in C_i = \left\{ c' \in C^i \,\middle|\, P^{-1}(P(c')) \setminus c' \neq \varnothing \right\}$, for $i = 1, ..., n$, and put more emphasis on objects called "information sets". These are introduced next.

Definition 3.5 *For an extensive form F, the system H_i of disjoint subsets of moves defined by $H_i = \{P(c)\}_{c \in C_i}$ is player i's **information partition**, for all $i = 1, ..., n$. An element $h \in H_i$ is one of player i's **information sets**. Moves in the set $X_i = \cup_{h \in H_i} h$ are player i's **decision points**, or the moves assigned to player i. The system $\mathcal{X} = (X_1, ..., X_n)$ is the **assignment** of decision points to (personal) players.*

That different information sets $h \in H_i$ do not overlap follows because C^i is a partition of Y, the set of nontrivial choices C_i is a subset of the set C^i of all choices, and condition (3.5) applies, for all $i = 1, ..., n$. Effectively, for every personal player i, the information partition H_i is a partition of the set X_i of i's decision points. The elements of this information partition, the information sets, express the indistinguishability of the (plays contained in the) moves contained in them. Note that for a personal player her decision points (or information sets) are only those moves (or sets of moves) where she has nontrivial choices, because otherwise there is no decision to be taken.[12]

The conventional formalism emphasizes information sets because these are easy to interpret as models of what a player knows. When the "state" is $x \in X$,

---

[12] This is not necessary. If trivial choices are specified more carefully, their images under P can be used to model what a player knows even when she has no decision to take.

then the plays that are regarded possible by player i are those passing through some move $x \in h$ in her information set $h \in H_i$. Among the states *in* this information set h, however, the player cannot distinguish.

Conventional extensive form representations also tend to give the information partitions and the assignment of decision points as separate objects besides choice partitions. Hence, you may find an extensive form being defined as a five-tuple $F = (\mathcal{T}, \mathcal{X}, H, C, p)$, where

- $\mathcal{T} = (W, N)$ is a rooted tree,

- $\mathcal{X} = (X_0, X_1, ..., X_n)$ is the assignment of decision points to personal players and chance (where $X_0 = X \setminus (\cup_{i=1}^{n} X_i)$ and $\mathcal{X}$ is a partition of X),

- $H = (H_i)_{i=1}^{n}$ is the collection of information partitions for personal players (each H_i is a partition of X_i),

- $C = (C_i)_{i=1}^{n}$ are the (nontrivial) choice partitions for personal players, and

- $p : P^{-1}(X_0) \to \mathbb{R}_{++}$ is an assignment of probabilities for chance moves.

To such an "extensive" definition, of course, consistency conditions must be added to meet conditions (3.5) and (3.6). Much of this is, however, redundant, because information sets can always be reconstructed from choices—but *not* vice versa.

One way of embedding the conventional definition of an extensive form into the present is as follows. Call an extensive form (Definition 3.4) *conventional* if no move is assigned to more than one player, i.e. if X_i and X_j are disjoint for all $j \neq i$ (for X_i and X_j as defined in Definition 3.5).

If for some extensive form every nontrivial choice (resp. every information set) contains precisely one node, i.e. if $\{y\} \in C_i$ for all $y \in P^{-1}(X_i)$ (resp. $\{x\} \in H_i$ for all $x \in X_i$) for all $i = 1, ..., n$, then this must be a conventional extensive form. But it is, moreover, one with *perfect information* (see Definition 3.2). The decision points in an extensive form with perfect information count as (degenerate) information sets which contain precisely one move. Thus, an extensive form with perfect information (Definition 3.2) is a *special case* of a conventional extensive form, which itself is a special case of an extensive form with imperfect information (Definition 3.4).

Imperfect Information Games As under perfect information, turning an extensive form into a game takes a specification of the (personal) players' preferences. This again requires an implicit appeal to von Neumann-Morgenstern decision theory as preferences over lotteries need to be representable as expected utilities over plays. (To avoid such an implicit appeal to decision theory, one would have to specify preferences on lotteries over plays instead of Bernoulli utility functions.)

Definition 3.6 *An **extensive form game** with n players is a pair $G = (F, v)$ where $F = (T, C, p)$ is an extensive form with n players and $v : W \rightarrow \mathbb{R}^n$ is the payoff function mapping each play $w \in W$ into a payoff vector $v(w) = (v_1(w), ..., v_n(w))$.*

The graphical representation is like that under perfect information, except that the representation of choices and/or information sets is now more complicated. Information sets are depicted by connecting moves in the same information set with a broken line (see Figure 3.3 for an example).

Nontrivial choices are usually represented by labeling all edges that lead to nodes in a given choice with the same symbol that identifies this particular choice. For example, in Figure 3.3 one could label the edges leading from x_3 to x_5 and x_6 and the edges leading from x_4 to x_9 and x_{10} with the symbol $q_{2H}(H)$, meaning that the endpoints of these edges belong to the choice of $q_2 = q_{2H}$ after firm 2 has observed signal H (the second choice in C^2). With the present definition of an extensive form, this may require multiple labels on some edges, because endpoints of edges may belong to nontrivial choices of several players simultaneously.

Example 3.7 *Consider again several firms serving the same market, but now suppose that there is no leader. For simplicity, suppose there are three (profit maximizing) firms $i = 1, 2, 3$ each of which can choose either a low or a high output level, $0 < q_{iL} < q_{iH}$.*

But now none of the firms can observe a choice by any other firm (or a signal about the opponents' choices). The profit of any firm will depend on the decisions by all other firms through the market price (the "inverse demand function" which all firms understand). In oligopoly theory this is known as a "Cournot oligopoly", where firms choose quantities supplied.

Figure 3.4 gives two graphical representations of the extensive form. The upper left is a conventional extensive form with 15 nodes. The root in the upper left corner is controlled by firm $i = 1$ choosing between q_{1H} and q_{1L}, as expressed by the labeling of the edges. This is followed by two moves belonging to firm $i = 2$ which are contained in the same information set (depicted as a dotted line), expressing that firm $i = 2$ cannot distinguish between those two moves.

Finally, firm $i = 3$ chooses between its two output levels at an information set encompassing four moves between which it cannot distinguish. All plays pass through one of these four moves in firm 3's information set (again depicted by a dotted line). The choices are depicted by labeling the corresponding edges.

On the lower right of the figure the same interaction is depicted by a tree with nine nodes, where the top node is the root $x_0 = W$ and all other nodes

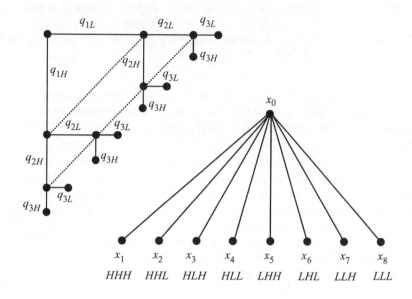

Figure 3.4: Two ways to model simultaneous decisions

are terminal. The choice sets here are as follows:

$$C_1 = C^1 = \{\{x_1, ..., x_4\}, \{x_5, ..., x_8\}\},$$
$$C_2 = C^2 = \{\{x_1, x_2, x_5, x_6\}, \{x_3, x_4, x_7, x_8\}\},$$
$$C_3 = C^3 = \{\{x_1, x_3, x_5, x_7\}, \{x_2, x_4, x_6, x_8\}\}.$$

This is depicted by associating with each terminal node a triplet of letters, where e.g. HLL means $q_1 = q_{1H}$, $q_2 = q_{2L}$, and $q_3 = q_{3L}$. Whether the players' ignorance about their opponents' choices is depicted by large information sets, as on the upper left, or by assigning the same move (the root) to all players, as on the lower right, does not matter. What is important is the hardwiring of choices that comes from the ignorance about the opponents' decisions.

Extensive forms with perfect information have the feature that they nicely decompose. Every move is the root of a subtree. Therefore, and because under perfect information there are no simultaneous moves, the analysis can be decomposed into a sequence of single person optimization problems. This is what Kuhn's algorithm exploits. Not surprisingly, therefore, Kuhn's algorithm fails for extensive form games with imperfect information.

Subgames   In perfect information games the decomposition into subtrees works, because both these objects have precisely the same structure as the original tree. Looking for objects with the same structure contained in an imperfect information game leads to the notion of a *subgame*. But while subtrees rise at every move in a tree, subgames turn out to be very rare.

Definition 3.7 *A **subgame** of an extensive form $F = (T, C, p)$ is an extensive form $F' = (T', C', p')$ where T' is a subtree of T, where C' consists of those nontrivial choices in C which are contained in the set of nodes of T' (together with a restriction of trivial choices to the set of nodes of T'), and where p' is the restriction of p to nodes in T'.*

The demanding piece of this definition is the combination of how C' is determined with the requirement that F' has to be an extensive form. The set C' is constructed by collecting, for each player $i = 1, ..., n$, those (nontrivial) choices from C^i that are fully contained in (the set of nodes of) T'. If this (together with an appropriate restriction of trivial choices) results in choice partitions, a subgame has been found. But, if for some $i = 1, ..., n$ there is some nontrivial choice $c \in C^i$ which has a nonempty intersection with both (the set of nodes of) T' and its complement (in the set of nodes of T), then C' can never assign a choice partition to i, simply because c will not be present in C' and, therefore, i's choice set under C' does not cover (the set of nodes except the root of) T'.

Roughly, a subgame consists of a subtree such that every nontrivial choice (or information set) of the original extensive form is either fully *contained in* or *disjoint from* the (set of nodes of) the subtree. This is a very demanding requirement. For example, the extensive form in Example 3.7 has no proper subgame; i.e., the only subgame is the full game itself. The extensive form in Figure 3.3 has only one proper subgame, starting at x_2.

But even if there are proper subgames, there is nothing in the definition of a subgame to ensure that subgames can be analyzed as single person decision problems. Hence, Kuhn's algorithm is bound to fail.

Summary

- An extensive form (with imperfect information) consists of a rooted tree, choice partitions for each player satisfying condition (3.5), and a probability assignment for chance moves satisfying condition (3.6). Choices are non-overlapping sets of nodes, expressing a hardwiring of decisions for informational reasons.

- The images of choices under the immediate predecessor function are called information sets and express the indistinguishability of moves to the players.

- An extensive form has perfect information if there are no simultaneous moves and all information sets contain precisely one move.

- A subgame consists of a subtree such that every nontrivial choice (or every information set) is either fully contained in or disjoint from the subtree.

Exercises

Exercise 3.11 Reconsider Example 3.5 with three firms, one leader and two followers, where each firm can choose only between two output levels. But now assume that, after the leader has chosen $q_1 = q_{1H}$, chance sends with probability $1 - \varepsilon > 0$ the signal $t = H$ and with probability $\varepsilon > 0$ the signal L, and after the leader has chosen $q_1 = q_{1L}$ chance sends $t = L$ with probability $1 - \varepsilon$ and $t = H$ with probability ε. Draw an extensive form. How many proper subgames does this extensive form have?

Exercise 3.12 Reconsider the 2-Marienbad game. How does the extensive form change when player 2, once it is her turn to move, learns only the total number of matches remaining on the table after 1's move, but not their arrangement?

Exercise 3.13 (Matching pennies) Two players each put one coin on the table but keep it hidden from the other player until she has put down her own coin. Once both have put down their coins, they reveal to each other the sides of the coins lying face up on the table. If the sides match, player 1 wins a dollar from player 2, if they don't, player 2 wins a dollar from player 1. Draw two different extensive form representations for this game. Try applying Kuhn's algorithm. Why does it not work?

Exercise 3.14 Modify the previous exercise (Matching pennies) as follows. Player 1 is first asked whether she wishes to participate in the game. If she does not, a terminal node is reached, where no player receives anything. If she accepts, player 2 is asked for her consent. If she declines, again no player receives anything; if she accepts, the Matching pennies game is played. Draw the extensive form and determine the number of subgames.

Exercise 3.15 Let the set of plays be $W = \{w_1, ..., w_{10}\}$ and

$$N = \left\{W, \{w_1, w_2\}, \{w_3, w_4\}, \{w_5, w_6\}, \{w_3, ..., w_{10}\}, (\{w\})_{w \in W}\right\}.$$

Verify that the pair $T = (W, N)$ is a rooted tree. Suppose the choices of player 1 are given by $c_1^1 = \{\{w_3, w_4\}, \{w_7\}, \{w_8\}\}$, $c_1^2 = \{\{w_5, w_6\}, \{w_9\}, \{w_{10}\}\}$, and $c_1^3 = \left\{\{w_1, w_2\}, \{w_3, ... w_{10}\}, (\{w_j\})_{j=1}^6\right\}$. Which of those are trivial? Is it possible that player 2's nontrivial choices are

$$c_2^1 = \{\{w_1\}, \{w_3\}, \{w_5\}, \{w_7\}, \{w_9\}\} \text{ and}$$
$$c_2^2 = \{\{w_2\}, \{w_4\}, \{w_6\}, \{w_8\}, \{w_{10}\}\}?$$

Exercise 3.16 Show that if, in an extensive form with perfect information (Definition 3.2), every immediate successor of a decision point is declared a (singleton) choice of the player who controls the decision point (and the union of all immediate successors of decision points belonging to other players or chance are declared one large trivial choice), then the resulting triplet (T, C, p) meets the definition of an extensive form with imperfect information (Definition 3.4) when p is extended to $Y = N \setminus \{W\}$ by setting $p(y) = 1$ for all $y \in Y \setminus P^{-1}(X_0)$.

Notes on the Literature: The conventional tree model of extensive forms with imperfect information originates in Kuhn [104] (see also Selten [163]), generalizing the concept by von Neumann and Morgenstern [134] and von Neumann [133]. The Stackelberg duopoly (see Example 3.5) was introduced by von Stackelberg [172]. An extensive account of models of oligopolistic competition, presenting more general versions of the examples in this section, can be found in Friedman ([62] and [63]). Noisy observations of the leader's choices in the Stackelberg game were discussed by Bagwell [10]; see also Hurkens and van Damme [91], and Güth, Kirchsteiger, and Ritzberger [74]. The notion of a subgame has been stressed by Selten [162]. A generalization of (conventional) extensive forms which allows for simultaneous decisions being represented by the same move, and for even more generality, has also been introduced by Selten [165].

3.3   Strategies

Pure Strategies   Up to this point, what players can do has been described locally. Players are asked at move $x \in X$ to choose among the choices $c \in C_i$ that are available there in the sense that $x \in P(c)$. But, of course, when players reason about the game they will take into account what they do at *all* moves where they have a choice. This leads to the concept of a "strategy". Such a "pure strategy" will assign to every move a choice and, thereby, a set of immediate successors of the move, selecting among the plays that pass through the move.

Definition 3.8 *For an extensive form F with n players a **pure strategy** for player i is a function $s_i : X \to C^i$ such that*

$$s_i^{-1}(c) = P(c), \text{ for all } c \in s_i(X) = \cup_{x \in X} s_i(x) \tag{3.7}$$

for all players $i = 1, ..., n$.

Condition (3.7) on a pure strategy has two aspects. First, it guarantees that s_i assigns to each move $x \in X$ some choice $c \in C^i$ which is *available* at x, because $s_i^{-1}(c) = \{x' \in X \,|\, s_i(x') = c\} \subseteq P(c)$ (where $P(c) = h$ is an information set for player i if c is nontrivial). Second, it guarantees that to each move x in an information set $h = P(c)$ the *same* choice gets assigned, because $P(c) \subseteq s_i^{-1}(c)$. In other words, condition (3.7) guarantees that s_i maps each move into choices *available* at that move and that s_i is *measurable* with respect to the information partition H_i; i.e., $s_i^{-1}(c) \in H_i$ for all nontrivial choices $c \in s_i(X)$. "Measurability" of s_i says that a pure strategy is a random variable which does not convey more information than what is embodied in the information partition H_i of player i.[13] Effectively, a pure strategy assigns the same available choice to all moves in one information set.

Since the images of choices under the immediate predecessor function P are information sets, a pure strategy can also be thought of as a function that assigns to every information set of a player one of the choices (of this player) that are available at this information set. Therefore, often the domain of a pure strategy s_i is taken to be the set H_i of information sets of player i, mapping information sets $h \in H_i$ into nontrivial choices $c \in C_i$ (available at h). A pure strategy, therefore, is like a "book" in which every "page" contains an instruction for one particular information set (and the book has precisely as many pages as the player has information sets). The instruction

---

[13]   Intuitively, a function is *measurable* with respect to a partition of its domain if it is constant on every element (viz. a subset of the domain) of the partition.

for a particular information set, of course, specifies one (but only one) of the choices available at this information set.

For a personal player i, denote by S_i the *set of all possible pure strategies*, i.e. the set of all functions that satisfy Definition 3.8, for all $i = 1, ..., n$. If a pure strategy is a "book", then the set S_i of all pure strategies is the "library" of all those books. If the tree is finite, then the "library" S_i will be finite, because, with finitely many "pages" and finitely many instructions for each "page", only finitely many "books" can be written.

How many pure strategies there are and which instructions they can contain is, in fact, completely determined by the underlying extensive form, for any personal player. Hence, pure strategies are functions *derived from* an extensive form and they are not primitives of the game. This is to say that, once you have specified an extensive form, you have no more choice on which strategies you allow for the players, because the extensive form determines uniquely a set of pure strategies for every player.

Example 3.8 *Reconsider the 2-Marienbad game (Example 3.1; see Figure 3.2). Since in this extensive form player 1 has four information sets, one with four choices and three with two choices available, she has $4 \cdot 2^3 = 32$ different pure strategies. Player 2 has three information sets, two with three choices and one with two choices available; therefore, player 2 has $2 \cdot 3^2 = 18$ pure strategies. In the noisy Stackelberg game (see Example 3.5) of Figure 3.3 firm $i = 1$ has two pure strategies, but the two followers, firms $i = 2, 3$, each have four pure strategies, because each of them controls two information sets with two choices each. In the two-player bargaining game of Example 3.2 each of the two players has infinitely many pure strategies (in fact, a continuum).*

This illustrates that many extensive form games have a large number of pure strategies for each player. The more information sets a player has, the more strategies she has. This is so because a pure strategy specifies a choice for *every* information set, even for those information sets that cannot be reached when the pure strategy at hand is actually implemented.

A list with one pure strategy for each player, $s = (s_1, ..., s_n)$, is known as a *pure strategy combination*. It specifies one "book" from each of the individual "libraries". If player i has K_i pure strategies, then there are $K = \prod_{i=1}^{n} K_i$ pure strategy combinations for the game. Denote by $S = S_1 \times ... \times S_n$ the Cartesian product of the individual pure strategy sets, viz. the set of all pure strategy combinations.

Any pure strategy combination $s \in S$ induces, together with the probability assignment p, a probability distribution $\pi_s : W \to \mathbb{R}_+$ with $\sum_{w \in W} \pi_s(w) = 1$ on the set W of plays as follows. For any node $y \in Y$, let the probability of y being reached from its (unique) immediate predecessor $x = P(y)$ under $s \in S$ be

$$\pi_s(y \,|\, x) = \begin{cases} p(y) & \text{if} & y \in \cap_{i=1}^{n} s_i(x) \\ 0 & \text{otherwise} \end{cases} \tag{3.8}$$

Then the probability that the play $w \in W$ will occur, if players use the pure strategy combination $s \in S$, is given by $\pi_s(w) = \prod_{w \in y} \pi_s(y|P(y))$, where $\pi_s(W|P(W)) = 1$ at the root, because $P(W) = W$.

With these probabilities over plays the Bernoulli utility function v can now be used to obtain payoffs from pure strategy combinations in the extensive form game $G = (F, v)$ as expected payoffs from the lottery over plays induced by a pure strategy combination. Applying the expected utility theorem, the *expected payoff function* of player i on pure strategy combinations, $u_i : S \to \mathbb{R}$, is

$$u_i(s) = \sum_{w \in W} \pi_s(w) v_i(w) \tag{3.9}$$

for all $s \in S$ and all $i = 1, ..., n$.[14] Again, the whole vector of payoffs is $u(s) = (u_1(s), ..., u_n(s))$, for all $s \in S$, and, thus, constitutes a function $u : S \to \mathbb{R}^n$.

Mixed Strategies   Depending on the application, one may want to allow players also to use other types of strategies. In particular, allowing players to use randomization devices may be of interest. Throwing dices, flipping coins, or spinning roulette wheels to decide on the pure strategy actually implemented by a player may be means of hiding the decision from others, or may simply be an expression of indifference. For example, in the Matching pennies game of Exercise 3.13 no player knows the decision of the other when putting down her coin. So, it may not be unreasonable to toss the coin rather than put it down deliberately. But tossing the coin means picking either one of the two pure strategies (choices) with a probability of $1/2$ (if the coin is fair).

A random device like a roulette wheel, which picks one pure strategy according to the probability distribution governing its outcomes, is known as a *mixed strategy*. It is as if pure strategies are "mixed" according to a probability distribution chosen by the player. Formally, a mixed strategy is a lottery over pure strategies, the probabilities (attached to particular pure strategies) of which are chosen by the player.

Definition 3.9 *For a given extensive form F with n players, a **mixed strategy** for player i is a probability distribution on pure strategies, i.e. a function $\sigma_i : S_i \to \mathbb{R}_+$ such that $\sum_{s_i \in S_i} \sigma_i(s_i) = 1$, for all $i = 1, ..., n$.*

Imagine a fair pointer spinning over a circular disk with as many different segments along its rim as the player has pure strategies ("books in her library"). Letting the player choose a mixed strategy is allowing the player to choose the length of the different segments, thus determining the probability that the pointer will come to a halt in a particular segment. The segment

---

[14]   If there are infinitely many plays, the sum would have to be replaced by an integral. But all the theory will be developed for finite trees, so sums are always sufficient.

where the pointer comes to a halt indicates the "book" from the "library" which will instruct the player as to what she will choose at (all of) her information sets. Notice that the player always may make one particular segment (pure strategy) the whole circumference, thus assigning probability zero to all other segments (pure strategies). This corresponds to choosing this particular pure strategy with probability 1. In this sense mixed strategies are generalizations of pure strategies. Any pure strategy can be identified with the mixed strategy that assigns probability 1 to it (and probability 0 to all other pure strategies).

It seems as if random selections of strategies are rare in the real world. But this may often be due to a deliberate decision not to use a random device, which nevertheless remains potentially available. Moreover, in some situations random devices are indeed used. Telecommunications companies often monitor their clients according to random schemes. In American football the offensive team has to decide whether to pass or run. Passing generally gains more yards, but choosing the action not expected by the other team is even more important. Teams decide to run on some occasions and to pass on others, which acts like a random device.

The latter observation, in fact, provides an alternative interpretation of mixed strategies. In games that are played very often against varying opponents, the *frequencies* of the opponents' pure strategies are equivalent to mixed strategies of the opponents. Hence, a mixed strategy can be thought of as a distribution of potential contestants in a large population, where each potential contestant is known to use a particular pure strategy. For a player, whose opponents are drawn at random from such populations, the frequencies of pure strategies "simulate" a mixed strategy of the opponent.

Admittedly, however, many applications forbid random decisions. Yet, mixed strategies still have significance. Under von Neumann-Morgenstern decision theory they represent assessments or "beliefs" about what other players will do. When a player reasons about what her opponents will do, she will assign probabilities to the various strategy combinations among the other players. Knowing that all players make their decisions independently, she will assign these probabilities independently for each opponent. This, formally, results in a mixed strategy combination over the strategy choices of other players. In this interpretation, mixed strategies model how players reason about their opponents in a way that is consistent with the rules of the game.

Whether players use random devices, or pure strategies occur with certain frequencies in large populations, or mixed strategies are viewed as beliefs about the others is largely a matter of the particular application. Yet, the last interpretation is the "safest", because in any case a mixed strategy represents what anybody else should *expect* about the player position under scrutiny. Since what players expect about each other ultimately drives any strategic consideration, all the above interpretations will do for the purpose of theory— and, indeed, they are formally equivalent.

The set of *all* mixed strategies of player i, viz. the set of all probability distributions on S_i, is denoted Δ_i and called the player's *mixed strategy set* or her *mixed strategy space*. It has a nice geometric structure. Since the tree is taken to be finite, the set of pure strategies of player i is a finite set, say, with $K_i > 1$ elements. For any fixed mixed strategy $\sigma_i \in \Delta_i$, number the pure strategies from 1 to K_i, so $S_i = \{s_i^1, ..., s_i^{K_i}\}$, and denote $\sigma_i(s_i^k) = \sigma_i^k$, for all $k = 1, ..., K_i$. Then the space of mixed strategies Δ_i can be identified with the $(K_i - 1)$-dimensional unit simplex, i.e. with $\Delta_i = \left\{ \sigma_i \in \mathbb{R}_+^{K_i} \,\middle|\, \sum_{k=1}^{K_i} \sigma_i^k = 1 \right\}$. This way the space of mixed strategies, instead of being a space of functions, becomes a compact convex subset of a Euclidean vector space. (For the geometry of a simplex, see Figure 2.1 in Section 2.2.)

A whole list (or vector) of mixed strategies, one for each player, $\sigma = (\sigma_1, ..., \sigma_n)$, is known as a *mixed strategy combination*. It is an element of the space of all mixed strategy combinations $\Theta = \Delta_1 \times ... \times \Delta_n$, the Cartesian product of the individual mixed strategy sets. Since Θ is a product of simplices, it is geometrically a compact convex polyhedron with dimension $\sum_{i=1}^n K_i - n$. For example, in a two-player game where each player has two pure strategies, each player's mixed strategy space is a one-dimensional simplex and the space Θ of mixed strategy combinations looks like the unit square.

Since a non-cooperative game is a game with complete rules, players have to take their decisions independently when they decide on mixed strategies. Any opportunity for coordination and correlation of strategy choices must already be covered by the rules of the game, otherwise it cannot have complete rules. But then, since strategies are derived from the rules of the game, decisions about strategy choices must be independent across players.

The same point can also be made by invoking the interpretation of mixed strategies as beliefs. Consider an outside observer who knows as much about the players as they know about each other (and the game). If this observer learns some player's, say, i's, strategy choice, this will not affect the observer's assessment of the other players' choices. This is so because, by the assumption of complete rules, player i cannot have more information than the observer. Indeed, they have identical information (and identical analyses of that information) with respect to the other players, so there is nothing that the observer can learn from player i. Therefore, for any strategy choice of player i, the conditional probability over the others' strategy choices is the same as the unconditional probability. It follows that the observer's assessment of the strategy choices of all players must be the *product* of her assessments of their individual choices.

This observation has an important consequence. It determines, for any given mixed strategy combination, the probability distribution on pure strategy *combinations*. In particular, by independence, the probability of any pure strategy combination $s = (s_1, ..., s_n) \in S$ is the product of its "marginals", i.e. the product of the probabilities assigned by the individual mixed strategies to

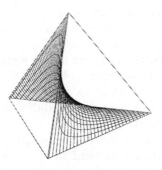

Figure 3.5: Product distributions

the components s_i, for each $i = 1, ..., n$. More formally, if $\sigma = (\sigma_1, ..., \sigma_n) \in \Theta$ is a mixed strategy combination and $s \in S$ a pure strategy combination, then the probability $\sigma(s)$ of $s \in S$ under $\sigma \in \Theta$ is given by[15]

$$\sigma(s) = \sigma(s_1, ..., s_n) = \prod_{i=1}^{n} \sigma_i(s_i), \text{ for all } s \in S. \qquad (3.10)$$

The product structure restricts the probability distributions that mixed strategy combinations can generate on the set of pure strategy combinations. In general there are many more probability distributions over pure strategy combinations than those that can be induced by the product distributions derived from mixed strategy combinations.

Recall that, for a two-player game with two pure strategies for each player (a "2 × 2 game"), the space of mixed strategies looks like the unit square and has, therefore, dimension 2. On the other hand, in 2 × 2 games, there are four pure strategy combinations (which in $\Theta = [0,1]^2$ are represented by the corners of the unit square). Hence, the set of probability distributions over pure strategy *combinations*, $\Delta(S) = \{\varphi : S \to \mathbb{R}_+ \,|\, \sum_{s \in S} \varphi(s) = 1\}$, can

---

[15] The symbol $\sigma(s)$ for the value of the mapping $s \mapsto \sigma(s)$ entails a slight abuse of notation, because "σ" is also used for the vector $\sigma = (\sigma_i)_{i=1}^{n} \in \Theta$. But, because we will think of $\sigma \in \Theta$ as a vector (of vectors) and of $\sigma(s)$ as (the value of) a function, usually no confusion can arise.

be identified with a three-dimensional simplex. Figure 3.5 illustrates how the probability distributions that can be induced by mixed strategy combinations are embedded in the higher-dimensional space $\Delta(S)$ of all probability distributions over pure strategy combinations.

The triangular pyramidal object is the three-dimensional unit simplex $\Delta(S)$ for a 2×2 game. Embedded in it is a twisted surface which depicts all points corresponding to probability distributions induced by mixed strategy combinations. It is the image of Θ under the mapping defined by (3.10). This visually reveals that the convexity of $\Theta = \times_{i=1}^{n}\Delta_i$ is enforced by the parametrization. If Θ, the space of mixed strategy combinations, is viewed as a subset of $\Delta(S)$, the space of probability distributions over pure strategy combinations, then Θ is not convex anymore.

Given that it is known which probability distributions on pure strategy combinations are induced by mixed strategy combinations, the payoff function can be extended to the space of mixed strategy combinations. Again, the expected utility theorem is invoked to obtain a function $U = (U_1, ..., U_n)$: $\Theta \rightarrow \mathbb{R}^n$ which represents the players' preferences on lotteries over pure strategy combinations. The *expected payoff* from the mixed strategy combination $\sigma \in \Theta$ for player i is given by

$$U_i(\sigma) = \sum_{s \in S} \sigma(s)u_i(s) = \sum_{s \in S} \sigma(s) \sum_{w \in W} \pi_s(w)v_i(w) \qquad (3.11)$$

for all $i = 1, ..., n$ where $\sigma(s)$ is defined in (3.10). Because $\sigma(s)$ is a product, for each $s \in S$, the expected payoff function of every player is *linear* in the mixed strategy of each player. This linearity makes expected utilities particularly simple to manipulate.

Behavior Strategies   Pure strategies are plans of action for all contingencies. Mixed strategies are probability distributions over the former. Both notions portray a player as making up her mind *prior* to an actual playing of the game. Once a pure strategy has been selected, its detailed instructions are sufficient to delegate the actual playing of the game to a team of agents which executes the player's strategy.

But for a rational player it should not make a difference whether she decides prior to or during the actual playing of a game. Suppose, for the moment, there is in fact an optimal way to play a game which a rational player can identify prior to any playing—say, by introspection. If the same rational player unexpectedly woke up in the middle of a play, one would think that her (rational) behavior during the playing would amount to an execution of the plan she would otherwise have selected ex ante. (Such a property is often referred to as "time consistency".) For less rational people the ex ante decision seems a bit demanding, and they will tend to take decisions stepwise as they go along.

This piecemeal tactics is captured by the notion of a *behavior strategy*. Like a mixed strategy, behavior strategies involve randomizations. But now the random device will not select a full course of action for all contingencies in one go: rather, many small random devices will select choices at information sets. Like with a pure strategy, a behavior strategy can be thought of as a "book", with each of its pages referring to one particular information set. But instead of an instruction on what to choose at the information set, this "book" gives a probability distribution over choices available at the information set on each of its pages.

Definition 3.10 *For a given extensive form, a* **behavior strategy** *for player i is a function ρ_i which assigns to every move $x \in X$ a probability distribution b over choices $c \in C^i$ such that*

$$\rho_i^{-1}(b) = \cap_{c \in \text{supp}(b)} P(c), \text{ for all } b \in \rho_i(X) = \cup_{x \in X} \rho_i(x). \tag{3.12}$$

Condition (3.12) is analogous to (3.7). If b is a probability distribution over choices for which two choices $c, c' \in \text{supp}(b)$ do *not* have precisely the same predecessors, $P(c) \neq P(c')$, then by (3.5) $P(c)$ and $P(c')$ are disjoint; $P(c) \cap P(c') = \emptyset$. Therefore,

$$\rho_i^{-1}(b) = \{x \in X \,|\, \rho_i(x) = b\} \subseteq \cap_{c \in \text{supp}(b)} P(c) = \emptyset$$

implies that *no* move $x \in X$ gets assigned such a distribution. Hence, a behavior strategy assigns distributions with supports in choices *available* at (the moves in) the same information set. Moreover, if $b \in \rho_i(X)$ and $b(c) > 0$, then, for any pair of moves $x, x' \in P(c)$,

$$x, x' \in \cap_{c' \in \text{supp}(b)} P(c') = P(c) \subseteq \rho_i^{-1}(b)$$

implies that $\rho_i(x) = b = \rho_i(x')$. Hence, a behavior strategy assigns the same distribution (over available choices) to all moves in an information set. In short, a behavior strategy is a random variable which assigns to every move a distribution over available choices such that it is measurable with respect to the information partition H_i.

Clearly, if $b, b' \in \rho_i(X)$ and $c \in \text{supp}(b) \cap \text{supp}(b')$, then $\rho_i^{-1}(b) = P(c) = \rho_i^{-1}(b')$ implies $b = b'$, because ρ_i is a function. Hence, a behavior strategy assigns to every choice $c \in C^i$ at most one positive number. This allows for a simplified definition. One may think of a behavior strategy for player i as a vector $b_i = (b_i(c))_{c \in C^i}$ of non-negative numbers which satisfies

$$\sum_{c' \subseteq P^{-1}(P(c))} b_i(c') = 1, \text{ for all } c \in C^i. \tag{3.13}$$

The component $b_i(c)$ of the vector b_i is the largest number that a fixed behavior strategy ρ_i assigns to choice $c \in C^i$. By the previous argument, this largest number is unique.

Condition (3.13) is an alternative definition of a behavior strategy. If $c \in C^i$ is trivial (viz. if $P^{-1}(P(c)) = c$), then at every move $x \in P(c)$ there is *precisely one available* choice, namely c itself. In this case the choice $c \in C^i$ must be assigned probability 1. If move x belongs to some information set $h \in H_i$ of player i, then, by condition (3.5), the choices $c \in C^i$ that are available at x are the same for all moves x' in the information set $h = P(c) \in H_i$. In this case a behavior strategy assigns the *same* probability distribution over choices to *all* moves *in* the information set h.

While Definition 3.10 is conceptually the correct definition, it is notationally more convenient to work with the representation by a vector b_i which satisfies (3.13). Therefore, this representation by b_i will be adopted from now on.

Hence, a behavior strategy assigns probability distributions over (available) choices to information sets. These probabilities are the *conditional* probabilities of choices available at the information set, given that the information set is reached. But, of course, a behavior strategy also assigns probabilities over choices to information sets which cannot be reached, given what the behavior strategy assigns to earlier information sets. Like a pure strategy, a behavior strategy assigns probabilities over choices to *all* information sets. Still, using behavior strategies often simplifies the analysis.

The *set of all behavior strategies* for player i is denoted by B_i, for all $i = 1, ..., n$. Again, B_i is the "library" of all "books" giving probability distributions for choices at information sets on each of its pages. Since these probability distributions may always assign 1 to one choice (and 0 to the others) at each information set, this "library" contains the original "library of pure strategies" S_i. So, behavior strategies are yet another generalization of pure strategies. By condition (3.13) the space B_i can geometrically be thought of as a product of simplices (of various dimension), one for each information set.

A vector $b = (b_1, ..., b_n)$ consisting of one behavior strategy for each player is called a *behavior strategy combination* or a behavior strategy profile. Hence, the Cartesian product $B = B_1 \times ... \times B_n$ is the set of all behavior strategy combinations.

One way to think of behavior strategies is to view a player not as a single person, but as a *team of agents*. An "agent" ih of player i is an artificial player who controls exactly one of the information sets $h \in H_i$ of player i (and has trivial choices at all other moves) and has the same incentives as player i. The latter is ensured by giving agent ih the same payoff function as player i, i.e. $v_{ih}(w) = v_i(w)$, for all $w \in W$. If all information sets of player i are assigned to separate agents, player i can be dispensed with and substituted for by her team of agents. This is a device to *decentralize* the player's decisions.

If the game is played by the team of agents for each player, then the space B of behavior strategy combinations (of the game played by the original players) becomes the space of mixed strategy combinations for the game played by the agents. The following example illustrates.

Example 3.9 *In the 2-Marienbad game (Example 3.1) player 1 has four information sets and, thus, can be substituted for by a team of four agents. The space of all behavior strategies of player 1 in this game can, therefore, be identified with $\Delta^3 \times \Delta^1 \times \Delta^1 \times \Delta^1$, the product of a three-dimensional simplex with the unit cube. It has, thus, dimension 6, while the space of mixed strategies of player 1 has dimension 31. Player 2, in the 2-Marienbad game, controls three information sets, so her team consists of three agents. The space of all her behavior strategies has dimension 5, while her mixed strategy space has dimension 17. This illustrates that using behavior strategies can drastically reduce the dimension of the problem.*

In the noisy Stackelberg game (Example 3.5) firm $i = 1$ controls only a single information set, but firms $i = 2$ and 3 each control two information sets (corresponding to the two signals), so they can be split into two agents each. In the Cournot oligopoly (Example 3.7) there is no difference between players and agents, because each player is assigned only a single information set. In the bargaining game (Example 3.2) each player can be split into a continuum of agents.

To define the payoff function on behavior strategies, proceed as follows. For a given behavior strategy combination $b \in B$ the probability that node $y \in Y$ will be reached from its immediate predecessor $x = P(y) \in X$ is

$$\pi_b(y\,|x) = p(y) \prod_{i=1}^{n} b_i(c_i) \tag{3.14}$$

where $(c_i)_{i=1}^n \in \times_{i=1}^n C^i$ is the (unique) choice combination such that $y \in \cap_{i=1}^n c_i$. The realization probability of a play $w \in W$ is, therefore, given by $\pi_b(w) = \prod_{w \in y} \pi_b(y\,|P(y))$, where again $\pi_b(W\,|P(W)) = 1$ for the root (because by definition $P(W) = W$).

With these probabilities specified, the *expected payoff* function on behavior strategies $U : B \to \mathbb{R}^n$ is given by

$$U_i(b) = \sum_{w \in W} \pi_b(w) v_i(w) \tag{3.15}$$

for all $i = 1, ..., n$.[16] Formally it resembles the expression for pure strategies, but it involves taking expectations with respect to randomized strategies, as under mixed strategies. Now, however, the randomization is local at each information set, in accordance with the piecemeal approach taken by behavior strategies.

---

[16] The same symbol $U(.)$ as under mixed strategies is used here to emphasize that randomization is involved, and because no confusion can arise if the argument of the function is specified.

Summary

- A pure strategy of a personal player assigns to every information set of this player precisely one of the choices available at (moves in) this information set.

- A mixed strategy for a personal player is a probability distribution over her pure strategies.

- A behavior strategy of a personal player assigns to every information set of this player a probability distribution over the choices available at (moves in) this information set.

- The set of mixed (behavior) strategies of a personal player can be identified with a compact convex subset of a Euclidean space. The same holds for the space of mixed (behavior) strategy combinations.

Exercises

Exercise 3.17 Reconsider the Centipede game from Exercise 3.9 with $K = 4$. Write down all pure strategies for player 1 (resp. 2). What is the dimension of the mixed strategy space and what is the dimension of the space of behavior strategies for player 1 (resp. 2)?

Exercise 3.18 If a player in an extensive form has k information sets and at information set h has $m_h > 1$ (nontrivial) choices available, how many pure strategies does she have?

Exercise 3.19 Consider two firms serving the same market represented by an inverse demand function $p(q) = \max\{0, 3 - q\}$, giving the market price as a function of aggregate supply $q = q_1 + q_2$. Firms try to maximize profits $\pi_i(q_1, q_2) = p(q)q_i - q_i$ where the cost function is $c_i(q_i) = q_i$, for $i = 1, 2$. Their quantity choices are constrained to integer quantities, i.e. $q_i \in \{0, 1, 2, 3\}$. Firm $i = 1$ is the leader and chooses q_1 first. Firm $i = 2$ learns the leader's choice and moves in response to it. Draw the extensive form and determine the number of pure strategies for firm $i = 2$.

Exercise 3.20 Consider a duel where two players are located 21 steps apart and each of them has a gun with one shot in it. They walk towards each other and at each step (steps are taken simultaneously) each of them may either shoot or wait (for the next step). The decisions on whether or not to shoot are taken simultaneously. If one shoots, her shot is wasted, but the opponent may be killed with probability $\rho(k)$ which depends on the contestants' remaining distance $k = 1, 3, ..., 21$. If one of the two is killed, the game terminates. If after ten steps by each neither of the two is dead, they both go home alive, but with the embarrassment of having been bad marksmen. The probability of hitting satisfies $0 \leq \rho(k+2) < \rho(k) \leq 1$ for all $k = 1, ..., 21$. Sketch the extensive form. How many pure strategies does each player have? What is the dimension of the space of behavior strategies for player 1 (resp. 2)? How would you modify the extensive form to account for the fact that player 1 is a better marksman than player 2? Let the payoff from being killed be -1, that from killing the opponent be $+1$, and that from both walking home alive after ten steps without anyone being hit be 0. Calculate the expected payoff for some behavior strategy combinations.

Exercise 3.21 Consider a two-player extensive form where each player controls one information set with two available choices. So, both players have two pure strategies and there are four pure strategy combinations $s = (s_1, s_2) \in S = S_1 \times S_2$. Can the probability distribution $\varphi(s_1^1, s_2^1) = 1/3$, $\varphi(s_1^1, s_2^2) = 1/9$, $\varphi(s_1^2, s_2^1) = 2/9$, and $\varphi(s_1^2, s_2^2) = 1/3$ on pure strategy combinations be induced by a mixed strategy combination?

Notes on the Literature: The idea of a pure strategy originates in Borel ([29]; see also von Neumann [133]). Pure and mixed strategies are systematically introduced by von Neumann and Morgenstern ([134], Chapters 11 and 17). The metaphor of "books" and "libraries" is taken from Luce and Raiffa ([105], p. 159) who attribute it to Harold Kuhn. The interpretation of mixed strategies as frequencies of pure strategies in large populations was first suggested by Nash [130] and has since inspired evolutionary game theory (see Ritzberger and Weibull [152]); for textbooks on evolutionary game theory, see Hofbauer and Sigmund [89], Weibull [179], Vega-Redondo [176], or Samuelson [158]. The interpretation of mixed strategies as beliefs is due to Aumann [7].

3.4   Perfect Recall

No-Absent-Mindedness   Though the extensive form is a model of rational decision making in strategic situations, it allows for a modeling of a much wider variety of phenomena. In particular, the extensive form in the following example is covered by the definition.

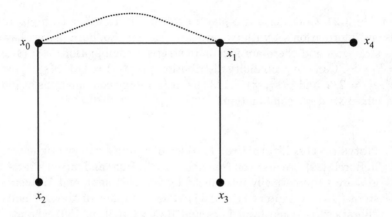

Figure 3.6: Drunken driver

Example 3.10 *Figure 3.6 illustrates the following single-player game. An individual is sitting in a bar planning her trip home (located at $x_3 = \{w_2\}$). To get home she has to take the highway and leave it at the second exit ($x_1 = \{w_2, w_3\}$). Turning at the first exit leads into a dangerous area ($x_2 = \{w_1\}$). If she continues beyond the second exit, she cannot return and will have to spend the night at a motel ($x_4 = \{w_3\}$). The individual is absent-minded and aware of this fact. At an intersection she cannot tell whether it is the first or the second, and she cannot remember how many intersections she has passed.*

Figure 3.6 depicts this situation by assigning to the (single) player an information set (represented by the dotted curve) which encompasses both moves of the tree. The two choices of the player are $\{x_1, x_4\}$ ("continue") and $\{x_2, x_3\}$ ("exit"). Clearly, this does not model a decision by a rational player. It models an "absent-minded" driver, who keeps forgetting her previous decisions.

This example is inconsistent with an expected utility decision theory. At an information set the moves (contained in it) count as "states". A fundamental presumption about states is that which state obtains does *not* depend on the decision maker's choice. But in Example 3.10 it does. By choosing "exit", the decision maker can prevent "state" x_1 from materializing.

Beyond illustrating that extensive forms can model decision problems of players who are not perfectly rational, this example raises the following issue. What precisely is required to make an extensive form represent decision problems of rational players?

The key to an answer is to recall what the tree structure under perfect information imposes on choices. Under perfect information every choice consists of precisely one node. Therefore, choices inherit important properties of the tree. In particular, under perfect information the collection

$$(W(c))_{c \in C_i} = (\{w \in W \mid \exists x \in c : w \in x\})_{c \in C_i}$$

of sets of plays passing through nontrivial choices is itself a *tree* for each player i. (More precisely, it is order isomorphic[17] to a tree, albeit to an unrooted tree.) Hence, any succession of choices constitutes a refinement of the selection among plays. Under the general definition of an extensive form with imperfect information (Definition 3.4), this is not necessarily so.

In particular, under imperfect information, there is no guarantee that choices are indeed (partially) ordered. To develop a notion of what a "succession of choices" may mean, a relation like "comes before" needs to be defined on *sets* of nodes. For a personal player i and two nontrivial choices $c, c' \in C_i$ (or two information sets $h, h' \in H_i$) of this player, say that c *comes before* c' (or h comes before h'), denoted $c <_i c'$ (or $h <_i h'$), if there are $x \in c$ and $x' \in c'$ (or $x \in h$ and $x' \in h'$) such that x comes before x'; i.e.,

$$c <_i c' \text{ if there are } x \in c \text{ and } x' \in c' \text{ such that } x' \subset x \qquad (3.16)$$

(and likewise for information sets). This constitutes a natural extension of set inclusion on nodes to sets of nodes (choices and information sets of a player). But with this definition in Example 3.10 the choice "continue" ($\{x_1, x_4\}$) comes before itself! Yet, for a relation to be a partial ordering it must be irreflexive; i.e., there must not be any element that "comes before itself". That a choice comes before itself means, roughly, that the player (to whom this choice belongs) is not aware that she is taking an action. It does *not* involve forgetting something, because "forgetting" means that something has disappeared from the memory that has been there previously. But if the term "previously" does not have a meaning because the relation $<_i$ is not a partial ordering, then "disappearance of *past* information" does not have a meaning. Indeed, in Figure 3.6 there is no change in the information state of the player between x_0 and x_1, because both moves belong to the same information set.

A choice that comes before itself is against the essence of an extensive form. After all, an extensive form uses the tree representation as a means to

---

[17] An order isomorphism between two partially ordered sets is a bijection (one-to-one and onto function) which respects the orderings on the two sets.

distinguish between past, present, and future. If this distinction is obscured by a choice that comes before itself, why then use a tree at all? Indeed, if a choice comes before itself, it can be shown that there is a play $w \in W$ which *cannot* be reached under *any* pure strategy combination.

Therefore, to model rational players, extensive forms like in Figure 3.6 must be ruled out. And indeed, such extensive forms have traditionally been ruled out by making the following assumption part of the definition of an extensive form.

Definition 3.11 *In an extensive form the set of nontrivial choices C_i (or the information partition H_i) of player i satisfies* **no-absent-mindedness** *if no play passes through a nontrivial choice (or an information set) more than once, i.e. if*

$$x, y \in c \ (or \ x, y \in h) \ and \ x \cap y \neq \emptyset \ imply \ x = y \qquad (3.17)$$

for all $c \in C_i$ (or for all $h \in H_i$).[18] The entire extensive form satisfies no-absent-mindedness if C_i (or H_i) satisfies no-absent-mindedness for all players $i = 1, ..., n$.

This condition ensures that all plays can be reached by some pure strategy combination. But it does more. It effectively ensures that mixed strategies do at least as well as behavior strategies. A proof of the following result is given in the appendix to this section.

Theorem 3.2 *An extensive form satisfies no-absent-mindedness if and only if for every probability distribution on plays that is induced by some behavior strategy combination there is a mixed strategy combination that induces the same distribution.*

Perfect Recall   Unfortunately, no-absent-mindedness is not enough to guarantee sensible choice sets. The following two examples illustrate phenomena which are allowed under Definition 3.4 and satisfy Definition 3.11, but are inconsistent with rational players.

Example 3.11 *Figure 3.7 depicts two single player extensive forms where the player is forgetful. (Moves are labeled with the identity of the player who controls them.) In part (a) the player ($i = 1$) controls all moves in the tree, but at the immediate successors of the root $x_0 = W$ she has forgotten what she has chosen at the root, depicted by a broken line connecting $x_1 = \{w_1, w_2\}$ and $x_2 = \{w_3, w_4\}$, which indicates an information set.*

Part (b) depicts a stylized version of bridge, where one party consisting of two "players" is treated as a single player ($i = 1$), and the other team is

---

[18]   Recall that by (3.1) $x \cap y \neq \emptyset$ implies either $x \subseteq y$ or $y \subset x$.

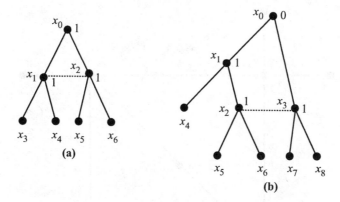

Figure 3.7: Forgetful decision makers

ignored in the representation. Bridge is often given as an example in which a player (viz. a team of two) is forgetful, because during the bidding process each of the two partners knows her own hand, but not the hand of the partner. Here, a very stylized version is given. The partner who opens the bidding can get only two possible hands (a chance move at the root $x_0 = W$). One of those hands does not leave her any choice but to pass (so she has no decision point after being allocated this hand). The other hand ($x_1 = \{w_1, w_2, w_3\}$) allows her either to pass ($x_2 = \{w_2, w_3\}$) or to bid a contract ($x_4 = \{w_1\}$). Her partner, who controls the information set depicted by the broken line, sees the bid (pass or contract), but not the hand on which the opening bid was based. If a contract was bid, she will have to pass, but if the opening bid was a pass, she may now either bid a contract ($\{x_5, x_7\}$) or also pass ($\{x_6, x_8\}$).

The two examples in Figure 3.7 differ with respect to what the player forgets. In part (a) she forgets what she *has done* at a previous move. In part (b) she forgets something she *has known* at a previous move. Both, however, involve a deficiency of memory, because information that the player possessed disappears along some play. Yet, in both cases no-absent-mindedness (Definition 3.11) is fulfilled, and information sets are partially ordered.

An even worse example is depicted in Figure 3.8. Again this is a single player game with a chance move at the root $x_0 = W$ (in the middle) and the two dotted lines indicate information sets h and h' of (the only personal) player 1. (Choices c_k and c'_k, for $k = 1, 2$, are indicated by labeling the edges leading towards them.) In this extensive form not only does the player forget something she knew, but on top of that the two information sets (and, thereby, choices) *cannot* be ordered by the relation $<_1$ from (3.16).

All these examples illustrate that the definition of choices needs to be restricted if rational players are to be modeled. The key to the appropriate restrictions is to recall the structure that choices inherit from the tree under

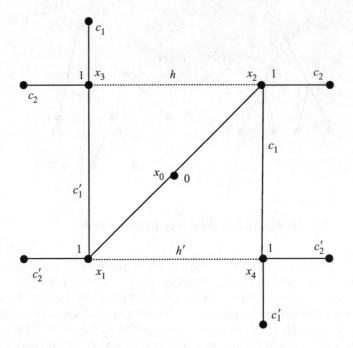

Figure 3.8: Unordered information sets

perfect information—that any succession of choices translates into a refinement of the selection among plays. After all, this is the reason to call them "choices".

Since imperfect information is meant to restrict what players know, but not to obscure what the choices are, the most straightforward way is to impose on choices under imperfect information the very same requirement. To do so, we need to look at the union of the nodes in a choice (viz. a set of plays), referred to as the set of plays *passing through* a choice,

$$W(c) = \{w \in W \,|\, \exists x \in c : w \in x\} = \cup_{x \in c} x, \qquad (3.18)$$

for all $c \in C_i$, for $i = 1, ..., n$, and require that those satisfy a tree property. This leads to a concept known as "perfect recall", which is almost invariably assumed in applications of game theory.

Definition 3.12 *In an extensive form the set of nontrivial choices C_i of player i satisfies* **perfect recall** *if every succession of choices translates into a refinement of the selection among plays, i.e. if*

$$c <_i c' \text{ implies } W(c') \subseteq W(c), \text{ for all } c, c' \in C_i. \qquad (3.19)$$

The extensive form satisfies perfect recall if the sets C_i satisfy perfect recall for all players $i = 1, ..., n$.

It is easy to see that this is equivalent to requiring that the system $(W(c))_{c \in C_i}$ satisfies (3.1) for each player $i = 1, ..., n$ (see Exercise 3.26 below); i.e., it is a tree,[19] though not a rooted one.

Clearly, perfect recall is stronger than no-absent-mindedness. For, if $x, x' \in c \in C_i$ and $x' \subseteq x$, then either $x = x'$ (i.e. no-absent-mindedness holds) or $x' \subset x$. In the latter case both $P(x)$ and $y' = P(x')$ are elements of the same information set $h = P(c)$. Because c is nontrivial, there is $c' \in C_i$, $c' \neq c$, available at h and there is $y \in c' \cap P^{-1}(y')$ such that $y \subset x$. But this means that $c <_i c'$ holds. Yet, there is a play which, whenever it intersects h, continues in c' and, therefore, never intersects c (because $C_i \subseteq C^i$ and C^i is a partition). Hence, $W(c') \subseteq W(c)$ cannot hold, despite $c <_i c'$, in violation of perfect recall.

On the other hand, the examples in Figures 3.8 and 3.9 satisfy no-absent-mindedness, but violate perfect recall. So, no-absent-mindedness is considerably weaker than perfect recall. Indeed, perfect recall rules out all examples in Figure 3.6 to 3.8. (You may wish to convince yourself about that.)

The name "perfect recall" may seem somewhat surprising, given that it is defined in terms of choices. Indeed, traditionally perfect recall is defined differently. The traditional definitions are rather complicated and will not be given here. They are usually argued in terms of a memory requirement, and this is where its name comes from. The usual explanation is that perfect recall

> ... is equivalent to the assertion that each player is allowed by the rules of the game to remember everything he knew at previous moves and all of his choices at those moves. ([104], p. 213)

Indeed, it can be shown that perfect recall implies that players never forget what they knew or what they did. But it does a little more. Consider the one-player extensive form in Figure 3.9. Again, it starts with a chance move at the root $x_0 = W$ (in the middle) and the single player 1 has two information sets ($\{x_2, x_3, x_4\}$ and $\{x_1, x_5, x_6\}$), each containing three moves, depicted by dotted lines.

The story that goes with Figure 3.9 is as follows. Player 1 is confronted with two panels, each with two buttons (c_1 and c_2 on the left and c'_1 and c'_2 on the right panel). The two panels get serially connected to an electric circuit by a chance move (at the root) which the player cannot observe. The player is then asked to press one button on each of the two panels. The result will depend on the ordering of the two panels in the serial connection and on which combination of buttons the player presses. For example, if the circuit is

---

[19] Again, more precisely, it is order isomorphic to a tree, because singleton sets of plays may not belong to it.

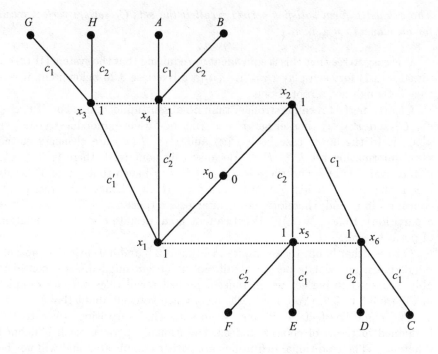

Figure 3.9: "Rational" failure of perfect recall

switched from the left to the right and the player presses c_1 on the left and c_2' on the right panel, the outcome will be D. If the circuit is from right to left, the same combination of buttons will result in A, and so on. The outcomes will be learned by the player only after she has pressed buttons on *both* panels. Hence, she possesses precisely the same information at both of her information sets.

This extensive form fails perfect recall. Yet, the story is consistent with the player being rational. Indeed, the player in this game forgets neither what she knew nor what she did. She simply learns the consequences of her actions when it is too late. What goes wrong in this example is that choices (and information sets) are *not* partially ordered by the relation $<_1$. Perfect recall rules that out, because it also implies that choices (and information sets) are partially ordered by the relation "comes before" on choices (and information sets).[20]

---

[20]   To be sure, *partial* ordering means that, if there are *no* two choices $c, c' \in C_i$ such that $c <_i c'$, then C_i is partially ordered. If between any two (distinct) elements the relation has to hold (one way or the other), then the ordering would be *complete* (see Chapter 2.1).

Kuhn's Theorem   Hence, perfect recall not only is a memory requirement but also supplies a notion of past, present, and future, which is necessary to make sense of something temporal like a memory. In fact, perfect recall can be characterized by the simultaneous satisfaction of the following three independent conditions:

(P1) Players never forget what they did.[21]

(P2) Players never forget what they knew.[22]

(P3) There is an unambiguous meaning to past, present, and future. (The relation $<_i$ is a partial ordering.)

Imposing perfect recall yields extensive forms which are consistent with the rationality of players and which lend themselves easily to applications. Working with such extensive forms has a remarkable consequence which supports the intuition that for a rational player it should not matter whether she plans her strategy in advance or executes it during a playing of the game. The following theorem is known as Kuhn's Theorem. (A proof is in the appendix to this section.)

Theorem 3.3 *An extensive form satisfies perfect recall if and only if, for every probability distribution on plays that is induced by some mixed strategy combination, there is a behavior strategy combination that induces the same distribution.*

The above is a slight understatement of what it really says. Since perfect recall implies no-absent-mindedness, under the former the (set of) probability distributions on plays induced by mixed strategies *coincides precisely* with those induced by behavior strategies. Kuhn's theorem effectively allows us to work in an analysis of perfect recall extensive form games with the team of agents rather than with players. Because the space of behavior strategies is lower-dimensional than the space of mixed strategies, this greatly simplifies the analysis.

Note the difference between Theorems 3.2 and 3.3. The first says that under no-absent-mindedness mixed strategies do at least as well as behavior strategies. The second says that under perfect recall behavior strategies do at least as well as mixed strategies. Both theorems, however, are characterizations in the sense of giving conditions that are simultaneously necessary and sufficient.

From now on perfect recall will always be assumed. One may think of it as a part of the definition of an extensive form game played by rational players.

---

[21]   The notion that a player never forgets what she did is traditionally formalized by the concept of "complete inflation", due to Dalkey [35].

[22]   Formally: $h <_i h'$ implies $W(h') \subseteq W(h)$, for all $h, h' \in H_i$ and all $i = 1, ..., n$ (see Ritzberger [151]).

Summary

- An extensive form satisfies no-absent-mindedness if every play passes through any choice (or information set) at most once.

- Mixed strategies do at least as well as behavior strategies if and only if the extensive form satisfies no-absent-mindedness.

- An extensive form satisfies perfect recall if every succession of (nontrivial) choices translates into a refinement of the selection among plays.

- Intuitively, perfect recall means that players never forget what they did, or what they knew, and that past, present, and future have an unambiguous meaning.

- Kuhn's theorem states that behavior strategies do at least as well as mixed strategies if and only if the extensive form satisfies perfect recall.

Appendix

To demonstrate Theorems 3.2 and 3.3, the following definitions are helpful. For a node $x \in N$, let $R_i(x)$ be the set of pure strategies $s_i \in S_i$ of player i such that there exists a pure strategy combination $s \in S$ with ith component s_i and $\sum_{w \in x} \pi_s(w) > 0$, i.e. those pure strategies of i under which x can be reached. For any choice $c \in C^i$ let $S_i(c) = \{s_i \in S_i \,|\, c \in s_i(X)\}$, i.e. those pure strategies of i which actually pick c. Then, given a mixed strategy $\sigma_i \in \Delta_i$, by Bayes' rule the conditional probability that choice $c \in C^i$ will materialize given that $x \in P(c)$ is reached is

$$a_i(c\,|\,x) = \frac{\sum_{s_i \in R_i(x) \cap S_i(c)} \sigma_i(s_i)}{\sum_{s_i \in R_i(x)} \sigma_i(s_i)} \qquad (3.20)$$

if this quantity is defined, and $a_i(c\,|\,x) = \sum_{s_i \in S_i(c)} \sigma_i(s_i)$ otherwise.

Proof of Theorem 3.2 ("if") Suppose (3.17) fails, i.e. that there is $c \in C_i$ such that $c <_i c$. Let $y, y' \in c$ be such that $y' \subset y$. Since $c \in C_i$ is nontrivial, there is $c' \neq c$, $c' \in C_i$, with $P(c') = P(c) \in H_i$. Let $w \in P(y') \cap W(c')$ be a play that passes through $P(y')$ and c'. Any behavior strategy combination $b \in B$ which satisfies $b_j(\widehat{c}) > 0$, for all $\widehat{c} \in C^j$ and all $j = 1, ..., n$, will induce a positive probability of w to materialize, $\pi_b(w) > 0$. But for every pure strategy combination $s \in S$, either $s_i(P(y')) = c$ (so every play in $P(y')$ will be in $W(c)$ rather than in $W(c')$) implies $\pi_s(w) = 0$, or $s_i(P(y)) \neq c$ (so $P(y')$ cannot be reached) implies $\pi_s(w) = 0$. (See Figure 3.10(a), where the dotted line indicates the choice c.) Since by (3.7) these are the only two possibilities,

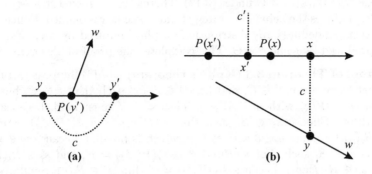

Figure 3.10: Illustrations of proofs

$\pi_s(w) = 0$, for all $s \in S$. It follows that $\pi_\sigma(w) = \sum_{s \in S} \sigma(s)\pi_s(w) = 0$, for all mixed strategy combinations $\sigma \in \Theta$. Hence, if every distribution on plays induced by behavior strategies can be reproduced by mixed strategies, no-absent-mindedness (3.17) must hold.

("only if") Suppose for some $c \in C_i$ there is $x \in P(c)$ such that $R_i(x) \cap S_i(c) = \emptyset$, i.e., $s_i(x) = c$ implies $s_i \notin R_i(x)$. This is possible only if there is $x' \in P(c)$ such that $x \subset x'$ and $c <_i c$, because otherwise one can always choose the choices on the path to x without violating the constraint $s_i(x') = c$, for all $x' \in P(c)$. Hence, (3.17) is violated. In other words, if no-absent-mindedness holds, then $R_i(x) \cap S_i(c)$ is always nonempty, for all $x \in P(c)$, and all $c \in C_i$.

Assume no-absent-mindedness. Given some behavior strategy $b_i \in B_i$, construct a mixed strategy $\overline{\sigma}_i \in \Delta_i$ from it as follows:

$$\overline{\sigma}_i(s_i) = \prod_{h \in H_i} b_i(s_i(h)), \text{ for all } s_i \in S_i \qquad (3.21)$$

where $s_i(h) = \cup_{x \in h} s_i(x)$ is a single choice by condition (3.7). Since $\overline{\sigma}_i$ is non-negative and

$$\sum_{s_i \in S_i} \overline{\sigma}_i(s_i) = \sum_{s_i \in S_i} \prod_{h \in H_i} b_i(s_i(h)) = \prod_{h \in H_i} \left(\sum_{c \subseteq P^{-1}(h)} b_i(c) \right) = 1$$

from condition (3.13), it does indeed define a mixed strategy. Substituting (3.21) into (3.20) yields, for $x \in P(c)$,

$$a_i(c|x) = \frac{b_i(c) \sum_{s_i \in R_i(x)} \prod_{h \in H_i \backslash P(c)} b_i(s_i(h))}{\sum_{c' \subseteq P^{-1}(P(c))} b_i(c') \sum_{s_i \in R_i(x)} \prod_{h \in H_i \backslash P(c)} b_i(s_i(h))} = b_i(c)$$

because $R_i(x) \cap S_i(c) \neq \emptyset$ under (3.17). Therefore, the mixed strategy defined in (3.21) induces the behavior strategy from which it was defined. Hence, under no-absent-mindedness any distribution on plays induced by some $b \in B$ can be reproduced by some $\sigma \in \Theta$. This completes the proof of Theorem 3.2.

Proof of Theorem 3.3 (Kuhn's theorem)   ("if") Suppose (3.19) fails, i.e. that there are $c, c' \in C_i$ and $(x, x') \in c \times c'$ such that $x \subset x'$, but there also is $w \in W(c)$ with $w \notin W(c')$. Then $c \neq c'$ must hold, because C^i is a partition. Choose $s_i \in S_i$ such that $s_i(P(x)) = c$, $s_i(P(x')) \neq c'$, and $s_i \in R_i(y)$ for some $y \in c$ with $w \in y$, which is possible because $w \notin W(c')$. Choose $s_i' \in S_i$ such that $s_i'(P(x)) \neq c$, $s_i'(P(x')) = c'$, and $s_i' \in R_i(P(x))$. Then $s_i \notin R_i(P(x))$, because $s_i(P(x')) \neq c'$, but $s_i \in S_i(c)$. Similarly, $s_i' \in R_i(P(x))$, because $s_i'(P(x')) = c'$, and $s_i' \notin S_i(c)$. (See Figure 3.10(b), where the dotted lines indicate the choices c and c'.) Choose $\sigma_i \in \Delta_i$ such that $0 < \sigma_i(s_i) = 1 - \sigma_i(s_i') < 1$.

Then, using (3.20), $a_i\,(c\,|P(x)\,) = 0$ follows from $s_i \notin R_i(P(x))$, $s_i' \notin S_i(c)$, and $s_i' \in R_i(P(x))$. Since $s_i(P(x)) = c$ implies $s_i(P(y)) = c$ and $s_i'(P(x)) \neq c$ implies $s_i'(P(y)) \neq c$ from condition (3.7), $s_i \in R_i(y)$ and $s_i \in S_i(c)$ imply $a_i\,(c\,|P(y)\,) > 0$. Since this probability distribution (induced by $\sigma_i \in \Delta_i$ over choices available at $h = P(c) \in H_i$ via (3.20)) conditions on the particular moves in the information set h, it cannot be induced by a behavior strategy. Therefore, if any distribution on plays induced by mixed strategies can be duplicated by behavior strategies, perfect recall must hold.

("only if") Assume perfect recall. Then $c' <_i c$ implies $W(c) \subseteq W(c')$, for all $c, c' \in C_i$. Suppose there is $y \in c$ such that there is *no* $y' \in c'$ with $y \subset y'$. Then from $y \subseteq W(c')$ it follows that there is $y' \in c'$ with $y' \subset y$, because $y' = y$ is impossible, since perfect recall implies no-absent-mindedness. Since $c' \in C_i$, there is $c'' \in C_i$, $c'' \neq c'$, available at $P(y')$ and a play $w \in y \cap P(y') \cap W(c'')$, hence, $w \in W(c)$, but $w \notin W(c')$. Since this contradicts $W(c) \subseteq W(c')$, it follows that for *all* $y \in c$ there is $y' \in c'$ such that $y \subset y'$.

Now consider two moves $x, x' \in P(c)$ for $c \in C_i$. If there is $s_i \in R_i(x)$ with $s_i \notin R_i(x')$, then there must be some $c' \in C_i$ such that $c' <_i c$ and $c' \notin s_i(X)$. Then perfect recall implies that $W(c) \subseteq W(c')$ and the above argument implies that there is $y' \in c'$ such that $x \subseteq y'$. But the latter implies, from $c' \notin s_i(X)$, that $s_i \notin R_i(x)$, a contradiction. Thus, $R_i(x) \subseteq R_i(x')$ and, reversing the roles of x and x', $R_i(x) = R_i(x')$, for all $x, x' \in P(c)$. It follows from (3.20) that under perfect recall $a_i\,(c\,|x\,) = a_i\,(c\,|x'\,)$, for all $x, x' \in P(c)$, and, hence, a_i is a behavior strategy. This completes the proof of Kuhn's theorem.

Exercises

Exercise 3.22 Show that the extensive form in Figure 3.9 fails perfect recall, but satisfies no-absent-mindedness.

Exercise 3.23 Show that any extensive form with perfect information satisfies perfect recall.

Exercise 3.24 Show that the set of nontrivial choices C_i of player i in an extensive form satisfies perfect recall if and only if, for every move $x \in h$ in an information set $h \in H_i$ of player i, the ordered set of information sets and choices preceding move x (in the sense of $<_i$) is the same as for any other move $x' \in h$, for all $h \in H_i$.

Exercise 3.25 Show that the set of nontrivial choices C_i of player i in an extensive form satisfies perfect recall if and only if, for all moves $x \in h \in H_i$, for all choices $c \in C_i$ available at x, and for all $y, y' \in X_i$,

$$\text{if } y, y' \in h' \text{ and there is } z \in c \cap P^{-1}(x) \text{ such that } y \subseteq z,$$
$$\text{then there is } x' \in h \text{ and } z' \in c \cap P^{-1}(x') \text{ such that } y' \subseteq z'$$

(i.e. it is, in fact, Selten's [163] definition of perfect recall).

Exercise 3.26 Show that, if a (finite) extensive form satisfies perfect recall, then for each player i the collection $(W(c))_{c \in C_i}$ satisfies (3.1) and vice versa.

Notes on the Literature: Example 3.10 is taken from Piccione and Rubinstein [141], who also introduced the term "no-absent-mindedness". The condition itself was part of the definition of an extensive form in Kuhn's [104] original paper. Theorem 3.2 is essentially new, though a closely related statement appears in Piccione and Rubinstein ([141], Proposition 1). The traditional definitions of perfect recall are by Kuhn [104], where also the original proof of Kuhn's theorem appears ([104], Theorem 4), and later by Selten [163]. That the current definition of perfect recall is equivalent to those is shown by Ritzberger [151], where also the decomposition into (P1)-(P3) is demonstrated. That complete inflation ("players do not forget what they did") is implied by perfect recall is noted by Dalkey [35]. Okada [136] has shown that for a certain class of games complete inflation even characterizes perfect recall.

3.5   Incomplete Information

The maintained presumption of non-cooperative game theory is that it studies exclusively games of *complete information*. This means that the extensive
form (the rules of the game) and the payoff function (the incentives of players) are *common knowledge*: everybody knows this data, everybody knows
that everybody knows, everybody knows that everybody knows that everybody knows, and so on ad infinitum. This assumption appears acceptable for
the extensive form (the rules), but demanding when it comes to the payoff
function (the incentives). With respect to the latter, realism suggests incomplete information.

But if the payoff function is not known, how will a player evaluate what
her opponents will do? It is not enough to form beliefs about the opponents'
incentives, because what the opponents will do depends on what they expect
about the incentives of the first player. This requires beliefs about beliefs and
seems to lead into an infinite regress.

Harsanyi Transformation   The extensive form, however, provides ways to
cut the Gordian knot. Harsanyi [79] has introduced a device which translates
situations of incomplete information into games of complete, though imperfect, information. The idea is as simple as it is ingenious. List all games, viz.
payoff functions,[23] that players consider possible. Let an initial chance move
select one of those. Then every player learns her own (component of the) payoff function, but not those of the opponents. Hence, a larger "meta-game" has
been constructed which reflects the players' uncertainty about the incentives
of other players. This meta-game is now assumed to be common knowledge
and, therefore, is a game with complete, but imperfect, information.

This transformation is known as the *Harsanyi transformation*. Sometimes,
the complete information game *after* the Harsanyi transformation is referred
to as a "game of incomplete information", if it needs to be emphasized that
one tries to model incomplete information. One should not, however, get con-
fused by such a liberal use of language: The Harsanyi transformation yields a
complete information game with imperfect information.

Example 3.12 *Consider the following court case. A plaintiff alleges that the
defendant was negligent in providing adequate safety at a chemical plant, a
charge which the plaintiff estimates is true with some probability p. The plaintiff files suit, but the case is not decided immediately. In the meantime, the
defendant and the plaintiff can settle out of court.*

---

[23]   Harsanyi [79] has also described a device which translates incomplete information
about the rules (viz. the tree) into uncertainty about the payoff function. For a formal
execution of Harsanyi's program see Mertens and Zamir [116].

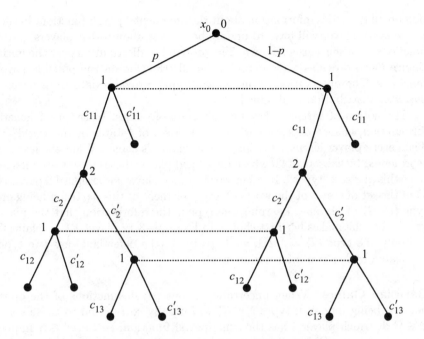

Figure 3.11: Court case

This appears as a situation of incomplete information, because the plaintiff does not know whether the defendant is liable or blameless. On the other hand, the defendant faces no such uncertainty, because she knows her own safety provisions and she knows the damages (and possible gains from winning the case) for the plaintiff. Figure 3.11 depicts the extensive form that models this. At the root ($x_0 = W$) a chance move chooses whether the defendant is liable or not. The plaintiff (player 1) does not learn that, hence the information set depicted by the dotted line. In ignorance about the chance move, the plaintiff then decides whether to sue (c_{11}) or take no action at all (c'_{11}). In the latter case the game ends. If she sues, the defendant (player 2), who knows about the chance move, can offer an out of court settlement (c_2) or hold out (c'_2). In the first case the plaintiff, still ignorant about the chance move, can settle (c_{12}) or refuse (c'_{12}); in the second case she can try (c_{13}) or drop the case (c'_{13}). The three (nondegenerate) information sets of the plaintiff represent her uncertainty about whether she is playing "the left" or "the right" part of the tree.

This example is particularly simple, because the issue of beliefs about beliefs does not show up if only one of the players faces uncertainty. If several players face incomplete information, the observation of one's own payoff function possibly carries information about the opponents' payoff function. In such a case every player will have to update her beliefs about other players' payoff functions once she learns her own. The probability distribution over the initial chance move provides a (common) *prior* distribution on the possible payoff functions. The observation of one's own "type" provides data from which a *posterior* distribution is obtained.

The notion of a *type* is key for this class of games. In the most general formulation, chance chooses initially from a set of "states of the world" Ω. Then every player learns something about chance's choice. What she learns is represented by subsets of Ω, where for a fixed player these subsets are referred to as this player's "types". In other words, every player i is assigned a partition T_i of the set of states of the world Ω, every element of this partition being one type $t_i \in T_i$ of player i. Learning the type is the information that the player gets before she makes her first choice. In Example 3.12 player 1 (the plaintiff) has only one type ($T_1 = \{\Omega\}$), while player 2 (the defendant) has two types (indicated by her two information sets).

Bayesian Games When uncertainty about payoff functions of the opponents is being modeled, types $t_i \in T_i$ will usually correspond to those states $\omega \in \Omega$ at which player i has the same payoff function; i.e., $\omega, \omega' \in t_i$ implies $v_i(w, \omega) = v_i(w, \omega')$, for all $w \in W$ and all $t_i \in T_i$. Since a type of player i is a subset of Ω, it constitutes an event, and the player can update the probability of any other event, in particular that her opponents' payoff functions will take a certain value. But, of course, types can be specified more generally.

Because players update their prior beliefs by Bayes' rule once they learn their types, it has become customary to refer to the "meta-games" that result from the Harsanyi transformation as *Bayesian games*. The following definition is a bit of a hairsplitting, because it is already covered by the definition of an extensive form (Definition 3.4).

Definition 3.13 *An n-player **Bayesian game**, or an n-player incomplete information game after the Harsanyi transformation, is a four-tuple (F, T, v, μ) where*

- *F is an **extensive form**,*

- *$T = (T_1, ..., T_n)$ consists of n partitions T_i, the **type-sets**, of a common set Ω of **states**,*

- *$v : W \times \Omega \to \mathbb{R}^n$ is the **payoff function** depending on plays $w \in W$ and on states $\omega \in \Omega$, and*

- *$\mu : \Omega \to \mathbb{R}_{++}$ is a probability distribution, $\sum_{\omega \in \Omega} \mu(\omega) = 1$, on states, the system of **prior beliefs**.*

The interpretation of a Bayesian game is as follows. Initially (at the root of the "meta-tree") chance will choose a state $\omega \in \Omega$. Then each player learns her type $t_i(\omega) = \{t_i \in T_i \mid \omega \in t_i\}$; i.e., she reaches an information set which consists of the states in $t_i(\omega)$. Thus, states correspond to immediate successors of the root (of the "meta-tree") and types to sets of those. At each such state the tree of F rises. Choices of player i in the new extensive form are the unions of the choices in F over states in t_i, for all $t_i \in T_i$ and all $i = 1, ..., n$. (At the root of the enlarged extensive form all players have a trivial choice encompassing all states.) Therefore, in general, a Bayesian game has no proper subgames.

Payoffs in a Bayesian game depend on states. This offers considerable flexibility to model uncertainty not only about the motivation of opponents, but also about the rules of the game. Indeed, if you wish to model a situation where some player believes that she has a choice (at some information set in F) when she actually does not, then all you need to do is to make the payoff function for those states where she actually has no choice constant in the choices which she only believes she has. The tree here serves as a model of how players *perceive* the situation, while the payoff function captures what actually matters.

Since each T_i is a partition of Ω, realizations of different *type combinations* $t = (t_1, ..., t_n) \in T = T_1 \times ... \times T_n$ are mutually exclusive events. Hence, upon observing her own type, a player can update her beliefs about the opponents' types by Bayes' rule.[24] Let the vector of types of all players except player i be denoted by $t_{-i} = (t_1, ..., t_{i-1}, t_{i+1}, ..., t_n) \in T_{-i} = \times_{j \neq i} T_j$. When i observes t_i, application of Bayes' rule yields the conditional probability of the opponents' types given i's type, i's *posterior belief* μ_i, by

$$\mu_i(t_{-i} \mid t_i) = \frac{\text{Prob}(t_{-i}, t_i)}{\sum_{\tau_{-i} \in T_{-i}} \text{Prob}(\tau_{-i}, t_i)} = \frac{\sum_{\omega \in \cap_{j=1}^n t_j} \mu(\omega)}{\sum_{\omega \in t_i} \mu(\omega)} \tag{3.22}$$

for all $t_{-i} \in T_{-i}$ and all $t_i \in T_i$. In Example 3.12 this updating is trivial, because player 1 has only a single type, and player 2 knows the opponent's type. Hence, player 1's posterior equals the prior, and player 2's posterior is either zero or one.

In fact, the posterior belief is always trivial to derive when the distribution over type combinations induced by μ satisfies independence across players, i.e. when the joint distribution of type combinations is a product distribution of individual types (which, in general, it need not be, of course).

---

[24] Bayes' rule asserts that the *conditional* probability of an event B_h, out of mutually exclusive events $B_1, ..., B_k$, given another event A, $\text{Prob}(B_h \mid A)$, is the joint probability that both A and B_h occur, $\text{Prob}(A \cap B_h) = \text{Prob}(A \mid B_h) \text{Prob}(B_h)$, divided by the marginal probability that A occurs, $\text{Prob}(A) = \sum_{l=1}^k \text{Prob}(A \cap B_l)$, i.e., $\text{Prob}(B_h \mid A) = \text{Prob}(A \cap B_h) / \text{Prob}(A)$ (see Section 2.4).

Example 3.13 *Consider an auction of a work of art for which every bidder knows how she values it, but not how other bidders value it. Assume, moreover, that all bidders value the object purely for consumption purposes (hanging the painting on the wall) and do not consider resale opportunities. In this case the valuations are* private *and can be modeled as independent draws from a (cumulative) probability distribution* $\Phi : [\underline{t}, \overline{t}] \to \mathbb{R}_+$. *Assume, for simplicity, that* Φ *has a density function* $\phi : [\underline{t}, \overline{t}] \to \mathbb{R}_+$.[25] *Let there be* n *bidders, each of whom obtains payoff* $t_i - q$ *in the event that her valuation is* $t_i \in [\underline{t}, \overline{t}]$ *and she buys the object for a price of* $q \geq 0$, *and payoff* 0 *if she does not get the object.*

The rules for the auction are as follows. All bidders $i = 1, ..., n$ *simultaneously submit bids* b_i *and the object will be assigned to the highest bidder at a price equal to, say, the second highest bid (a "private-value sealed-bid second-price auction"). Since bidders do not know the valuations of others, this requires the Harsanyi transformation. More precisely, the set of states* Ω *is the set of possible* n-*tuples* $t = (t_1, ..., t_n) \in [\underline{t}, \overline{t}]^n = \Omega$ *of valuations for the* n *bidders. The type-set for bidder* i *is the set of states where* t_i *takes a particular value,* $T_i = \left\{ \{ t \in [\underline{t}, \overline{t}]^n \,|\, t_i = \tau \} \right\}_{\tau \in [\underline{t}, \overline{t}]}$, *and the system of prior beliefs is given by* $\mu(t) = \mu(t_1, ..., t_n) = \prod_{i=1}^{n} \phi(t_i)$. *(For a change, the set of states here is a continuum.)*

As for payoff functions, player i*'s payoff function assigns to any* n-*tuple* $b = (b_1, ..., b_n)$ *of bids (the choices) and to any state* $t = (t_1, ..., t_n)$ *the payoff* 0 *if some other bid exceeds* i*'s bid* b_i *and payoff* $t_i - \max_{j \neq i} b_j$ *if* b_i *is higher than any other bid. (Ties are broken randomly.) Since the opponents' bids will depend on their types, the player needs to form beliefs about those types. With the independence assumption Bayes' rule yields the simple formula* $\mu_i(t_{-i} | t_i) = \prod_{j \neq i} \phi(t_j)$, *for all* $t \in T$ *and all* $i = 1, ..., n$. *Hence the posterior equals the prior distribution. Indeed, the probability that player* i*'s type* t_i *will have the highest valuation among all bidders is simply* $\Phi(t_i)^{n-1}$, *the probability that* $n - 1$ *independent draws from* Φ *lie below* t_i.

According to Definition 3.13, a Bayesian game is played by the original players who entertain some uncertainty about their opponents. But no player takes any action before she has learned her type. So, in fact, one can let the types play the game, rather than the original players. Indeed, this is the formulation that Harsanyi originally adopted. What Definition 3.13 actually describes is sometimes referred to as the *Selten form*, because Harsanyi credits Selten with it. Under the Selten form the original players will play the "meta-game" generated by nature's choice of a state and the type partitions. Harsanyi's own formulation lets the *types* play the game, after they have updated their beliefs. If one thinks of F as an extensive form where all players

---

[25]   The function $\Phi : [\underline{t}, \overline{t}] \to \mathbb{R}_+$ is a (cumulative) *probability distribution* if it is nondecreasing, continuous from the right, and satisfies $\Phi(\underline{t}) = 0$ and $\Phi(\overline{t}) = 1$. It has a *density* if there is a function ϕ such that the integral of ϕ from $\underline{t}$ to any value $\tau \in [\underline{t}, \overline{t}]$ equals $\Phi(\tau)$.

control only one and the same information set (i.e. they decide once and for all simultaneously), then types correspond precisely to the *agents* of a player in the Bayesian game. In this sense, the *Harsanyi form* is played by the agents (types), and the Selten form is played by the original players.

Common Prior There is one restriction in the formalization of a Bayesian game which deserves discussion. What the types believe about other players is obtained by their updating from a *common prior*, modeled as a chance move with commonly known probabilities. Would it not be more "realistic" to allow players to start out with different priors? Nothing forbids a generalization of the definition which allows each player i to entertain her own individual prior belief about the initial chance move. But then one needs to go further. An individual prior about the chance move is not enough for the player to predict the behavior of others. For this she needs a prior belief about the prior beliefs of other players. Because other players too will have to hold such prior beliefs about i's prior beliefs, even this is not enough. Beliefs about beliefs about beliefs are required, and we are back to the infinite regress. A Bayesian game with individual priors would therefore, in general, have to come with an infinite hierarchy of beliefs for each player, unless at some level of believing the priors become common knowledge.

Even then, there remains a dubious smell around such a construction. After all, this stipulates that players can disagree about the probabilities of one and the same lottery. If I believe that a coin will fall heads with probability 3/4 and you believe that it will fall tails with probability 2/3, then we have a good bargain in prospect. You pay me one dollar if the coin comes up heads and I pay you one dollar if the coin comes up tails. Our payments will always sum to zero, but my expected profit is 1/2 and yours is 1/3, so we both expect a positive profit. Our expectations do not sum to zero.[26]

But can rational players "agree to disagree" in this manner? If you are rational, you will hold your beliefs for rational reasons. If your beliefs differ from mine, and I have evaluated all my data correctly, then your data is probably different from mine. But then, as soon as your beliefs are revealed to me, I have access to your data, at least insofar as it is relevant for my beliefs. Hence, once I learn that you are willing to bet against me, I should revise my beliefs to take account of this new information. And you should do the same. Aumann [8] shows that, under an appropriate specification, these revisions will ultimately lead the two of us to hold the same beliefs. Hence the assumption of a commonly known common prior μ.

In other words, in a Bayesian game, types (of different rational players) will hold *consistent* beliefs in the sense that all these beliefs are derivable from a common prior. Moreover, this common prior is commonly known, enabling

---

[26] That this entails an inconsistency, of course, relies on the assumption that the underlying random experiment can be replicated arbitrarily often.

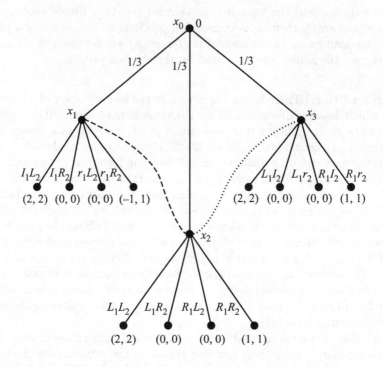

Figure 3.12: Player 1 does not know that player 2 knows

types to reconstruct any level of beliefs about beliefs for any other type or player. This is often referred to as the *Harsanyi doctrine*. Roughly, it holds that any two rational decision makers who have access to identical information will come up with identical conclusions.[27]

With this proviso explained, extensive forms used to model incomplete information turn out to be a surprisingly powerful tool. For example, Bayesian games can be used to model situations where some players are uncertain about what other players *know*, even though they know the other players' payoff functions.

Example 3.14 *Consider the two-player game in Figure 3.12. There are three states, corresponding to* $\omega_1 = x_1$, $\omega_2 = x_2$, *and* $\omega_3 = x_3$, *among which chance chooses at the root with prior probabilities 1/3 each. (Personal players 1 and 2 both have a trivial choice* $\{x_1, x_2, x_3\}$ *at the root* $x_0 = W$.)

---

[27]   Counterarguments to the Harsanyi doctrine may hold that what is uncertain may be the outcomes of games. Short of a theory yielding unique solutions for every game, different individuals may bet on different solutions. Such counterarguments, however, do not offer a *consistent* alternative.

Player 1 *learns whether state* ω_1 *has occurred or not. Thus, her type-set is* $T_1 = \{\{\omega_1\}, \{\omega_2, \omega_3\}\}$. *This is indicated by the dotted line connecting* x_2 *and* x_3 *which depicts player* 1*'s nondegenerate information set. Her other information set is the singleton set* $\{x_1\}$. *Player* 2 *learns whether state* ω_3 *has occurred or not. Thus, her type-set is* $T_2 = \{\{\omega_1, \omega_2\}, \{\omega_3\}\}$. *This is indicated by the broken line that connects* x_1 *and* x_2 *which depicts player* 2*'s nondegenerate information set. Her other information set is* $\{x_3\}$. *Both players have non-trivial choices available at the moves* x_1, x_2, *and* x_3. *Choices are indicated by labeling the corresponding edges. The letters* l_i *(resp.* L_i*) and* r_i *(resp.* R_i*) refer to player* i*'s choices at her singleton (resp. nondegenerate) information set,* $i = 1, 2$; *e.g.,* $r_1 L_2$ *denotes player* 1 *choosing* r_1 *and player* 2 *choosing* L_2. *(Check that player* i *has the same choices available at any two moves in the same information set if it belongs to her.) The values of the payoff function are recorded at the terminal nodes of the tree, where the first entry is player* 1*'s and the second, player* 2*'s payoff.*

Player 1 *knows player* 2*'s payoff function at all states, because player* 2 *has the same payoff function at all states. Hence,* 2 *also knows that* 1 *knows* 2*'s payoff function at all states, and so on. In short,* 2*'s payoff function is* common knowledge *at all states. If* ω_3 *obtains, player* 2 *also knows player* 1*'s payoff function, because in this case* 2 *regards only state* ω_3 *as possible. And the event that* 1 *has the payoff function that she has in* ω_3 *is* $\{\omega_2, \omega_3\}$, *a superset of* $\{\omega_3\}$. *But if* ω_1 *or* ω_2 *obtains, player* 2*'s type (information set) is* $\{\omega_1, \omega_2\}$, *an event over which* 1*'s payoff function is not constant. Hence, in those two states* 2 *does* not *know* 1*'s payoff function. Therefore, the event "player* 2 *knows player* 1*'s payoff function" is* $\{\omega_3\}$. *But if* ω_3 *obtains, player* 1*'s type (information set) is* $\{\omega_2, \omega_3\}$, *an event that is* not *a subset of* $\{\omega_3\}$. *Consequently, at* ω_3 *player* 1 *does* not *know that* 2 *knows* 1*'s payoff function, despite the fact that this is the only state at which* 2 *does indeed know* 1*'s payoff function.*

The last example illustrates how many different varieties of uncertainty a Bayesian game can model when payoff functions are appropriately chosen. In this sense, this class of games provides a rich environment. But, of course, the gain in generality comes at an expense. The cost is that either the number of players becomes very large—in the Harsanyi form—or the number of pure strategies becomes very large—in the Selten form.

Summary

- The Harsanyi transformation translates situations of incomplete information into imperfect information extensive form games with complete information, known as Bayesian games.

- In a Bayesian game, "types" represent the information that a player may have about the state of the world.

- The Harsanyi doctrine holds that the beliefs that various types entertain will be derivable as posteriors from a commonly known common prior distribution, because rational players cannot "agree to disagree".

- In the Harsanyi form the Bayesian game is played by the types; in the Selten form it is played by the original players.

Exercises

Exercise 3.27 (Swing voter's curse) Consider an election race between two candidates each of whom is supported by almost half the electorate. Whether candidate 1 or 2 wins depends on the votes of two citizens. There are two states of the world. The citizens agree that in state ω_1 candidate 1 is best, and in state ω_2 candidate 2 is best. Both citizens obtain payoff 1 if the best candidate for the state wins and payoff 0 otherwise. If the candidates tie, both citizens obtain $1/2$.

Citizen 1 knows the state, while citizen 2 estimates that the state is ω_1 with probability 0.9 and ω_2 with probability 0.1. Each citizen can vote for candidate 1, or for candidate 2, or abstain. They do this simultaneously. Formulate this as a Bayesian game (give the extensive form). Can it be an optimal choice for citizen 2 to abstain?

Exercise 3.28 Each of two individuals is assigned randomly and independently a ticket showing an integer between 1 and 5, indicating the size of a prize she may receive. Then each individual is asked (independently and simultaneously) whether she wishes to exchange her prize for the other individual's prize. If both agree, the prizes are exchanged; otherwise each individual receives her own prize. Both individuals wish to maximize their expected monetary payoffs.

Formulate this as a Bayesian game. Do you think that rational individuals will want to exchange their tickets if they were assigned one with a value of at least 2?

Exercise 3.29 Calculate the (posterior) beliefs about the other player's type for the four types from Example 3.14. Use these to calculate the payoff function for each type separately under the assumption that the opponent will always choose L_i (resp. l_i) at her two information sets. What is the best choice for the type under consideration? Now suppose that both players learn that the state is ω_3 and the opponent chooses R_1 (resp. r_2). What is now the optimal choice?

Exercise 3.30 Consider a two player Bayesian game where each player $i = 1, 2$ can be of two types, t_{i1} and t_{i2}, for $i = 1, 2$. There are four states, $\Omega = \{\omega_1, ..., \omega_4\}$, and $t_{11} = \{\omega_1, \omega_2\}$, $t_{12} = \{\omega_3, \omega_4\}$ are player 1's types, while $t_{21} = \{\omega_1, \omega_3\}$, $t_{22} = \{\omega_2, \omega_4\}$ are player 2's types. Let the system of prior beliefs be given by $\mu(\omega_1) = 0.01$, $\mu(\omega_2) = 0$, $\mu(\omega_3) = 0.09$, and $\mu(\omega_4) = 0.9$. Calculate the posterior beliefs for both types of player 1 and 2. What does type t_{11} know about the opponent's type when she is called upon to move? How does this compare with what type t_{21} believes?

Notes on the Literature: Example 3.12 is based on Png [142]. Bayesian games were introduced by Harsanyi [79]. (Exercise 3.30 is due to him.) Vickrey [177] initiated the study of auctions, an instance of which is in Example 3.13 (see also Wilson [180], [182], and Milgrom and Weber [120]; for surveys see McAfee and McMillan [110] and Milgrom [117]; Myerson [127] considers the design of optimal auctions). The Harsanyi doctrine is argued by Aumann [8]. Example 3.14 is based on Morris, Rob, and Shin [122]. Exercise 3.27 is based on Feddersen and Pesendorfer [58].

4

Games in Normal Form

This chapter deals with an alternative representation of games, the normal (or strategic) form. The emphasis is on the relation of the normal form to the extensive form, where the latter is seen as the basic representation of games. Though the normal form is a somewhat coarser representation of a game, it lends itself quite naturally to solution theory, because of its formal simplicity.

The basic material of this chapter is collected in the first section. Later sections constitute advanced reading and can be skipped if less emphasis on theoretical issues is desired. The second section discusses which class of extensive form games are represented by a given normal form. It shows that a normal form is not simply a trivial extensive form. This material originates in the early days of game theory. The third section introduces more recent results on the same issue. The final section prepares the ground for later discussions on the robustness of solution concepts.

4.1   Normal Form Games

Normal Form   The notion of a pure strategy is suggestive. Recall that a pure strategy (in an extensive form game) assigns a choice to each and every information set of a player. Given a pure strategy for each opponent, therefore, a player can figure out what she should do in all contingencies that may arise in the course of the game. If all players have done this, each of them has chosen a pure strategy and these pin down which plays may materialize (up to chance moves). Hence, after all players have decided on a pure strategy, they may well submit those to an umpire, who can then uniquely reconstruct what will happen in the game if it were actually played. This provides an alternative way to ascertain the outcome of a game, without having the players play it.

More precisely, in an n-player game every n-tuple $s = (s_1, ..., s_n) \in S = S_1 \times ... \times S_n$ of pure strategies generates a unique probability distribution π_s

over plays $w \in W$. Applying expected utility theory, each such distribution π_s translates uniquely into expected payoffs $u(s) = (u_1(s), ..., u_n(s))$, for every pure strategy combination $s \in S$ (see (3.8) and (3.9) in Section 3.3). Therefore, the game might well be represented in the following form.

Definition 4.1 *An n-player **normal form game** Γ is a pair $\Gamma = (S, u)$ where*

- $S = S_1 \times ... \times S_n$ *is the Cartesian product of the n sets S_i of **pure strategies** $s_i \in S_i$ of all players $i = 1, ..., n$ and*

- $u = (u_1, ..., u_n) : S \to \mathbb{R}^n$ *is the (expected) **payoff function** which assigns for every player i a payoff $u_i(s)$ to every pure strategy combination $s \in S$, for all $i = 1, ..., n$.*

The advantage of the normal form over the extensive form is its simplicity. By looking at a normal form game the analysis of strategic interaction boils down to studying a real-valued function on a product set.

An explicit specification of chance moves is absent from a normal form game. Rather, chance moves are "built into" the payoff function by taking expectations with respect to the probability distribution over plays induced by pure strategy combinations—which, of course, constitutes an implicit appeal to the expected utility hypothesis. In principle, chance could be represented as an extra player with a fixed mixed strategy (and without a payoff) in a normal form. But this is usually not done.

For the larger part of the theory, the set S of pure strategy combinations will be assumed to be finite. In this case Γ is a *finite* normal form game. In an *infinite* normal form game at least one of the sets S_i is infinite.

One may wonder whether there is, in analogy with an extensive form, a normal form representation of a game which does not specify preferences or payoffs. And indeed, there is. It is a triplet (S, W, Π) where S is the (product) set of pure strategies, as above, W is a set of plays (or consequences), presumably coming from an extensive form, and $\Pi : S \to \Delta(W)$ is a function which assigns to every pure strategy combination a probability distribution $\Pi(s) = \pi_s \in \Delta(W)$ over plays.[1] Such a *normal form* (or *normal game form*) is a pure representation of the rules of the game, without a specification of preferences. Given preferences over plays for each player (which conform to the expected utility hypothesis), the normal form contains all the information that is required to calculate the payoff function on pure strategy combinations in the corresponding normal form game.[2]

---

[1]   Notation is suggestive here: $\Delta(W) = \{\pi : W \to \mathbb{R}_+ \,|\, \sum_{w \in W} \pi(w) = 1\}$ denotes the set of all probability distributions on the (finite) set W.

[2]   In the older literature a normal form game is sometimes referred to as a *game in strategic form*.

For games with only two players, the normal form can be very conveniently represented by two matrices, A and B, of the same dimension. The idea is as follows. Player 1's pure strategies are identified with the rows of the matrices (player 1 is the "row player") and player 2's pure strategies are identified with the columns of the matrices (player 2 is the "column player"). The entries in matrix A (resp. B) are the player 1's (resp. 2's) payoff from the strategy combination associated with the corresponding row and column. This is often called a "bimatrix game" and is fully specified by a pair of matrices (A, B).

Example 4.1 *Consider the extensive form game in Figure 3.12. This is a two-player game where each player controls two information sets with two choices each, so each player has four pure strategies. If pure strategies of player 1 are identified with rows and pure strategies of player 2 are identified with columns, the bimatrix representation is*

$$
A = \begin{pmatrix} 2 & 2/3 & 4/3 & 0 \\ 2/3 & 1/3 & 1 & 2/3 \\ 4/3 & 1/3 & 2/3 & -1/3 \\ 0 & 0 & 1/3 & 1/3 \end{pmatrix}, \; B = \begin{pmatrix} 2 & 2/3 & 4/3 & 0 \\ 2/3 & 1/3 & 1 & 2/3 \\ 4/3 & 1 & 2/3 & 1/3 \\ 0 & 2/3 & 1/3 & 1 \end{pmatrix}.
$$

The entries in matrix A (resp. B) are player 1's (resp. 2's) payoffs; e.g., the third row corresponds to player 1 choosing r_1 at her information set $\{x_1\}$ and L_1 at her information set $\{x_2, x_3\}$. The second column corresponds to player 2 choosing l_2 at her information set $\{x_3\}$ and R_2 at her information set $\{x_1, x_2\}$. Accordingly, player 1 obtains $\frac{1}{3} \cdot (-1) + \frac{1}{3} \cdot 0 + \frac{1}{3} \cdot 2 = \frac{1}{3}$ and player 2 obtains $\frac{1}{3} \cdot 1 + \frac{1}{3} \cdot 0 + \frac{1}{3} \cdot 2 = 1$. Note that the initial chance move has been eliminated from the picture by taking expectations with respect to its probabilities. More compactly, the two matrices A and B are often written as a single matrix, the entries of which are vectors (u_1, u_2) giving the payoffs to both players.

That a pure strategy specifies a choice for each and every move (resp. information set) has also a downside. In particular, a pure strategy specifies choices also at moves that cannot be reached under this pure strategy. Two pure strategies (of the same player) which differ only at such moves (which cannot be reached under either of them) will, therefore, induce the same (probability distribution over) plays. Hence, they will have the same payoffs, for each given strategy combination of the opponents. This will introduce payoff duplications in the normal form. Whether or not these duplications are accounted for distinguishes various types of normal form games.

Suppose we start from an extensive form game G and derive *all* pure strategies (for all players $i = 1, ..., n$). The normal form game thus derived will be referred to as the *unreduced* normal form associated with G. It will, in general, be a huge normal form game with many payoff ties.

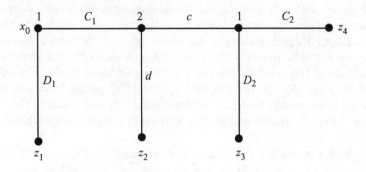

Figure 4.1: Mini-centipede

Similarly, players could be split into agents. An agent ih (of a player i) is an artificial player, who controls a single information set $h \in H_i$ and has the same payoff function as the player to whom she belongs, $u_{ih} = u_i$. The set of pure strategies S_{ih} of agent ih would then be the set of choices available at (any of the moves in) her information set $h \in H_i$. The normal form game

$$(\times_{i=1}^{n} \times_{h \in H_i} S_{ih}, \times_{i=1}^{n} \times_{h \in H_i} u_{ih})$$

played between the agents of all players is called the *agent normal form*. This representation entails that agents make their decisions independently, even if they belong to the same player; i.e., the team of agents (of a given player) is coordinated only by payoffs, but not by any other coordination device, like pre-play communication or the like.

The latter, one may view as a misspecified model. Yet, one may wish to retain the information contained in the definition of agents. The compromise is not to view agents as independent, but to allow them implicitly to coordinate by making mixed strategies of player i distributions on the product set $\times_{h \in H_i} S_{ih}$ (which amounts to allowing "correlated" strategies among agents of the same player). This representation is known as the *standard form*, as introduced by Harsanyi and Selten ([82], pp. 30).

Reduced Normal Forms The unreduced normal form, the agent normal form, and the standard form all retain the information that is explicit in the extensive form. So, these three normal forms are the most "extensive" normal form representations of an extensive form.

Example 4.2 *Consider the extensive form in Figure 4.1. (Moves are labeled with the identity of the player who has nontrivial choices available, and nontrivial choices are depicted by labeling the corresponding edges.) Player 1's pure strategy set in the unreduced normal form consists of the strategies*

(D_1, D_2), (D_1, C_2), (C_1, D_2), and (C_1, C_2). *The first two of those induce the same play: termination of the game after the first move. Recording (D_1, D_2) and (D_1, C_2) as separate strategies, therefore, represents a clear-cut redundancy. It may be preferable to identify those two strategies to a single one which prescribes termination at the first move.*

The observation that two different pure strategies (of the same player) may always (i.e. for all strategy combinations among the others) induce the same (probability distributions over) plays, because they differ only at moves which cannot be reached, suggests a lower-dimensional normal form representation. In terms of the normal form (S, W, Π) (viz. without payoff specification), the following holds. (A proof is in the appendix to this section.)

Proposition 4.1 *Two strategies $s_i, s_i' \in S_i$ of a player i in an extensive form differ only at moves that cannot be reached under either of these strategies if and only if the equation $\pi_s = \pi_{s'}$ holds for all $s_{-i} = (s_1, ..., s_{i-1}, s_{i+1}, ..., s_n) \in S_{-i} = \times_{j \neq i} S_j$, where $s = (s_{-i}, s_i) \in S$ and $s' = (s_{-i}, s_i') \in S$.*

Proposition 4.1 implies that two strategies $s_i, s_i' \in S_i$ which differ only at moves that cannot be reached will, for each given strategy combination $s_{-i} \in S_{-i}$ among the other players, always give the same payoffs (to all players); i.e., $u(s_{-i}, s_i) = u(s_{-i}, s_i')$. Therefore, when player i decides between s_i and s_i' she will always find herself indifferent. This justifies the following definition.

Definition 4.2 *Two strategies $s_i, s_i' \in S_i$ of player i are called **strategically equivalent** if $u(s_{-i}, s_i) = u(s_{-i}, s_i')$, for all $s_{-i} \in S_{-i}$, where $u = (u_1, ..., u_n)$.*

Since it does not matter for any player which of two strategically equivalent strategies $s_i, s_i' \in S_i$ player i picks, the presence of both these strategies in the normal form seems like redundant information.

Hence, strategically equivalent strategies may be identified and replaced by a single representative (one for each equivalence class). This leads to the *semi-reduced* normal form or *pure-strategy reduced* normal form. Hence, a normal form game is in semi-reduced normal form (or pure-strategy reduced normal form) if for each player i the mapping $s_i \mapsto (u(s_{-i}, s_i))_{s_{-i} \in S_{-i}}$ is one-to-one; i.e., if $u(s_{-i}, s_i) = u(s_{-i}, s_i')$, for all $s_{-i} \in S_{-i}$, implies $s_i = s_i'$.

If there is no risk of misunderstanding, the semi-reduced normal form will often be referred to simply as "the normal form". This is because it entails a considerable simplification, as the example below illustrates.

Example 4.3 *Consider a duel where the two contestants hold one-bullet pistols and get two chances to shoot at each other. Suppose that when they have the initial opportunity to shoot (simultaneously), the probability of hitting the opponent is $1/2$. If both survive this first round, they both take one step towards*

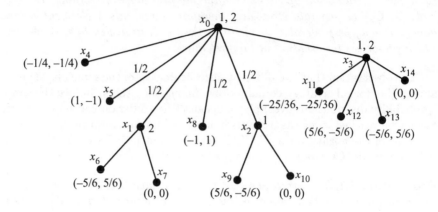

Figure 4.2: Two-step duel

each other and get a second opportunity to shoot (if they have not yet fired at
the first try). In this second round the probability of hitting the opponent is
5/6, say, because they are now closer.

Each contestant values highest the outcome where he himself survives and
the opponent gets killed (payoff 1). That both stay alive is the second-best
outcome for both of them (payoff 0). A contestant who dies obtains payoff −1,
irrespective of whether the opponent also dies or survives. Figure 4.2 depicts
the extensive form.

Nodes x_1 and x_5 corresponds to player 1 shooting first and player 2 waiting
(with a chance move deciding whether player 2 is hit); x_2 and x_8 corresponds
to player 2 shooting first and player 1 waiting (again with chance deciding
between x_2 and x_8); x_3 corresponds to both waiting for the second round; x_4
corresponds to both shooting at the first instance; and so forth.

Chance moves that lead only to terminal nodes have been replaced by the
expected payoffs from the lottery induced by the chance moves. (In Exercise
4.3 you are asked to verify that this does not change the normal form.) Moves
are labeled with the players who decide there, and in payoff vectors the first
entry is 1's and the second 2's payoff, as usual.

The nontrivial choices for players 1 and 2 associated with this extensive
form are given by

$$C_1 = \{\{x_1, x_4, x_5\}, \{x_2, x_3, x_8\}, \{x_9\}, \{x_{10}\}, \{x_{11}, x_{12}\}, \{x_{13}, x_{14}\}\},$$
$$C_2 = \{\{x_2, x_4, x_8\}, \{x_1, x_3, x_5\}, \{x_6\}, \{x_7\}, \{x_{11}, x_{13}\}, \{x_{12}, x_{14}\}\}.$$

Hence, each player has three information sets with two choices each; i.e., each *player has* $2^3 = 8$ *pure strategies. The unreduced normal form game is given (in bimatrix form) by*

$$
A = \begin{pmatrix}
-1/4 & -1/4 & -1/4 & -1/4 & 1/2 & 1/12 & 1/2 & 1/12 \\
-1/4 & -1/4 & -1/4 & -1/4 & 1/2 & 1/12 & 1/2 & 1/12 \\
-1/4 & -1/4 & -1/4 & -1/4 & 1/2 & 1/12 & 1/2 & 1/12 \\
-1/4 & -1/4 & -1/4 & -1/4 & 1/2 & 1/12 & 1/2 & 1/12 \\
-1/2 & -1/2 & -1/2 & -1/2 & 0 & 0 & -5/6 & -5/6 \\
-1/12 & -1/12 & -1/12 & -1/12 & 0 & 0 & -5/6 & -5/6 \\
-1/2 & -1/2 & -1/2 & -1/2 & 5/6 & 5/6 & -25/36 & -25/36 \\
-1/12 & -1/12 & -1/12 & -1/12 & 5/6 & 5/6 & -25/36 & -25/36
\end{pmatrix}
$$

and

$$
B = \begin{pmatrix}
-1/4 & -1/4 & -1/4 & -1/4 & -1/2 & -1/12 & -1/2 & -1/12 \\
-1/4 & -1/4 & -1/4 & -1/4 & -1/2 & -1/12 & -1/2 & -1/12 \\
-1/4 & -1/4 & -1/4 & -1/4 & -1/2 & -1/12 & -1/2 & -1/12 \\
-1/4 & -1/4 & -1/4 & -1/4 & -1/2 & -1/12 & -1/2 & -1/12 \\
1/2 & 1/2 & 1/2 & 1/2 & 0 & 0 & 5/6 & 5/6 \\
1/12 & 1/12 & 1/12 & 1/12 & 0 & 0 & 5/6 & 5/6 \\
1/2 & 1/2 & 1/2 & 1/2 & -5/6 & -5/6 & -25/36 & -25/36 \\
1/12 & 1/12 & 1/12 & 1/12 & -5/6 & -5/6 & -25/36 & -25/36
\end{pmatrix}
$$

Here the first four rows (resp. columns) correspond to strategies that prescribe shooting at the first instance. The fifth row (resp. column) is the strategy which prescribes not shooting at any of the three information sets; the last row (resp. column) is the strategy not to shoot initially, but then shooting at both information sets reached in the second round. The sixth and seventh rows (resp. columns) represent strategies that prescribe not shooting initially and shooting in the second round only at one of the two information sets.

The first four strategies (rows) (resp. columns) for each player prescribe shooting initially, and then various combinations of shooting and not shooting at the two information sets reached in the second round.

This is obscure, because once a contestant has shot in the first round he has wasted his only bullet and a prescription to shoot at any later information set is vacuous, because it is physically impossible. Yet, such prescriptions show up as strategies in the unreduced normal form.

Since the only physically possible one of the first four strategies is "shoot initially and then never again" and all of the first four strategies are strategically equivalent, they might well be identified as a single strategy ("shoot at your first opportunity"). This identification results in the semi-reduced normal form given below. (Again, rows correspond to player 1's strategies and columns to player 2's strategies. The upper left entry is player 1's payoff, the lower right player 2's.)

Here each contestant has five pure strategies, the first one of which represents shooting at the start of the duel.

Yet, nothing that is strategically relevant to the situation at hand has been lost by going from the unreduced to the semi-reduced normal form. Moreover, the remaining payoff ties still indicate extensive form structures.

$-\frac{1}{4}$ $\quad$ $-\frac{1}{4}$	$\frac{1}{2}$ $\quad$ $-\frac{1}{2}$	$\frac{1}{12}$ $\quad$ $-\frac{1}{12}$	$\frac{1}{2}$ $\quad$ $-\frac{1}{2}$	$\frac{1}{12}$ $\quad$ $-\frac{1}{12}$
$-\frac{1}{2}$ $\quad$ $\frac{1}{2}$	0 $\quad$ 0	0 $\quad$ 0	$-\frac{5}{6}$ $\quad$ $\frac{5}{6}$	$-\frac{5}{6}$ $\quad$ $\frac{5}{6}$
$-\frac{1}{12}$ $\quad$ $\frac{1}{12}$	0 $\quad$ 0	0 $\quad$ 0	$-\frac{5}{6}$ $\quad$ $\frac{5}{6}$	$-\frac{5}{6}$ $\quad$ $\frac{5}{6}$
$-\frac{1}{2}$ $\quad$ $\frac{1}{2}$	$\frac{5}{6}$ $\quad$ $-\frac{5}{6}$	$\frac{5}{6}$ $\quad$ $-\frac{5}{6}$	$-\frac{25}{36}$ $\quad$ $-\frac{25}{36}$	$-\frac{25}{36}$ $\quad$ $-\frac{25}{36}$
$-\frac{1}{12}$ $\quad$ $\frac{1}{12}$	$\frac{5}{6}$ $\quad$ $-\frac{5}{6}$	$\frac{5}{6}$ $\quad$ $-\frac{5}{6}$	$-\frac{25}{36}$ $\quad$ $-\frac{25}{36}$	$-\frac{25}{36}$ $\quad$ $-\frac{25}{36}$

That two strategies are strategically equivalent is a condition on the pay-offs, and not on the induced probability distributions on plays. In coincidental cases this may lead to identifying two pure strategies as strategically equivalent even though they may induce different probability distributions on plays for some strategy combinations of the opponents. Yet, these are exceptional cases. Indeed, Proposition 4.1 can be used to show that for *almost all* payoff functions strategic equivalence is precisely the same as "inducing the same probability distributions on plays".[3] Therefore, the step to the semi-reduced normal form simply eliminates strategy duplications arising from variations of behavior in circumstances that cannot arise (under these strategies).

But one may go even further. It could happen that in a normal form game some pure strategy $s_i \in S_i$ of player i can be reproduced by taking an appropriate randomization over the remaining strategies in $S_i \setminus \{s_i\}$ in the sense that there is $\sigma_i \in \Delta_i = \left\{ \sigma_i : S_i \to \mathbb{R}_+ \left| \sum_{s_i \in S_i} \sigma_i(s_i) = 1 \right. \right\}$ such that

$$u(s_{-i}, s_i) = \sum_{s_i' \in S_i \setminus \{s_i\}} \sigma_i(s_i')\, u(s_{-i}, s_i'), \text{ for all } s_{-i} \in S_{-i}, \qquad (4.1)$$

where $u = (u_1, ..., u_n)$, $\sigma_i(s_i') \geq 0$, for all $s_i' \in S_i \setminus \{s_i\}$, and $\sum_{s_i' \in S_i \setminus \{s_i\}} \sigma_i(s_i') = 1$. If this is the case, all players are again indifferent as to whether player i uses s_i or the mixed strategy σ_i. Since the mixed strategy σ_i perfectly substitutes for s_i, one may well delete s_i from the set of player i's pure strategies.

This leads to the notion of a reduced (or mixed-strategy reduced) normal form. A given normal form game $\Gamma = (S, u)$ is in (mixed-strategy) *reduced*

---

[3] "Almost all" here means, roughly, that if the property under scrutiny does not apply, it will apply after an arbitrarily small perturbation of the payoff function.

normal form whenever

$$u(s_{-i}, s_i) = \sum_{s_i' \in S_i} \sigma_i(s_i') \, u(s_{-i}, s_i'), \text{ for all } s_{-i} \in S_{-i}, \qquad (4.2)$$

implies $\sigma_i(s_i) = 1$, for all $s_i \in S_i$ and for all players $i = 1, ..., n$. It follows that in a reduced normal form game Γ the vectors $(u(s_{-i}, s_i))_{s_{-i} \in S_{-i}}$ are linearly independent across the pure strategies $s_i \in S_i$ of player i for all i.

Hence, the following hierarchy of normal form representations derives from an extensive form game:

Unreduced normal form   Agent normal form   Standard form
Semi-reduced normal form
Reduced normal form

The top row lists the most precise normal form representations. Going downwards eliminates stepwise information on the extensive form. Which extensive form information can still be recovered in reduced (normal) forms will be discussed later.

Still, of course, for any given normal form game there is an extensive form game which has the original normal form game as "its" normal form representation: namely, the extensive form where all players choose simultaneously between their pure strategies once and for all at the root of the tree. Therefore, while conceptually the normal form is a derived representation of a game, formally it may well be thought of as a game representation of its own. The distinction between extensive and normal form then is as follows.

- In an extensive form the rules of the game are the primitive notion, and strategies are derived objects.

- In a normal form game strategies are primitives, and the rules of the game are kept implicit.

Since strategies are the primitives of normal form games, an issue arises about randomized strategies. In an extensive form game mixed and behavior strategies were introduced simply by allowing players to use randomization devices (either globally or locally). In a normal form this would change the game. Hence, in order to allow for randomized strategies, the definition of a new game is required to introduce mixed strategies.

Definition 4.3 *The* **mixed extension** *of a normal form game* $\Gamma = (S, u)$ *is the infinite normal form game* (Θ, U), *where* $\Theta = \times_{i=1}^n \Delta_i$ *and*

$$U_i(\sigma) = \sum_{s \in S} \sigma(s) \, u_i(s), \text{ for all } \sigma \in \Theta,$$

where $\sigma(s) = \sigma(s_1, ..., s_n) = \prod_{i=1}^n \sigma_i(s_i)$, *for all* $s \in S$, *for all* $i = 1, ..., n$.

Recall that the joint distribution of pure strategies is a product distribution, because of the non-cooperative framework, where players make their decisions independently in a game with complete rules.

Behavior strategies do not exist for normal form games. On the other hand, the agent normal form can be used to mimic behavior strategies. Mixed strategies in the agent normal form are probability distributions over the agents' strategies. Since every agent controls only a single information set, an agent's strategy set is precisely the set of (nontrivial) choices available at (every move in) this information set. Therefore, *mixed* strategies in the agent normal form correspond to *behavior* strategies in the extensive form.

Summary

- A normal form game consists of a set of pure strategies for every player and a payoff function which assigns to every strategy combination a payoff for every player.

- If strategically equivalent strategies are identified, the (unreduced) normal form is turned into the semi-reduced normal form. If all strategies that can be duplicated by mixtures over other strategies are deleted, the reduced normal form is obtained.

- The mixed extension of a normal form game is another normal form game where strategies are probability distributions over the original pure strategies and payoffs are derived by taking expectations.

Appendix

Proof of Proposition 4.1   If the two strategies $s_i, s_i' \in S_i$ differ only at moves that cannot be reached under either of them (i.e. if $s_i(x) \neq s_i'(x)$ implies $\pi_s(w) = 0 = \pi_{s'}(w)$, for all $w \in x$ and all $s_{-i} \in S_{-i}$), then clearly $\pi_s(w) = \pi_{s'}(w)$, for all $w \in W$ and all $s_{-i} \in S_{-i}$, where $s = (s_{-i}, s_i)$ and $s' = (s_{-i}, s_i')$.

Conversely, if there is some $x \in X$ such that $s_i(x) \neq s_i'(x)$ and $\pi_s(w) + \pi_{s'}(w) > 0$, for some $w \in x$ and some $s_{-i} \in S_{-i}$, then there is $y \in P^{-1}(x)$ and $w' \in y$ such that either $\pi_s(w') = 0 < \pi_{s'}(w')$ or $\pi_s(w') > 0 = \pi_{s'}(w')$ (where again $s = (s_{-i}, s_i)$ and $s' = (s_{-i}, s_i')$). This completes the proof of the proposition.

Exercises

Exercise 4.1 The tree of a centipede game is given by $W = \{w_1, ..., w_5\}$ and

$$N = \left\{ W, \{w_2, ..., w_5\}, \{w_3, ..., w_5\}, \{w_4, w_5\}, (\{w\})_{w \in W} \right\}.$$

Nontrivial choices of player 1 are given by

$$C_1 = \{\{\{w_1\}\}, \{\{w_2, ... w_5\}\}, \{\{w_3\}\}, \{\{w_4, w_5\}\}\}$$

and for player 2 by

$$C_2 = \{\{\{w_2\}\}, \{\{w_3, ..., w_5\}\}, \{\{w_4\}\}, \{\{w_5\}\}\}.$$

There are no chance moves. Payoff vectors are $v(w_1) = (2,0)$, $v(w_2) = (1,2)$, $v(w_3) = (4,1)$, $v(w_4) = (3,4)$, and $v(w_5) = (5,3)$, where the first entry is player 1's payoff and the second is 2's. Derive the unreduced and the semi-reduced normal form game.

Exercise 4.2 In the representation of Example 4.3 in Figure 4.2, chance moves that lead exclusively to terminal nodes have been pruned and replaced by (a terminal node with) the expected payoffs from the lottery induced by the chance moves. Derive an alternative extensive form representation (to Figure 4.2) by reinserting explicitly all chance moves (that lead exclusively to terminal nodes). For the new extensive form game derive the semi-reduced normal form. Is it different from the one derived in Example 4.3? Is the game in Example 4.3 a game of perfect information?

Exercise 4.3 Verify generally (for a finite extensive form game) that replacing chance moves that lead only to terminal nodes by (terminal nodes with) the expected payoffs from the associated lottery does not change the normal form of the game.

Exercise 4.4 Show that, in a normal form game $\Gamma = (S, u)$, for some $s_i \in S_i$ the equations

$$u(s_{-i}, s_i) = \sum_{s_i' \in S_i} \sigma_i(s_i') \, u(s_{-i}, s_i'), \text{ for all } s_{-i} \in S_{-i},$$

hold for some $\sigma_i \in \Delta_i$ with $\sigma_i(s_i) < 1$ if and only if condition (4.1) holds (i.e., the support of σ_i need not be restricted to $S_i \setminus \{s_i\}$).

Exercise 4.5 Show that, in the mixed extension (Θ, U) of a normal form game $\Gamma = (S, u)$, the payoff function U_i is linear in player i's strategy; i.e., for all $\sigma_i, \sigma_i' \in \Delta_i$ and any λ with $0 \leq \lambda \leq 1$ one has $U_i(\sigma_{-i}, \lambda \sigma_i + (1 - \lambda)\sigma_i') = \lambda U_i(\sigma_{-i}, \sigma_i) + (1 - \lambda)U_i(\sigma_{-i}, \sigma_i')$, for all $\sigma_{-i} \in \Theta_{-i} = \times_{j \neq i} \Delta_j$.

Exercise 4.6 Consider the mixed extension of a finite normal form game $\Gamma = (S, u)$. Define the mixed extension of the mixed extension, under the expected utility hypothesis. Is there an essential difference from the (original) mixed extension of Γ?

> **Notes on the Literature:** The concept of the normal form is due to von Neumann and Morgenstern ([134], Chapter 11), who draw on the contributions by Borel [29] and von Neumann [133]. Example 4.3 is borrowed from Luce and Raiffa [105]. The agent normal form has been introduced by Selten [163] and the standard form by Harsanyi and Selten [82]. The significance of the semi-reduced normal form is discussed by Thompson [175] (see the next section) and the reduced normal form is advocated by Kohlberg and Mertens [97] in the context of strategic stability.

4.2   Thompson Transformations

Often there are different extensive forms that appear to represent the same strategic situation. For instance, a single player's decision problem between three alternatives, a_1, a_2, and a_3, say, could be represented either as a simultaneous decision among the three alternatives, or as a sequential decision where the decision maker first decides whether or not to take a_1, and then, if she decided not to take a_1, decides between a_2 and a_3. (Draw the trees!) While the extensive forms of these two (single-player) games differ, they both have the same normal form. Likewise, irrespective of whether simultaneous decisions by two players are represented by a succession of information sets (with the same plays passing through) or by the two players deciding at the same moves, the normal form stays the same. Surprisingly, this observation generalizes.

Three Transformations   I shall now introduce three transformations which do not appear to change any "essential" features of the extensive form game. And I will argue that the common characteristics of these "inessential" transformations is that they do not affect the semi-reduced normal form of the game. Moreover, the converse is also true. Two extensive form games that have the same semi-reduced normal form are related by iterative applications of those "inessential" transformations (or their inverses). Thus, the (semi-reduced) normal form preserves the information that is "essential" about an extensive form game. Any solution concept that does not depend solely on the normal form, therefore, has to argue why (at least) one of the transformations is inappropriate.

Since the three "inessential" transformations *characterize* the set of extensive form games (with perfect recall) with the same (semi-reduced) normal form, they provide an answer to the question of which information on extensive form structures is lost in going to the normal form. They do not, however, pin down how extensive form structures are reflected in the normal form—the topic of the next section.

The transformations are originally due to Thompson [175]. What is presented below, however, are modifications of three of Thompson's four transformations. The modifications are designed to preserve perfect recall.

In what follows it is useful to introduce the following mapping from sets of plays to sets of nodes. For a fixed tree $\mathcal{T} = (W, N)$, let $W^{-1} : 2^W \to 2^N$ be defined by

$$W^{-1}(a) = \{x \in N \,|\, x \subseteq a, \text{ and there is no } y \in N : x \subset y \subseteq a\} \qquad (4.3)$$

for all sets of plays $a \subseteq W$.[4] The notation W^{-1} is no coincidence, because it can be shown that, for any rooted tree, this mapping is a function that provides an inverse to the function $W(m) = \{w \in W \mid \exists x \in m : w \in x\}$ on the (set of) sets of nodes $m \subseteq N$ which satisfy that there is no $y \in N$ with $x \subset y \subseteq W(m)$, for some $x \in m$ (see the appendix to this section). Effectively, $W^{-1}(a)$ is the coarsest partition of a set of plays $a \subseteq W$ by elements of N.[5]

The first transformation turns simultaneous decisions that are originally represented by a succession of information sets into truly simultaneous moves. It is a slight modification of what Thompson [175] called the "Interchange of moves". If any information set either is disjoint from or contains (the set of) all immediate successors of a particular move, then the successors may well be removed without affecting any player's position.

The inverse of the first transformation allows any extensive form to be turned into one that is *conventional*, i.e. such that at every move at most one player has nontrivial choices available. In what follows let $G = (F, v)$ and $\hat{G} = (\hat{F}, \hat{v})$ be two extensive form games with the same number ($n \geq 1$) of players. (For notation see Sections 3.1 and 3.2. In particular, recall that $P(x)$ denotes the immediate predecessor of node x in a tree.)

(I) (Interchanging simultaneous decisions) $\hat{G}$ is an **I-transform** of G if there is a move $\bar{x} \in N$ (in G) such that $P^{-1}(\bar{x}) \subseteq X$ and

$$\text{if } P^{-1}(\bar{x}) \cap P(c) \neq \emptyset, \text{ then } P^{-1}(\bar{x}) \subseteq P(c), \qquad (4.4)$$

for all $c \in C_i$ and all $i = 1, ..., n$, and $\hat{G}$ satisfies $\hat{T} = (W, N \setminus P^{-1}(\bar{x}))$, $\hat{C}_i = \left\{\hat{W}^{-1}(W(c))\right\}_{c \in C_i}$, for all $i = 1, ..., n$,

$$\hat{p}(x) = \begin{cases} p(x)p(P(x)) & \text{if} \qquad P(x) \in P^{-1}(\bar{x}) \\ p(x) & \text{otherwise} \end{cases}$$

for all $x \in N \setminus P^{-1}(\bar{x})$, and $v(w) = \hat{v}(w)$, for all $w \in W$.[6]

Hence, if no player who has (nontrivial) choices available at (immediate) successors of $\bar{x}$ can distinguish between those successors, then the successors are removed (from the set of moves) and these players get to decide simultaneously at $\bar{x}$. For a perfect recall game (I) produces a choice partition, because no player for whom $P^{-1}(\bar{x})$ overlaps with an information set can have nontrivial choices available at $\bar{x}$, by perfect recall and (4.4). So, the images of (nontrivial) choices available at $\bar{x}$ under $\hat{W}^{-1} \circ W$ do not intersect with other

---

[4] The function W^{-1} is easy to "compute". For every $w \in a$ find the largest node $x \in N$ which contains w, is contained in a, and satisfies that $P(x)$ is not contained in a. The (disjoint) union over all $w \in a$ of such nodes is $W^{-1}(a)$.

[5] A partition is coarser than another partition if every element of the second is contained in some element of the first.

[6] The notation $\hat{W}^{-1}$ indicates that this function is computed for the tree $\hat{T}$.

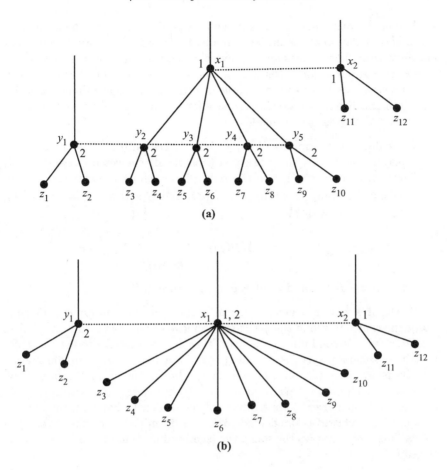

(a)

(b)

Figure 4.3: Interchanging simultaneous decisions

(nontrivial) choices (for which $\hat{W}^{-1} \circ W$ is the identity). Also, and most importantly, transformation (I) cannot change the normal form, because neither the number of nontrivial choices nor their ordering is affected by (I).

Example 4.4 *Figure 4.3 gives an example where the (immediate) successors of move $x_1 = \bar{x}$ (viz. $\{y_2, ..., y_5\}$) are removed when going from part (a) to part (b). Moves are labeled with the players who have nontrivial choices available, and dotted lines depict information sets. Choices available at information set $h = \{x_1, x_2\}$ are $c_1 = \{y_2, y_3, z_{11}\}$ and $c_2 = \{y_4, y_5, z_{12}\}$ where $p(y_2) = p(y_4) = 2/3$ and $p(y_3) = p(y_5) = 1/3$ in part (a). The images of these choices after transformation (I) are $\hat{c}_1 = \{z_3, ..., z_6, z_{11}\}$ and $\hat{c}_2 = \{z_7, ..., z_{10}, z_{12}\}$ where $\hat{p}(z_l) = 2/3$, for $l = 3, 4, 7, 8$, and $\hat{p}(z_l) = 1/3$, for $l = 5, 6, 9, 10$, in part (b).*

The next transformation is as in Thompson [175], save for the formalism. It is a generalization of the initial example. For a rational player it should not matter whether she decides among several choices at once, or splits them into consecutive sub-decisions. Iterative applications of this transformation (or, more precisely, its inverse) are capable of turning any conventional extensive form into another conventional extensive form, where at any information set precisely *two* choices are available.

(C) (Coalescing information sets) $\hat{G}$ is a **C-transform** of G if for some player $i = 1, ..., n$ there are two (nontrivial) choices $c, c' \in C_i$ (in G) such that $c = P(c')$, i is the only player who has nontrivial choices available at (any move in) $h' = P(c')$, and $\hat{G}$ satisfies $\widehat{T} = (W, N \setminus c)$,

$$\hat{C}_j = \left\{\hat{W}^{-1}(W(c))\right\}_{c \in C_j}, \text{ for all } j \neq i, \; \hat{C}_i = \left\{\hat{W}^{-1}(W(c))\right\}_{c \in C_i \setminus \{c\}},$$

$$\hat{p}(x) = \begin{cases} p(x)p(P(x)) & \text{if} \qquad P(x) \in c \\ p(x) & \text{otherwise} \end{cases}$$

for all $x \in N \setminus c$, and $v(w) = \hat{v}(w)$, for all $w \in W$.

Hence, if a player decides twice in a row (without improving information), the choice to refine the decision is removed and replaced by the choices available at the second information set. Transformation (C) produces a choice partition, because player i is the only player for whom $c = h'$ is an information set. Transformation (C) removes one choice, but in the normal form of G all strategies which do not take c (at $h = P(c)$), but differ only with respect to choices that come after c, are strategically equivalent. Therefore, the number of equivalence classes is unaffected by C-transforming. Since the ordering of (remaining) choices stays the same, the (semi-reduced) normal form also stays the same.

Example 4.5 *Figure 4.4 illustrates. In part (a) player $i = 1$ makes two successive decisions without learning anything in between. At $h = \{x_1, x_2, x_3\}$ she decides between $c_1^1 = \{y_3, y_4, y_7, y_8, y_{10}\}$ and $c_1^2 = \{y_1, y_2, y_5, y_6, y_9\}$, then at $h' = c_1^1$ between $c_1^3 = \{z_2, z_4, z_6, z_8, z_{10}\}$ and $c_1^4 = \{z_1, z_3, z_5, z_7, z_9\}$. (At the subset $\{x_1, x_2\}$ of h player $i = 2$ decides between $c_2^1 = \{y_1, y_3, y_5, y_7\}$ and $c_2^2 = \{y_2, y_4, y_6, y_8\}$.) In part (b) she decides only once, but among the three choices, $\hat{c}_1^1 = c_1^2$, $\hat{c}_1^2 = c_1^3$, and $\hat{c}_1^3 = c_1^4$, because $c = c_1^1$ has been removed. (Player $i = 2$ decides between $\hat{c}_2^1 = \{y_1, z_1, z_2, y_5, z_5, z_6\}$ and $\hat{c}_2^2 = \{y_2, z_3, z_4, y_6, z_7, z_8\}$ at the moves x_1 and x_2.)*

The third transformation, traditionally called "Addition of superfluous moves", is the most cumbersome. It is more transparent to split its definition into two parts, defining its inverse. The first part pins down when two choices can be considered irrelevant at some of the moves where they are available. Obviously, such irrelevance requires that no player cares and no player (except the player who takes these choices) learns about them.

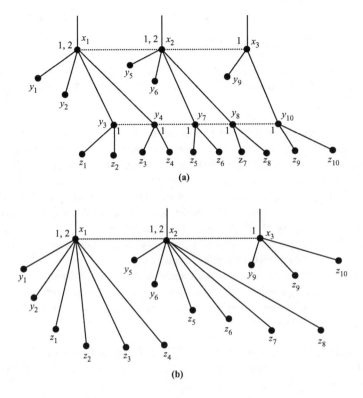

Figure 4.4: Coalescing information sets

Definition 4.4 *Let $G = (F, v)$ be an extensive form game with perfect recall and suppose that, for some player $i = 1, ..., n$, the two (distinct nontrivial) choices $c_1, c_2 \in C_i$ satisfy $P(c_1) = P(c_2) = h$. Let $h' \subseteq h$ be a subset of $h \in H_i$ and define*

$$N_k = \{x \in N \,|\, x \subseteq W(c_k) \cap W(h')\}$$

*as the nodes met after c_k along plays passing through h', for $k = 1, 2$. The choices $c_1, c_2 \in C_i$ are called **irrelevant at** (any move in) h' if i is the only player who has nontrivial choices available at (any move in) h', $c \cap N_1 \neq \emptyset$ implies $c \in \{c_1, c_2\}$, for all $c \in C_i$, and $\psi : N_1 \rightarrow N_2$ is a bijection[7] which satisfies*

(a) $y \subset x$ if and only if $\psi(y) \subset \psi(x)$, and $p(x) = p(\psi(x))$, for all $x, y \in N_1$,

---

[7]   A bijection ψ is a function that is (a) *onto* (or surjective; i.e., for all y in the range there is some x in the domain such that $\psi(x) = y$) and (b) *one-to-one* (or injective; i.e., $\psi(x) = \psi(x')$ implies $x = x'$).

(b) $x \in c$ if and only if $\psi(x) \in c$, for all $c \in C_j$, all $x \in N_1$, and all $j \neq i$,

(c) $N_1 \supseteq c \in C_i$ if and only if $N_2 \supseteq \psi(c) \equiv \cup_{x \in c} \psi(x) \in C_i$, for all $c \in C_i$,

(d) $v(w) = v(\psi(w))$, where $\psi(w) \in W(c_2) \cap W(h')$ denotes the play that ends at the image of the terminal node $\{w\} \in N_1$ under ψ, for all $\{w\} \in N_1$.

Condition (a), together with the assumption that ψ is a bijection, means that the tree through c_2 and h' is a *precise copy* of the tree through c_1 and h'. Condition (b) says that no player other than i ever *learns* whether c_1 or c_2 was taken at (some move in) h', because their information sets "go across" N_1 and N_2. Condition (c) says that, for any choice of i after c_1 at h', there is precisely one *corresponding choice* after c_2 at h'. Condition (d) requires that the choices c_1 and c_2 at h' do *not affect payoffs* $v = (v_1, ..., v_n)$ over plays through h', so no player cares whether c_1 or c_2 is chosen at h'. Note that Definition 4.4 applies only to perfect recall games.

With the notion of irrelevant choices at hand, (the inverse of) a finite repetition of Thompson's [175] "addition" and an ensuing refinement of information sets (to preserve perfect recall), as introduced by Elmes and Reny [55], can be defined as follows.

(D) (Deletion of irrelevant choices) $\hat{G}$ is a **D-transform** of G if for some player $i = 1, ..., n$ there are two different choices $c_1, c_2 \in C_i$ (in G) such that

    (a) at any move in $h = P(c_1) = P(c_2)$ player i is the *only* player who has nontrivial choices available and c_1 and c_2 are the *only* choices available to i at h,

    (b) c_1 and c_2 are *irrelevant at* the proper subset h' of h,

and $\hat{G}$ satisfies $\hat{T} = \left(\hat{W}, N \setminus (N_2 \cup h') \right)$, with $\hat{W} = W \setminus (W(c_2) \cap W(h'))$,

$$\hat{C}_j = \left\{ \hat{W}^{-1} \left(W(c) \cap \hat{W} \right) \right\}_{c \in C_j}, \text{ for all } j \neq i,$$

$$\hat{C}_i = \left\{ \left(\hat{W}^{-1} \left(W(c) \cap \hat{W} \right) \right)_{c \in C_i \setminus \{c_1\}}, c_1 \setminus P^{-1}(h') \right\}, \text{ and}$$

$$\hat{p}(x) = \begin{cases} p(x)p\left(P(x)\right) & \text{if} & P(x) \in h' \\ p(x) & \text{otherwise} \end{cases}$$

for all $x \in N \setminus (N_2 \cup h')$, and $v(w) = \hat{v}(w)$, for all $w \in \hat{W}$.

Transformation (D) reduces two choices that are irrelevant at a subset of the information set (where they are available) to their "relevant" parts.

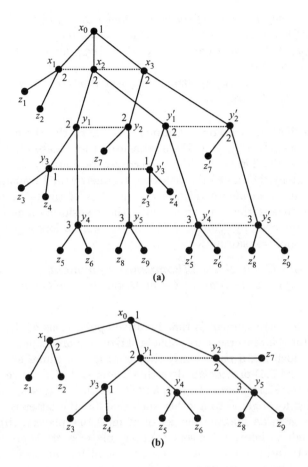

Figure 4.5: Deletion of irrelevant choices

Thereby, (D) eliminates the "duplicate" part of the tree. Elmes and Reny ([55], Lemma 5.2, using a result by Thompson [175]) have shown that (D) does not affect the (semi-reduced) normal form of the game. The role of the D-transformation (or, rather, its inverse) is to order originally unordered information sets.

Example 4.6 *Consider the extensive form in Figure 4.5(a). The nontrivial choices are as follows. We have $C_1 = \{\{x_1\}, \{x_2\}, \{x_3\}, \{z_3, z_3'\}, \{z_4, z_4'\}\}$ for player 1; for player 2 we have*

$$C_2 = \{\{z_1, y_1, y_2\}, \{z_2, y_1', y_2'\}, \{y_3, z_7\}, \{y_4, y_5\}, \{y_3', z_7'\}, \{y_4', y_5'\}\} ;$$

and for player 3 we have $C_3 = \{\{z_5, z_8, z_5', z_8'\}, \{z_6, z_9, z_6', z_9'\}\}$. The primed nodes are the images of the unprimed ones under the bijection ψ. Assume that $v(w_k) = v(w'_k)$, for all $w_k \in W$ which end at some terminal node $z_k = \{w_k\}$ and for all $w'_k \in W$ which end at some $z'_k = \{w'_k\}$ for $k = 3, ..., 9$. Then it is easy to verify that $c_1 = \{z_1, y_1, y_2\} \in C_2$ and $c_2 = \{z_2, y'_1, y'_2\} \in C_2$ are irrelevant at $h' = \{x_2, x_3\}$. Applying transformation (D), therefore, yields the extensive form in Figure 4.5(b).

Equivalent Extensive Form Games As far as the three "elementary" transformations, (I), (C), and (D), are believed to be inessential, they define when two extensive forms should be considered "essentially the same". Since being "essentially the same" is a reflexive, symmetric, and transitive relation, they define an *equivalence relation* on the class of (finite) extensive form games. This means that, by transitivity, the transformations can be used iteratively. Moreover, by symmetry, inverse transformations can be used along with the original transformations.

Definition 4.5 *G and $\hat{G}$ are **T(hompson)-equivalent** if G is an I-, C-, or D-transform of $\hat{G}$ (or $\hat{G}$ is an I-, C-, or D-transform of G) up to a renaming of players.* [8]

The key to the argument is that iterative applications of the three transformations (or their inverses) are capable of reducing any extensive form game to another which has a particularly simple structure. Elmes and Reny ([55], Lemmas 5.3 and 5.4) provide an algorithm, composed of iterative applications of the three transformations (I), (C), and (D) or their inverses, which reduces any extensive form game to another such that (a) the latter is *conventional* (every move is in an information set of at most one player), (b) *every* play passes through *all* information sets, (c) every player controls precisely *one* information set, and (d) information sets are completely *ordered*. On this game (I) is used (sufficiently often) to reduce it to a game $\tilde{G}$ which is even simpler and meets the following definition.

Definition 4.6 *An extensive form game $\tilde{G} = (\tilde{F}, \tilde{v})$ is called **one-shot** if the nodes are only the root and terminal nodes, i.e., $\tilde{N} = \left(\tilde{W}, (\{w\})_{w \in \tilde{W}}\right)$, and*

$$\sum_{\{w\} \in c \cap [\cap_{j \neq i} c_j]} \tilde{p}(\{w\}) \tilde{v}(w) = \sum_{\{w\} \in c' \cap [\cap_{j \neq i} c_j]} \tilde{p}(\{w\}) \tilde{v}(w) \qquad (4.5)$$

for all $(c_j)_{j \neq i} \in \times_{j \neq i} \tilde{C}^j$ implies $c = c'$, for all $c, c' \in \tilde{C}_i$, for all $i = 1, ..., n$.

In a one-shot game all players decide once and for all simultaneously at the root of the tree (i.e., the root is the only move). Since the reduction algorithm is rather cumbersome, its definition is avoided here (see Elmes and Reny [55]). But an example will illustrate how it works.

[8] A *renaming of players* is a bijection from the set $\{1, ..., n\}$ of players' names to itself.

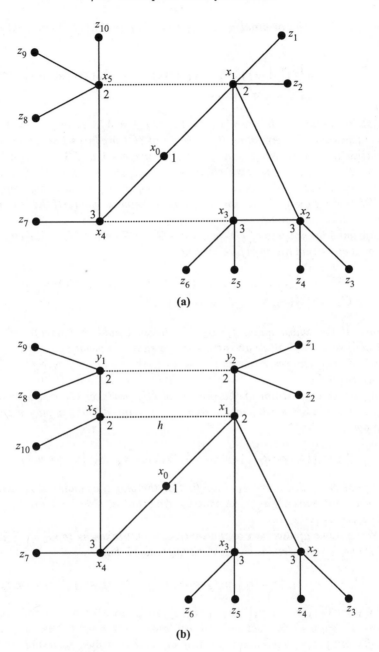

(a)

(b)

Figure 4.6: Applying the inverse of (C)

Example 4.7 *The original game G_1 is in Figure 4.6(a). Nontrivial choices are*

$$C_1 = \{\{x_1\}, \{x_4\}\}, C_2 = \{\{z_1, z_8\}, \{z_2, z_9\}, \{x_2, x_3, z_{10}\}\}\ \ and$$
$$C_3 = \{\{z_3, z_5, z_7\}, \{z_4, z_6, x_5\}\}\ .$$

Probability assignments are $p(x_3) = 1 - p(x_2) = 1/3$ (and $p(x) = 1$ for all other nodes $x \in Y$). Applying the inverse of (C) produces the extensive form G_2 in Figure 4.6(b). Nontrivial choices for players 1 and 3 are as in G_1, but for player 2 there is an extra choice $c = \{y_1, y_2\}$ on top of those from G_1. This is the choice that gets removed according to (C) when going from G_2 to G_1. Probability assignments are as before. (Convince yourself that the normal form remains unchanged.)

Applying the inverse of (D) generates G_3 in Figure 4.7(c). Player 1's nontrivial choices remain as before, but now

$$C_2 = \{\{z_1, z_8\}, \{z_2, z_9\}, \{x_2, x_3, z_7, z_{10}\}, \{y_1, y_2, z_7'\}\}\ \ and$$
$$C_3 = \{\{z_3, z_5, y_3\}, \{z_4, z_6, x_5\}\}\ ,$$

because of the added move y_3. By (D) it must hold that payoffs (for plays that end) at z_7 and at z_7' are precisely the same. Probability assignments are as before. The choices irrelevant at $h' = \{y_3\}$ are $\{x_2, x_3, z_7, z_{10}\} \in C_2$ and $\{y_1, y_2, z_7'\} \in C_2$.

A further application of the inverse of (D) produces G_4 in Figure 4.7(d). Players 1's and 3's nontrivial choices are unchanged, but for player 2 choices are now

$$C_2 = \{\{z_1, z_8, z_7'\}, \{z_2, z_9, z_7''\}, \{x_2, x_3, z_7, z_{10}\}, \{y_1, y_2, y_4\}\}\ ,$$

because of the added move y_4. Payoffs (for the play that ends) at z_7'' are as for (the play that ends at) z_7. The choices irrelevant at $h' = \{y_4\}$ are $\{z_1, z_8, z_7'\}$ and $\{z_2, z_9, z_7''\}$.

Yet another application of the inverse of (D) takes us to G_5 in Figure 4.8 (e). Player 1's nontrivial choices are unchanged, but

$$C_2 = \{\{z_1, z_1', z_8, z_7'\}, \{z_2, z_2', z_9, z_7''\}, \{x_2, x_3, z_7, z_{10}\}, \{y_1, y_4, y_5\}\}$$

and $C_3 = \{\{z_3, z_5, y_2, y_3\}, \{z_4, z_6, x_5, y_2'\}\}$, because of the added move y_5. Player 3's information set h_3 now contains the moves x_2, x_3, x_4, and y_5. Payoffs for plays that end at primed or double-primed terminal nodes are as for those that end at the unprimed versions. Probability assignments are as above. The choices irrelevant at $h' = \{y_5\}$ are $\{z_3, z_5, y_2, y_3\} \in C_3$ and $\{z_4, z_6, x_5, y_2'\} \in C_3$.

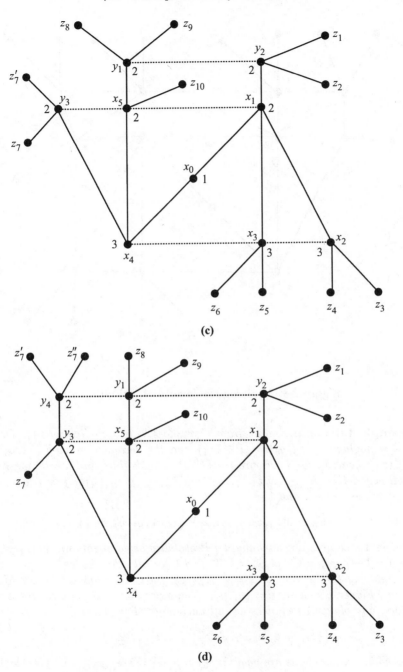

Figure 4.7: Applying the inverse of (D) twice

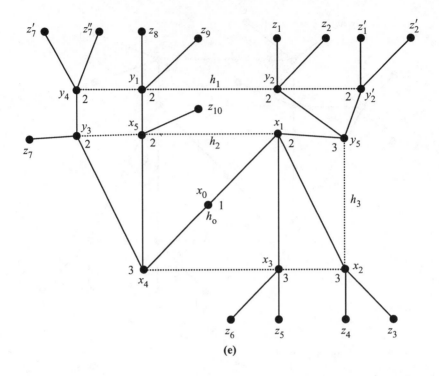

(e)

Figure 4.8: Applying the inverse of (D) once more

Example 4.8 *Continuing on from Example 4.7, two applications of (I), the first removing* $\{x_2, x_3, y_5\} = P^{-1}(x_1)$ *and the second removing* $\{x_5, y_3\} = P^{-1}(x_4)$, *generate* G_6 *in Figure 4.9(f).[9] Player 2's third and fourth nontrivial choices now become* $\{z_3, ..., z_6, z_7, z_{10}\}$ *and* $\{y_1, y_2, y_2', y_4\}$ *(the first two remain the same as in* G_5*), and*

$$C_3 = \{\{z_3, z_5, y_2, y_4, z_7\}, \{z_4, z_6, y_1, y_2', z_{10}\}\}.$$

(Player 1's choices are unchanged.) Probability assignments are now $p(z_5) = p(z_6) = 1 - p(z_3) = 1 - p(z_4) = 1/3$ *(and 1 for all other nodes).*

Applying (C) to remove $c = \{y_1, y_2, y_2', y_4\} \in C_2$ produces G_7 in Figure 4.9(g). Player 2's choice $c = \{y_1, y_2\}$ disappears, her other choices remain as before, and player 1's choices are all unchanged. For player 3 now

$$C_3 = \{\{z_1, z_2, z_3, z_5, z_7, z_7', z_7''\}, \{z_1', z_2', z_4, z_6, z_8, z_9, z_{10}\}\}.$$

Probability assignments are again $p(z_5) = p(z_6) = 1 - p(z_3) = 1 - p(z_4) = 1/3$. *(Would you have believed that* G_7 *is "essentially" the same game as* G_1*?)*

---

[9]   It is also feasible to apply (C) first to give player 2 only one information set, and then to proceed with applications of (I).

A final application of (I), removing $\{x_1, x_4\} = P^{-1}(x_0)$, now generates a one-shot game with 14 terminal nodes, where for players 2 and 3 nontrivial choices are as in G_7, but $C_1 = \{\{z_1, z_2, z_1', z_2', z_3, ..., z_6\}, \{z_7, z_7', z_7'', z_8, z_9, z_{10}\}\}$, and probability assignments are as in G_7. Payoffs coincide for (plays that end at) z_k and z_k' for $k = 1, 2$ and for (plays that end at) z_7, z_7', z_7''. Yet, by construction, these payoff ties do not induce (strategically) equivalent strategies in the normal form. (Construct the normal form!)

Once an extensive form game has been reduced to a one-shot game, things become transparent. In a one-shot game (nontrivial) choices coincide with pure strategies and there are no duplicate strategies by (4.5). It follows that *two* one-shot games will have identical normal forms if and only if they are the same.

This is easily extended. Say that two one-shot games are *isomorphic* if they are the same up to a renaming of players.[10] Likewise, say that two extensive form games have the *same normal form* (instead of "identical normal form") if the two (semi-reduced) normal forms are the same up to a renaming of players and a relabeling of pure strategies.[11] Then, the conclusion is that two one-shot extensive form games have the same normal form *if and only if* they are isomorphic.

Up to this point it has been argued that, if two extensive form games are T-equivalent, then they have the same normal form. Now consider two extensive form games, G and $\hat{G}$, that have the same (semi-reduced) normal form. Reduce G to a T-equivalent one shot-game $\tilde{G}_1$ by the procedure illustrated above. Likewise, reduce $\hat{G}$ to a one-shot game $\tilde{G}_2$. Because G and $\hat{G}$ have the same normal form, $\tilde{G}_1$ and $\tilde{G}_2$ must be isomorphic. Hence there is a renaming of players that takes $\tilde{G}_1$ into $\tilde{G}_2$. Inverting the original process, one can now go from $\tilde{G}_2$ back to $\hat{G}$. Hence, G and $\hat{G}$ are T-equivalent. This demonstrates (a version of) Thompson's theorem.

Theorem 4.1 *Two extensive form games G and $\hat{G}$ with perfect recall have the same normal form if and only if they are T-equivalent.*

Thompson's original setup allowed one extra transformation, called "Inflation/Deflation", and another version of (D), both of which were capable of destroying perfect recall (see [55] and [151]). The theorem still goes through in his original setup, but without the qualification that G and $\hat{G}$ satisfy perfect recall. The advantage of the present version is that the above three transformations preserve perfect recall. Hence, there is less doubt as to whether

---

[10]   More precisely, there is (a) a renaming ι of players, (b) a bijection ϕ on nodes such that the choices of $\iota(i)$ are the images (under ϕ) of i's choices, and (c) the payoff to $\iota(i)$ at the image (under ϕ) of a play equals the payoff to i from the play, for all plays and all i's.

[11]   A relabeling of strategies is, for each i, a bijection from i's pure strategies (in the semi-reduced normal form) to $\iota(i)$'s pure strategies (in the semi-reduced normal form) which preserves the payoff function accordingly.

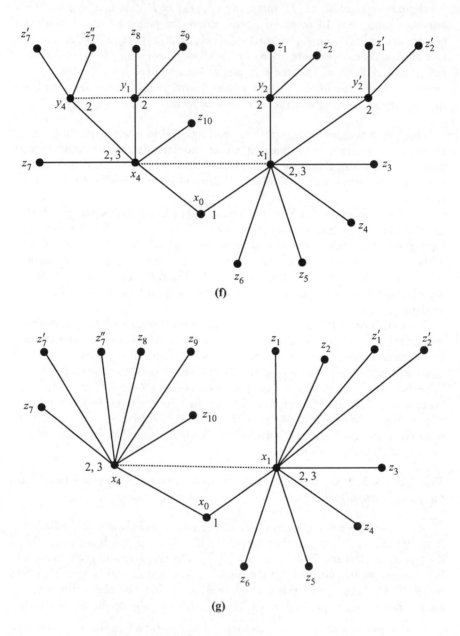

Figure 4.9: Applying (I) twice and (C) once

the present transformations are truly "inessential". (A proof of the following proposition is given in the appendix to this section.)

Proposition 4.2 *If G and $\hat{G}$ are T-equivalent, then G satisfies perfect recall if and only if $\hat{G}$ satisfies perfect recall.*

Of course, T-equivalence is based on a concept of rationality which ignores "framing effects" (effects induced by the way a problem is presented). After all, the Thompson transformations describe essentially different "framings" of the same strategic situation. But the underlying decision theory also ignores framing effects—at least if one adopts something like an independence axiom. Indeed, how far would a theory take us that assumes behavior driven by framing effects? In fact, a theory that does not depend on the fine details of the extensive form is probably more robust to misspecifications.

Summary

- "Interchanging simultaneous decisions", "Coalescing information sets", and "Deletion of irrelevant choices" characterize the equivalence class of extensive form games with perfect recall which have the same (semi-reduced) normal form.

- Therefore, any normal form game represents a class of (T-equivalent) extensive form games.

- Any solution concept that refers to information not contained in the normal form must give a plausible explanation for why at least one of the above three transformations is inappropriate.

Appendix

Proof of Proposition 4.2   Transformation (I) does not affect the number of choices, their ordering, or the plays passing through them. Hence, if G satisfies perfect recall, any I-transform $\hat{G}$ also does. Similarly, if G violates perfect recall, this violation must carry over to an I-transform $\hat{G}$.

Transformation (C) removes one choice, but otherwise preserves the ordering of choices and the plays passing through them. Hence, if G satisfies perfect recall, so will any C-transform $\hat{G}$. Now suppose a C-transform $\hat{G}$ satisfies perfect recall. The original game G has only one choice more and the set of plays passing through this extra choice $c \in C_i$ is the union of plays passing through choices available at the *same* information set in $\hat{G}$. These latter choices, moreover, come after the added choice c in G. Hence, if the C-transform $\hat{G}$ satisfies perfect recall, then G will also.

As for transformation (D), its domain are perfect recall games, so the game G that contains the two choices $c_1, c_2 \in C_i$, which are irrelevant at a subset of their information set, must have perfect recall. It remains to show that the image $\hat{G}$ also satisfies perfect recall, if G does. The ordering of all choices other than c_1 and c_2 is preserved, because for any play through N_2 there is a "copy" through N_1 which does not get removed by (D). Since (D) effectively removes plays, the sets of plays passing through all these choices are reduced by the same overall set of plays (those through h' and c_2). Hence, any violation of perfect recall in $\hat{G}$ must involve c_1 and/or c_2. But condition (c) in Definition 4.4 ensures that choices that come after c_1 or c_2 at h' in G do *not* come after c_1 or c_2 in $\hat{G}$. Hence, if G satisfies perfect recall, any D-transform $\hat{G}$ also will. This completes the proof of Proposition 4.2.

The rest of this appendix is devoted to showing that the inverse of the mapping $W(m) = \{w \in W \mid \exists x \in m : w \in x\}$ from 2^N to 2^W, restricted to the class $\mathcal{M}$ of sets of nodes below, is given by W^{-1} as defined in (4.3). Let $\mathcal{M}$ be the class of all sets of nodes $m \subseteq N$ for which there is *no* node $y \in N$ such that

$$x \subset y \subseteq W(m) \text{ for some } x \in m. \qquad (4.6)$$

The class $\mathcal{M} \subseteq 2^N$ consists of disjoint unions of nodes which provide the coarsest partition of the (set of) plays passing through them. This is what statement (c) of the next result says. The most prominent examples of sets in $\mathcal{M}$ are the nontrivial *choices* of players in a perfect recall game.

Lemma 4.1 *For a set $m \subseteq N$ the following three statements are equivalent:*
   (a) $m \in \mathcal{M}$;
   (b) either $m = \{W\}$ or m satisfies $W \notin m$ and $P(x) \setminus W(m)$ is nonempty for all $x \in m$;
   (c) if $x, y \in m$ and $x \cap y \neq \emptyset$, then $x = y$, and if $W(m) \supseteq y \in N$, then there is $x \in m$ such that $y \subseteq x$.

Proof We show that (b) implies (a), that (a) implies (c), and that (c) implies (b). To see that (b) implies (a), observe that the empty set $\emptyset$ and the singleton set $\{W\}$ consisting of the root only belong to $\mathcal{M}$. If $m \subseteq N$ is nonempty and $W \notin m$, then, for every $x \in m$ and $y \in N$ with $x \subset y$, the fact that $P(x) \subseteq y$ implies $y \setminus W(m) \neq \emptyset$, because $P(x) \setminus W(m)$ is nonempty for all $x \in m$ by hypothesis.

To show that (a) implies (c), let $x \in m \in \mathcal{M}$. If $x \subset y$, then $y \notin m$, because otherwise (4.6) fails. Hence, $x, y \in m$ and $x \subseteq y$ imply $x = y$. If $y \subseteq W(m)$, then for all plays $w \in y$ there is $x_w \in m$ such that $w \in x_w$. If for some $w \in y$ the inclusion $x_w \subset y$ holds, then $W(m) \supseteq y \supset x_w \in m$ violates (4.6). Hence, $y \subseteq x_w \in m$ for all $w \in y$, viz. (c).

Finally, to see that (c) implies (b), consider first the case $W \in m$. Then $m = \{W\}$, from (c), and $x \cap W \neq \emptyset$, for all $x \in N$, verify (b). If m is empty,

(b) holds trivially. If $m \subseteq N$ is nonempty, if it satisfies (c), and if $W \notin m$, then (b) must hold for all $x \in m$, because otherwise $P(x) \subseteq W(m)$ for some $x \in m$ implies that there is $x' \in m$ such that $P(x) \subseteq x'$ by the hypothesis, so $x \subset x'$ and $x, x' \in m$ contradict the first part of (c). This completes the proof of Lemma 4.1.

The main theorem shows that $W : \mathcal{M} \to 2^W$ is onto (surjective) and one-to-one (injective) viz. bijective. Intuitively, the theorem says that, in a perfect recall game, every set of plays can be selected by an appropriate assignment of nontrivial choices for the players.

Theorem 4.2 *The mapping $W : \mathcal{M} \to 2^W$ defined by*

$$W(m) = \{w \in W \mid \exists x \in m : w \in x\}$$

for all $m \in \mathcal{M}$ is a bijection (onto and one-to-one).

The proof proceeds in three steps.

Claim 1 $W^{-1}(a) \in \mathcal{M}$ for all sets of plays $a \in 2^W$.

Proof If $a = W$ (resp. $a = \emptyset$), then $W^{-1}(W) = \{W\} \in \mathcal{M}$ (resp. $W^{-1}(\emptyset) = \emptyset \in \mathcal{M}$). Next, let $x \in W^{-1}(a)$ for some nonempty $a \subset W$ and $x \subset y \in N$. Then $x \in W^{-1}(a)$ implies that $y \setminus a$ is nonempty by (4.3). If $y \subseteq W\left(W^{-1}(a)\right)$, then for every $w \in y$ there is $x_w \in W^{-1}(a)$ such that $w \in x_w$. But then $x_w \subseteq a$ for all $w \in y$ implies $y \subseteq a$, a contradiction. Hence, $y \setminus W\left(W^{-1}(a)\right)$ is nonempty. Since $x \in W^{-1}(a)$ and $y \in N$ are arbitrary, $W^{-1}(a) \in \mathcal{M}$. This completes the proof of Claim 1.

Claim 2 $W^{-1}\left(W\left(m\right)\right) = m$ for all $m \in \mathcal{M}$.

Proof If m is empty, then $W^{-1}\left(W\left(\emptyset\right)\right) = \emptyset$, and if $W \in m$, then by Lemma 4.1(b) $m = \{W\}$, so $W^{-1}\left(W\left(\{W\}\right)\right) = \{W\}$ as required. Hence, consider a nonempty set $m \in \mathcal{M}$ with $W \notin m$.

For any $x \in m$ we have $x \subseteq W(m)$ and $P(x) \setminus W(m) \neq \emptyset$ from Lemma 4.1(b). Hence, $x \in W^{-1}\left(W\left(m\right)\right)$ for all $x \in m$ implies $m \subseteq W^{-1}\left(W\left(m\right)\right)$.

If $x \in W^{-1}\left(W\left(m\right)\right)$, then $x \subseteq W(m)$ and $P(x) \setminus W(m) \neq \emptyset$, from $W \notin m$ and (4.3). Therefore, for every $w \in x$ there is $x_w \in m$ such that $w \in x_w$. By (4.6) it cannot be the case that $x_w \subset x$ for some $w \in x$, because $x \subseteq W(m)$. Hence, $x \subseteq x_w$ for all $w \in x$. But then, from the definition of a tree and (4.6), there is a single $x' \in m$ such that $x \subseteq x'$. If $x \subset x'$, then $x' \subseteq W(m)$ would imply the contradiction $x \notin W^{-1}\left(W\left(m\right)\right)$. Thus $x = x'$ and, therefore, $x \in m$. Hence, $W^{-1}\left(W\left(m\right)\right) \subseteq m$ completes the proof of Claim 2.

Claim 3 $W\left(W^{-1}\left(a\right)\right) = a$ for all $a \in 2^W$.

Proof If $a = W$, then $W^{-1}(a) = \{W\}$ and $W\left(W^{-1}(a)\right) = W$. Likewise, if a is empty, then $W\left(W^{-1}(a)\right) = \emptyset$. Now assume $W \neq W \setminus a \neq \emptyset$.

If $w \in a$, then there is $x \in W^{-1}(a)$ such that $w \in x$. For, let $N(w)$ denote (the set of) all nodes $y \in N$ with $w \in y$. By the definition of a tree, $N(w)$ is a completely ordered (by set inclusion) chain which contains the terminal node $\{w\}$ and the root W. For the latter two, $\{w\} \subseteq a$ and $P(W) \setminus a = W \setminus a \neq \emptyset$ hold. By Zorn's lemma, there exists a largest (with respect to set inclusion) element $x \in N(w)$ which satisfies $x \subseteq a$. This element must satisfy $P(x) \setminus a \neq \emptyset$ by maximality. Hence, $w \in x \in W^{-1}(a)$ by (4.3). Therefore, $w \in a$ implies $w \in W\left(W^{-1}(a)\right)$, for short, $a \subseteq W\left(W^{-1}(a)\right)$.

If $w \in W\left(W^{-1}(a)\right)$, then there is $x \in W^{-1}(a)$ such that $w \in x$. Hence, $w \in x \subseteq a$, because $x \in W^{-1}(a)$. Therefore, $W\left(W^{-1}(a)\right) \subseteq a$ completes the proof of Claim 3.

By Claim 2, $W : \mathcal{M} \to 2^W$ is one-to-one. By Claim 3 it is onto. So, the combination of the three steps verifies Theorem 4.2. (Note that if nontrivial choices are required to be elements of $\mathcal{M}$, this implies no-absent-mindedness, as specified in Definition 3.11.)

Exercises

Exercise 4.7 For each reduction step described in Examples 4.7 and 4.8 (Figures 4.6 to 4.9) determine the semi-reduced normal form in order to verify that (I), (C), and (D) do indeed leave the semi-reduced normal form unchanged.

Exercise 4.8 Return to Exercise 4.1 and reduce the extensive form game described there to a one-shot game by applying the three transformations (I), (C), and (D), or their inverses, iteratively.

Exercise 4.9 Consider the following extensive form game. Player 1 can go L ("left") or R ("right"). Player 2 learns what player 1 has done and then goes either T ("top") or B ("bottom"). Try to reduce this game to a one-shot game by using (I) and (C) only—but not (D). What is the problem?

Exercise 4.10 Give an example of an extensive form game where an application of (C) leads to the disappearance of a (proper) subgame. What is the crucial property in the construction of such an example?

Exercise 4.11 Verify that the following transformation also does not affect the semi-reduced normal form. If all immediate successors of a move are terminal nodes and are contained in a single choice for each player (i.e. so that which terminal node is reached depends solely on chance), then the immediate successors of this move are erased and the move becomes a terminal node at which the expected payoffs from the chance move accrue to the players.

Notes on the Literature: Thompson [175] introduced four transformations, of which the above three are variants (see also Dalkey [35] and Krentel, McKinsey, and Quine [99]). That (two of) Thompson's (four) transformations may destroy perfect recall was pointed out by Elmes and Reny [55]; see also Ritzberger [151], who introduced the three variants of Thompson's transformations presented here. Kohlberg and Mertens [97] state an additional transformation ("Addition of a superfluous mixture") which, if added to those by Thompson, yields the equivalence class of extensive form games with the same mixed-strategy reduced normal form. Theorem 4.2 is a new result.

4.3   Normal Form Information Sets

Payoff Ties   Thompson's theorem characterizes which extensive form games have the same (semi-reduced) normal form. They do not pin down which structures in the normal form "reflect" various extensive form concepts, like choices or information sets. Yet, it is pretty obvious that information sets, for example, are reflected in the normal form by *payoff ties*.

Example 4.9 *Consider again the extensive form in Figure 4.1. Player 1 opens and can either terminate immediately (D_1) or give the move to player 2 (C_1). Player 2 then can also terminate (d) or give the move back to 1 (c). Finally, player 1, if reached, can pick one of two choices (D_2 or C_2). The (semi-reduced) normal form (with plays/resp. terminal nodes standing in for payoffs) is given by (player 1 plays rows and player 2, columns)*

$$
\begin{array}{c|c|c|}
 & d & c \\
\hline
D_1 & z_1 & z_1 \\
\hline
C_1 D_2 & z_2 & z_3 \\
\hline
C_1 C_2 & z_2 & z_4 \\
\hline
\end{array}
$$

Hence, the extensive form is reflected by the fact that the strategy combinations (D_1, d) and (D_1, c) give the same payoffs (result in the same play) and the strategy combinations $(C_1 D_2, d)$ and $(C_1 C_2, d)$ give the same payoffs (play). Observe that the decision between the strategies $(C_1 D_2)$ and $(C_1 C_2)$ of player 1 makes a difference only if player 2 plays c. Likewise, 2's decision between c and d matters only if player 1 picks a strategy in the subset $\{(C_1 D_2), (C_1 C_2)\}$.

The key is the following observation. The choice that a player makes at one of her information sets matters (for the ultimate distribution on plays) if and only if this information set is reached. If a player intends to reach one of her information sets h, she can find herself in one of two situations. Either she will find herself at the information set h, and will have to decide how to continue, or she will find herself at some other information set h', which cannot be reached if h can, and will have to decide how to continue. These decisions can be made *independently*. The player's choice at h does not affect her available options at information sets that cannot be reached if h is reached.

This independence is captured in the normal form. Two strategies that (potentially reach h and) differ with respect to what they prescribe at h, but not otherwise, yield different payoffs if the other players play strategy combinations that reach h, and the same payoffs if the others play strategy combinations that do not reach h. Hence, the decision among strategies that prescribe different choices at an information set has an effect only if the opponents play within a particular subset of their strategy combinations.

Normal Form Information Sets To formalize this idea, fix a (semi-reduced) normal form game $\Gamma = (S, u)$. Two strategies $s_i, s_i' \in S_i$ of player i *agree on a strategy subset* $R_{-i} \subseteq S_{-i} = \times_{j \neq i} S_j$ if

$$u(s_{-i}, s_i) = u(s_{-i}, s_i'), \text{ for all } s_{-i} \in R_{-i}, \tag{4.7}$$

where $u = (u_1, ..., u_n)$. The most radical form of agreement is strategic equivalence (Definition 4.2): two strategies are strategically equivalent if and only if they agree on the full set S_{-i}. Hence, in a semi-reduced normal form, two strategies of player i can only agree on proper subsets of S_{-i}.

Definition 4.7 *Let Γ be a semi-reduced normal form game. The strategy subset $R \subseteq S$ is a **normal form information set** for player i if*

(a) *it has a product structure with respect to i's strategy component, i.e. if $R = R_{-i} \times R_i$ where $R_{-i} \subseteq S_{-i}$ and $R_i \subseteq S_i$, and*

(b) *for every (ordered) pair $(s_i, s_i') \in R_i \times R_i$ there is some $\bar{s}_i \in R_i$ which agrees with s_i on R_{-i} and agrees with s_i' on the complement $S_{-i} \setminus R_{-i}$.*

In the previous example $\{C_1 D_2, C_1 C_2\} \times S_2$ and $\{C_1 D_2, C_1 C_2\} \times \{c\}$ are normal form information sets for both players: for any pair (s_i, s_i') (from one of those two sets) let $\bar{s}_i = s_i$, for $i = 1, 2$. Indeed, these two correspond to the two information sets reached in the extensive form when player 1 does not terminate the game at the root.

Two points are worth noting. First, if the strategy subset R is a normal form information set for player i, then so is $(S_{-i} \setminus R_{-i}) \times R_i$. Second, while a normal form information set R for player i has a product structure with respect to i's strategy component (Definition 4.7(a)), the set R_{-i} need *not* have a product structure. Example 4.10 illustrates this.

Example 4.10 *(Selten's horse-game) Player 1 can initially give the move either to player 2 (A_1) or to player 3 (D_1). If player 2 gets to choose, she can either terminate the game (A_2) or give the move to player 3 (D_2). If player 3 gets to choose, she does not know which player gave her the move, and can choose either L or R. The normal form (with terminal nodes standing in for payoffs) is*

	A_2	D_2
A_1	z_5	z_3
D_1	z_1	z_1

L

	A_2	D_2
A_1	z_5	z_4
D_1	z_2	z_2

R

where player 1 plays rows, player 2 columns, and player 3 chooses between matrices. The set $R = [(\{D_1\} \times S_2) \cup \{(A_1, D_2)\}] \times S_3$ is a normal form information set for player 3. (For every (s_3, s_3') let $\bar{s}_3 = s_3$.) But the set $R_{-3} = [(\{D_1\} \times S_2) \cup \{(A_1, D_2)\}]$ is not a product (though its complement

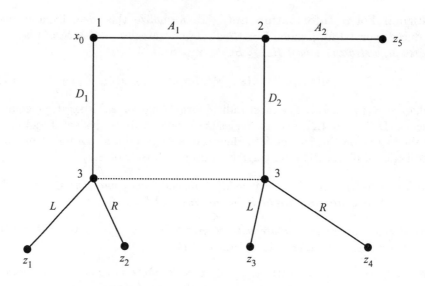

Figure 4.10: Selten's horse game

$S_{-3} \setminus R_{-3} = \{A_1\} \times \{A_2\}$ *is). Also, the set* $\{A_1\} \times S_{-1}$ *is a normal form information set for player 2 (again, for every* (s_2, s_2') *let* $\bar{s}_2 = s_2$*), but not for player 3. (If* $(s_3, s_3') = (L, R)$*, that* $\bar{s}_3$ *agrees with* $s_3 = L$ *on* $\{A_1\} \times S_2$ *implies* $\bar{s}_3 = s_3 = L$*, but* $\bar{s}_3 = L$ *does not agree with* $s_3' = R$ *on* $\{D_1\} \times S_2$*.) However, the set* $\{A_1\} \times S_{-1}$ *has a product structure.*

Normal form information sets "reflect" extensive form information sets in a precise sense. For an extensive form game $G = (F, v)$ and an information set $h \in H_i$ of player i in $F = (\mathcal{T}, C, p)$, let $S(h)$ denote the set of (pure) strategy combinations that reach h; i.e.,

$$S(h) = \left\{ s \in S \;\middle|\; \sum_{w \in W(h)} \pi_s(w) > 0 \right\}$$

where π_s denotes the probability distribution on plays induced by $s \in S$ (see (3.8) in Section 3.3) and $W(h) = \{w \in W \mid \exists x \in h : w \in x\}$. Let Γ_G denote the semi-reduced normal form of G. Mailath, Samuelson, and Swinkels [108] prove the following theorem.

Theorem 4.3 *The strategy subset* $R \subseteq S$ *of the semi-reduced normal form game* $\Gamma = (S, u)$ *is a normal form information set for player i if and only if there exists an extensive form game G without chance moves, with perfect recall, with semi-reduced normal form* $\Gamma_G = \Gamma$*, and with an information set* $h \in H_i$ *for player i such that* $S(h) = R$*.*

The proof of the "if" part relies on perfect recall. The latter implies that two strategies of player i that both reach h must prescribe the same choices at all information sets of player i that come before h, independently of what the other players do. Hence, $S(h) = S_{-i}(h) \times S_i(h)$. Furthermore, for a given pair $(s_i, s_i') \in S_i(h) \times S_i(h)$ the strategy $\bar{s}_i \in S_i(h)$ is constructed by specifying the same choices as s_i at all information sets of i that come before or (weakly) after h, and the same choices as s_i' at all information sets of i that come neither before nor after h. Since this construction ensures that $\bar{s}_i$ agrees with s_i on $S_{-i}(h)$ and with s_i' on $S_{-i} \setminus S_{-i}(h)$, the strategy subset $S(h)$ meets Definition 4.7. And this structure is preserved when passing to the semi-reduced normal form.

The proof of the "only if" portion is constructive. Given a normal form information set $R \subseteq S$ for player i in a (semi-reduced) normal form Γ, an extensive form with two "stages" is specified. First, players other than i choose among all of their strategies and player i chooses either "in" or some strategy in $S_i \setminus R_i$ (simultaneously). Then, after "in" by player i and $s_{-i} \in R_{-i}$ by the other players, player i reaches an information set h. After "in" by player i and $s_{-i} \in S_{-i} \setminus R_{-i}$ by the others, player i reaches an information set h'. Player i's choices at h are given by the equivalence classes of strategies induced by agreement on R_{-i}, and her choices at h' are given by the equivalence classes induced by agreement on $S_{-i} \setminus R_{-i}$. Since in the semi-reduced normal form all strategically equivalent strategies are identified to a single representative, this yields Γ as the associated normal form.

Theorem 4.3 can be extended to (extensive form) games *with* chance moves at the expense of a more involved formulation. Probably the easiest way to account for chance moves is by introducing an extra player ($i = 0$) in the normal form, who has a fixed mixed strategy, but no payoffs. To such a normal form game with "nature", the theorem applies directly. Otherwise, a more complicated formulation is required (see [108], Theorem 2).

Normal Form Subgames An extension of the logic of Theorem 4.3 is to define normal form subgames. Recall that a *subgame* of an extensive form is given by a subtree such that every nontrivial choice (or every information set) is either completely contained in or disjoint from the subtree.

Definition 4.8 *Let Γ be a semi-reduced normal form game. The strategy subset $R \subseteq S$ is a **normal form subgame** if R is a normal form information set for every player.*

A direct consequence of this definition is that a normal form subgame has a product structure, $R = \times_{i=1}^n R_i$. A statement analogous to Theorem 4.3 holds for normal form subgames. That is, R is a normal form subgame if and only if there is an extensive form game (without chance moves, with perfect recall, and with the original semi-reduced normal form) with a subgame that is reached precisely when a strategy combination in R is played.

An issue arises concerning whether the *mixed-strategy* reduced normal form also preserves the extensive form information identified by normal form information sets (or subgames). It turns out that this is not the case.

Example 4.11 *Consider the semi-reduced normal form game below. Player 1 plays rows and 2, columns. Payoffs are such that $u_1(s) = u_2(s)$, for all $s \in S$, so only a single number is given for each strategy combination.*

	s_2^1	s_2^2	s_2^3	s_2^4	s_2^5
s_1^1	2	5	2	5	8
s_1^2	4	6	4	6	8
s_1^3	0	0	4	4	8

The strategy subset $R = \{s_1^1, s_1^2\} \times \{s_2^1, ..., s_2^4\}$ is a normal form subgame. But player 2's strategy s_2^4 gives precisely the same payoffs as a uniform randomization between s_2^1 and s_2^5; i.e., $u(s_1, s_2^4) = \frac{1}{2}u(s_1, s_2^1) + \frac{1}{2}u(s_1, s_2^5)$, for all $s_1 \in S_1$. Eliminating s_2^4 yields the mixed-strategy reduced normal form

	s_2^1	s_2^2	s_2^3	s_2^5
s_1^1	2	5	2	8
s_1^2	4	6	4	8
s_1^3	0	0	4	8

where now the strategy subset $\{s_1^1, s_1^2\} \times \{s_2^1, s_2^2, s_2^3\}$ fails to be a normal form subgame: If $(s_2, s_2') = (s_2^2, s_2^3)$, then agreement on $\{s_1^1, s_1^2\}$ requires $\bar{s}_2 = s_2^2$; but s_2^2 does not agree with s_2^3 on $S_1 \setminus \{s_1^1, s_1^2\} = \{s_1^3\}$.

Hence, passing to the mixed-strategy reduced normal form may eliminate normal form information sets and thereby destroy extensive form information. Possibly even more surprising is the fact that passing to the mixed-strategy reduced normal form may also *create* artificial normal form information sets.

Example 4.12 *Consider the normal form game given by the left matrix below, where again $u_1(s) = u_2(s)$, for all $s \in S$, so only a single payoff number is specified. (Again, player 1 plays rows and 2, columns.) The right matrix is its mixed-strategy reduced normal form.*

	s_2^1	s_2^2	s_2^3
s_1^1	2	5	8
s_1^2	4	6	8
s_1^3	2	4	6
s_1^4	4	5	6

	s_2^1	s_2^3
s_1^1	2	8
s_1^2	4	8
s_1^3	2	6
s_1^4	4	6

Before the reduction (on the left) there is no normal form information set for player 1 with $R_1 = S_1$ and $S_2 \setminus R_2 \neq \emptyset$, because of the presence of player 2's

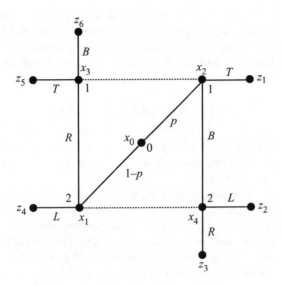

Figure 4.11: Representing both normal form information sets

second strategy. But after the reduction (on the right) both strategy subsets of the form $R = S_1 \times \{s_2\}$, with $s_2 \in S_2$, are normal form information sets for player 1. For example, if (s_1, s_1') is the first and fourth row, then $\bar{s}_1$ is the third row; if (s_1, s_1') is the fourth and the first row, then $\bar{s}_1$ is the second row; and so on.

While Theorem 4.3 identifies the normal form reflections of single information sets from an extensive form, it remains silent on how *all* normal form information sets (in a given normal form) can be represented in a *single* extensive form. Indeed, it turns out that this is not always possible. In the game

	L	R
T	a	a
B	a	b

(where 1 plays rows and 2 columns and a and b denote different payoff vectors) both players have a nontrivial normal form information set. Each of those can be represented by an obvious extensive form game. Either player 1 starts with terminating (payoffs a) or giving the move to 2, who then chooses between a or b; or player 2 starts with terminating (payoffs a) or giving the move to 1, who then chooses between a and b. But there is no single extensive form game where *both* players have an information set that corresponds to knowing that the opponent has given them the move.

However, there is an extensive form game with a chance move (at the root $x_0 = W$), where both players have an information set that corresponds to the *possibility* that the opponent may have given them the move, in Figure 4.11. (Nontrivial choices are identified by labeling the corresponding edges.) If the payoffs that accrue at z_1 and at z_2 are equal, and the payoffs are equal at z_4 and z_5,

$$v(z_1) = v(z_2) \text{ and } v(z_4) = v(z_5),$$

then a is the expected (with respect to the probabilities of the initial chance move) payoff vector from reaching either z_1, z_2, or z_4, z_5, and b is the expected payoff vector from reaching either z_3 or z_6. (Convince yourself by going through the calculations.)

Hence, in the extensive form of Figure 4.11 the normal form information sets of both players get represented simultaneously. But now the information sets do not correspond to "knowing" that the opponent has given up the chance to terminate, but instead correspond to the "possibility" that the opponent has passed on the decision. (Moreover, the extensive form in Figure 4.11 is degenerate in the sense that there are payoff ties across plays.)

Hence, representing all normal form information sets in a single extensive form may come at the cost of changing the interpretation of an "information set". But this is an issue for future research.

Summary

- Extensive form structures, like information sets, are reflected in the normal form by payoff ties.

- Payoff ties, which make the decision between certain strategies relevant only when the others use particular strategies, occur precisely if these strategies reach an information set in an extensive form game. This leads to normal form information sets.

- Extensive form information, like information sets and subgames, is preserved by the (semi-reduced) normal form, but not necessarily by the mixed-strategy reduced normal form.

Exercises

Exercise 4.12 In the normal form below, show that $R = \left\{ s_1^1, ..., s_1^4 \right\} \times \left\{ s_2^1, s_2^2 \right\}$ is a normal form information set for player 1 (the row player).

	s_2^1	s_2^2	s_2^3
s_1^1	a	c	e
s_1^2	a	c	f
s_1^3	b	d	e
s_1^4	b	d	f
s_1^5	g	g	g

Here a, b, c, d, e, and f denote different payoff vectors (u_1, u_2). Can you find an extensive form where R is precisely the set of strategy combinations that reach an information set h?

Exercise 4.13 In the normal form below, show that the strategy subset $R = \{s_1^1, s_1^2\} \times S_2$ is a normal form subgame.

	s_2^1	s_2^2	s_2^3	s_2^4
s_1^1	a	a	b	b
s_1^2	c	c	c	c
s_1^3	d	e	d	e

Again, a, b, c, d, and e denote payoff vectors (u_1, u_2). Is the strategy subset $R' = \{s_1^1\} \times \{s_2^2, s_2^3, s_2^4\}$ also a normal form subgame? Is the strategy subset $R' = \{s_1^1\} \times \{s_2^2, s_2^3, s_2^4\}$ a normal form subgame of the normal form $\Gamma' = (R, u)$?

Exercise 4.14 Find all normal form subgames of the game

	s_2^1	s_2^2	s_2^3
s_1^1	2 2	3 0	3 0
s_1^2	2 2	2 1	5 3
s_1^3	2 2	1 5	6 1

where the upper left number in each entry is player 1's (the row player's) payoff and the lower right is player 2's (the column player's).

Exercise 4.15 In the game from Example 4.10, show that the strategy subset $R = \{A_1\} \times S_2 \times S_3$ is a normal form information set for player 2, but not for players 1 and 3.

Exercise 4.16 In the semi-reduced normal form of the game from Exercise 4.1, find the normal form subgames that correspond to the subgames in the extensive form.

Notes on the Literature: The material in this section is based on Mailath, Samuelson, and Swinkels [108]. (Exercises 4.12, 4.13, and 4.14 are also taken from this source.) Example 4.10 is taken from Selten [163]. The last few remarks are inspired by Mailath, Samuelson, and Swinkels [109] and by an example that Abreu and Pearce [1] have given in a different context.

4.4   The Space of Games

Payoff Variations   When a game is formalized in a non-cooperative approach, the basic ingredients are complete rules, as captured by the extensive form. The very fact of a non-cooperative approach requires a precise understanding of the rules of the game. Yet, there may still be considerable uncertainty about the motives and incentives of players.

Therefore, it is often desirable to obtain results not only for a fixed game, but for a whole class of games that differ only with respect to the players' payoff functions, but not with respect to the rules. The normal form provides a convenient framework to formalize this idea.

In particular, for a fixed number of players and fixed pure strategy sets $S_1, ..., S_n$, the *space of all games* with these strategy sets can be defined as the set of all normal form games generated by varying the payoff function $u : S \rightarrow \mathbb{R}^n$. In general this is a functional space. But for finite S the set of all functions $u : S \rightarrow \mathbb{R}^n$ can be identified with a Euclidean vector space.

The payoff function u of a fixed finite normal form game can be written as a vector

$$u = (u(s))_{s \in S} = ((u_i(s))_{i=1}^n)_{s \in S}$$

with dimension nK, where $K = \prod_{i=1}^n |S_i|$ is the number of pure strategy combinations and $|S_i|$ denotes the number of elements in S_i. Then the space of all games is given by the set of all such vectors, viz. $\mathbb{R}^{nK}$. More precisely, the space of all games (with fixed strategy sets) can be identified with a Euclidean vector space of dimension nK.

The advantage of this identification is that in a Euclidean space there is a natural way to measure distance. Two (normal form) games $\Gamma = (S, u)$ and $\Gamma' = (S, u')$ may be called "similar" or "nearby" if the Euclidean distance $\|u, u'\| = \|u - u'\|$ between their payoff vectors is small.[12] Likewise, there is a natural way to measure when a certain subset of the space of all games is large or small. This can be done, for instance, by determining the volume or Lebesgue measure of the associated set of payoff vectors in $\mathbb{R}^{nK}$.

But there is a caveat to this. Arbitrary variations of the payoff function on S do *not* respect the payoff ties that are characteristic of extensive form structures. Thus, they do not necessarily respect the rules of the game.

Extensive Form Structures   To obtain the space of all games with a fixed *extensive* form, the payoff function $v : W \rightarrow \mathbb{R}^n$ on *plays* has to be varied. The

---

[12]   The Euclidean distance between two vectors x and y is defined as $\|x, y\| = \sqrt{(x - y) \cdot (x - y)}$ where $x \cdot y$ denotes the inner product of x and y.

material in the last section shows how the space of v's is embedded in the space of u's: strategy combinations that induce the same (probability distribution on) plays need to give the same payoffs.

This boils down to a system of linear equations; i.e., if $\pi_s = \pi_{s'}$ then $u(s) - u(s') = 0$, for all $s, s' \in S$. In other words, the space of all games with a fixed extensive form is a *linear subspace* of the space of all games (with fixed strategy sets). Indeed, one can write all the payoff ties that a fixed extensive form induces in its normal form as a linear map (from $\mathbb{R}^{nK}$ to itself), such that the kernel of this map is precisely the set of payoff vectors consistent with the extensive form.[13]

Unless this map is identically zero, the dimension of its kernel is smaller than the dimension nK of the embedding space. But the map is identically zero precisely if the extensive form that it describes is a one-shot game. Therefore, any extensive form that is not as trivial as a one-shot game will give rise to a *lower-dimensional linear subspace*.

Example 4.13 *Consider a simple two-player extensive form game with perfect information where initially player 1 can either terminate immediately or give the move to player 2. If 2 learns that it is her turn to move, she has two choices available. This gives rise to a normal form (where $u = (u_1, u_2)$)*

	s_2^1	s_2^2
s_1^1	$u\left(s_1^1, s_2^1\right)$	$u\left(s_1^1, s_2^2\right)$
s_1^2	$u\left(s_1^2, s_2^1\right)$	$u\left(s_1^2, s_2^2\right)$

where $u_i(s_1^1, s_2^1) - u_i(s_1^1, s_2^2) = 0$ for $i = 1, 2$ if s_1^1 stands for immediate termination by player 1. These two equations reduce the dimension of the extensive form space from $nK = 2 \cdot 4 = 8$ (the dimension of the normal form space) to a six-dimensional linear subspace.

An immediate consequence of this observation is that the "size" of an extensive form space, as measured against its normal form space, is about as small as you can get.[14] Even more strikingly, arbitrarily close (with respect to Euclidean distance in $\mathbb{R}^{nK}$) to a given extensive form game are many other games with almost identical payoffs, but different extensive forms. This seems to suggest that the space of all normal form games (with fixed strategy sets) is too large, because its sheer dimension makes games look close which—in the sense of the extensive form—may be viewed as distant.

Yet, there is often good reason to treat games with radically different extensive forms as being close (or "similar") if their payoff functions are close. The next example illustrates this.

---

[13] The *kernel* of a linear map is the set of all vectors that map into the origin (zero).
[14] Technically, "size" is measured by Lebesgue measure on $\mathbb{R}^{nK}$ and any (nontrivial) extensive form space has measure zero, because it is lower-dimensional.

Example 4.14 *Consider a "leader-follower" situation where first a "leader" (or several "leaders") can choose, and then a group of "followers" learns the leader's choice and gets to choose (simultaneously). Since this gives the leader a commitment opportunity, it will often be advantageous to her. The Stackelberg duopoly is a typical instance of this.*

Now suppose that the observation of the leader's choice by followers is not perfect anymore; i.e., with high probability followers will see the leader's true action, but with a small probability they will erroneously perceive some other action. If this tiny noise in the observation technology does not exclude any of the possible observations, then it can be shown that the first-mover advantage disappears when players are restricted to pure strategies (see [10]).

To this argument one may object that the introduction of imperfect observability radically changes the extensive form. And indeed it does. While in the original game every choice of the leader initiated a proper subgame, the game with imperfect observability does not have any proper subgames. Moreover, the number of plays is substantially larger in the latter as compared with the former game. Thus, the two extensive forms do indeed seem incomparable.

On the other hand, the normal form (without payoffs) of the game with imperfect observability is precisely the same as with perfect observability. As the probability of observing the true choice of the leader goes to 1, the payoffs of the (normal form) game with imperfect observability converge to those of the (normal form) game with perfect observability. Thus, in this case measuring distance between games in normal form space—ignoring the qualitative differences between extensive forms—seems to be the more appropriate way. After all, slightly noisy observations are meant to model an approximation to perfectly reliable observations.

Indeed, as the discussion of the Thompson transformations has shown, often different extensive forms are only different "framings" of the same strategic situation. And when a little bit of uncertainty is involved, there is good reason to turn to the full space of normal form games, in particular when robustness issues are at stake.

Summary

- For finite games the space of all (normal form) games with fixed strategy sets can be identified with a Euclidean vector space.

- The space of all games with a fixed extensive form is a linear subspace of the normal form space.

- Games can be "close to each other" if their payoff functions are close, even if they have different extensive forms.

Exercises

Exercise 4.17 Consider an extensive form game of perfect information where player 1 starts with two choices; then player 2 learns which of her two choices player 1 has taken and also chooses between two actions. What is the dimension of the space of normal form games with these strategy sets, and what is the dimension of the space of games with this extensive form?

Exercise 4.18 Consider two extensive form games. In the first, player i learns which of her $L > 1$ choices the player choosing before her has taken, and then makes her choice. In the second, player i does not learn precisely what the player before her has done, but she receives one of L signals (which are correlated with the choice of her predecessor). This is the only difference between the two extensive form games. Argue that the number of pure strategies of player i is the same in the two extensive forms and that the payoff functions of the two normal forms are the closer the higher the correlation between signals and true actions is.

Exercise 4.19 Using the expected utility hypothesis payoffs could be normalized. This would reduce the dimension of the space of normal form games with fixed strategy sets. By how much at most can it be reduced?

Exercise 4.20 In the game from Example 4.10, how many equations define the subspace that corresponds to this extensive form?

Exercise 4.21 Consider the extensive form from Figure 4.11. Fix the payoffs for the various plays, but vary the probabilities p and $1-p$ of the initial chance move. Which dimension does the resulting space of games have?

Notes on the Literature: The material in this section has been pioneered by Harsanyi [81] (see also Harsanyi [80]). Example 4.14 is based on Bagwell [10] and Güth, Kirchsteiger, and Ritzberger [74]; see also Hurkens and van Damme [91].

Part III

Solutions of Games

5

Solving Games

One could argue that predicting the outcome of a game is a matter of understanding the particular game, its historical context, the relations between the players, their psychology, etc. All these concrete matters may be important for any given game. Hence, many of them get summarized by the payoff function and the rules of the game. Solution theory, however, abstracts from the specific circumstances and attempts to identify general principles which are independent of the particular game under scrutiny.

Abstractly, a *solution concept* is a mapping from the set of games to outcomes or strategy combinations. It assigns to every game one or several "solutions". There are various ways to interpret such a mapping. It can be thought of as a rule which gives normative *recommendations* to rational players on how to behave in any given game. It may also be viewed as the game theorist's *predictions* of what might happen in the various games if those were played by rational players. In practice, it is often a way of arguing why certain observed phenomena persist.

There are a few "minimal" requirements that one may expect to hold for any viable solution concept. In particular,

- there should not be any game to which the solution concept assigns no solution (*existence*), and

- at least for some games, the solution concept should not assign all possible outcomes (*selectiveness*).

The first requirement ("existence") asks for solutions to be nonempty for all games; i.e., *something* should be assigned to *all* games. The second ("selectiveness") requires that the solution concept be not entirely agnostic; i.e., it should *not always* assign *everything* that is possible.

The domain of solution concepts will be normal form games, unless otherwise stated. The simplicity of this representation makes it the perfect vehicle for solution theory. (Here, finally, the hard work in representation theory pays off.)

5.1   Dominance

Dominated Strategies   In certain cases finding a solution is fairly easy. The following example is probably the most prominent game ever. It illustrates that individual and social rationality may be two quite different things (see also Section 1.2).

Example 5.1 *(Prisoners' dilemma) The story goes as follows. Two suspects are taken into custody and separated. The attorney believes that they are guilty of a major crime, but he has no adequate evidence to convict them. So, each of the suspects is confronted with two choices: either to confess to the crime or not. If they both do not confess, the attorney will book them on some minor charge, such as illegal possession of a weapon. If they both confess, they will be prosecuted, but the attorney will recommend less than the most severe sentence. If one confesses and the other does not, then the confessor will receive lenient treatment for turning state's evidence, whereas the latter will receive maximum punishment. If years in prison represent appropriate payoffs, this results in the following game:*

	cooperate	defect
cooperate	-1 $\quad -1$	-10 $\quad 0$
defect	0 $\quad -10$	-8 $\quad -8$

(Player 1 controls rows and player 2, columns; the upper left entry is 1's payoff, the lower right 2's.) The strategy label "cooperate" refers to holding out, and "defect" refers to confessing (which are traditional labels). This game is easy to solve. Observing that for each player "defect" is better than "cooperate", no matter what the opponent does, the only sensible solution is both players defecting. Even if one player expects the other to cooperate, she can make herself better off by defecting. Thus, they will both end up with 8 years in prison.

This example is particularly easy to solve, because the optimal choice for each player does not depend on what the other does. This is unusual, but can happen. The first lesson drawn from it is that a rational player will never use a particular strategy when she has another which always (against all strategy combinations among opponents) does better than the first.

This is, in fact, an implication of rationality. If every player behaves rationally (i.e., payoff maximizing), no player can be expected to select a strategy that is "dominated" in the above sense. More formally:

Definition 5.1 *In a (normal form) game* $\Gamma = (S, u)$ *a strategy* $s_i \in S_i$ *of player* i $(= 1, ..., n)$ *is* **strictly dominated** *if there is some other strategy* $s_i' \in S_i$ *such that*

$$u_i(s_{-i}, s_i') > u_i(s_{-i}, s_i), \text{ for all } s_{-i} \in S_{-i}, \tag{5.1}$$

where $s_{-i} = (s_i, ..., s_{i-1}, s_{i+1}, ..., s_n) \in S_{-i} = \times_{j \neq i} S_j$. *It is* **weakly dominated** *if there is* $s_i' \in S_i$ *such that*

$$\begin{aligned} u_i(s_{-i}, s_i') &\geq u_i(s_{-i}, s_i), \text{ for all } s_{-i} \in S_{-i}, \text{ and} \\ u_i(s_{-i}, s_i') &> u_i(s_{-i}, s_i), \text{ for some } s_{-i} \in S_{-i}. \end{aligned} \tag{5.2}$$

A strategy $s_i \in S_i$ *is* **undominated** *if it is not strictly dominated, and it is* **admissible** *if it is not weakly dominated.*

In the prisoners' dilemma (Example 5.1) "cooperate" is strictly dominated for both players. Hence, rational players will always choose "defect", yielding bilateral defection as the solution of this game.

Definition 5.1 refers only to strategies $s_i \in S_i$ and not to mixtures over those. This is *not* a restriction in two ways, and it *is* restrictive in one respect.

First, it is not restrictive as far as the strategy combinations among opponents are concerned. One might argue that dominance should be defined against beliefs about what the opponents do, hence against *mixed* strategy combinations of the opponents. But, by the expected utility hypothesis, (5.1) and (5.2) hold *if and only if* analogous inequalities hold for all mixed strategy combinations $\sigma_{-i} \in \Theta_{-i} = \times_{j \neq i} \Delta_j$ among the opponents. Second, Definition 5.1 does not allow for a pure strategy to be dominated by a mixed strategy. But, on the other hand, it does not require the game Γ to be finite. Hence, it also applies to the mixed extension of a finite game. In this case, $S = \Theta$ and $u = U$, and Definition 5.1 specifies when a mixed strategy is dominated by some other mixed strategy.[1]

The definition is, however, restrictive in the sense that it compares only strategies of the same status, i.e. pure with pure strategies or mixed with mixed strategies. Therefore it rules out dominance relations between mixed and pure strategies. That this does indeed eliminate something from the picture is illustrated by the following example.

Example 5.2 *Consider the two two-player games below (where only the payoffs for the row player 1 are specified).*

	s_2^1	s_2^2
s_1^1	-4	0
s_1^2	0	-4
s_1^3	-3	-3

	s_2^1	s_2^2
s_1^1	4	0
s_1^2	0	4
s_1^3	3	3

---

[1] Recall that a pure strategy can be thought of as a degenerate mixed strategy. Therefore, when dominance is applied to the mixed extension, more strategies will be dominated.

In the game on the left the third row is dominated by a mixed strategy with probability 1/2 on the first two rows. In the game on the right the same mixed strategy is dominated by the third row.

Hence, a pure strategy can be strictly dominated by a mixed strategy, and a mixed strategy can be strictly dominated by a pure strategy. But dominance concerns what players will or will not choose. So, either they have random devices at their disposal, in which case dominance should be applied to the mixed extension. Or they do not. In the latter case comparisons between pure strategies suffice. After all, if players cannot choose randomized strategies, it would not make sense to eliminate a pure strategy from consideration, only because it is dominated by a (nondegenerate) mixture—which is not available to the player.

One may wonder what the relation between dominated pure and mixed strategies is. Consider a mixed strategy $\sigma_i \in \Delta_i$ the support of which contains a dominated pure strategy $s_i \in S_i$. Since $s_i \in \text{supp}\,(\sigma_i)$ is dominated, there is a (pure or mixed) strategy $\sigma_i' \in \Delta_i$ (with $\sigma_i'(s_i) = 0$, without loss of generality) such that $U_i(s_{-i}, \sigma_i') > U_i(s_{-i}, s_i)$, for all $s_{-i} \in S_{-i}$, and we can choose $\sigma_i'' \in \Delta_i$ such that $\sigma_i''(s_i) = 0$ and $\sigma_i''(s_i') = \sigma_i(s_i') + \sigma_i(s_i)\sigma_i'(s_i')$, for all $s_i' \in S_i \setminus \{s_i\}$. Then,

$$
\begin{aligned}
U_i(s_{-i}, \sigma_i'') &= \sum_{s_i' \in S_i \setminus \{s_i\}} [\sigma_i(s_i') + \sigma_i(s_i)\sigma_i'(s_i')]\, u_i(s_{-i}, s_i') \\
&= \sum_{s_i' \in S_i \setminus \{s_i\}} \sigma_i(s_i')u_i(s_{-i}, s_i') + \sigma_i(s_i)U_i(s_{-i}, \sigma_i') \\
&> U_i(s_{-i}, \sigma_i), \text{ for all } s_{-i} \in S_{-i}.
\end{aligned}
$$

The same construction (of σ_i'') works for weak dominance. This argument proves the following lemma.

Lemma 5.1 *If the support of a mixed strategy $\sigma_i \in \Delta_i$ contains a dominated pure strategy, then σ_i is dominated.*

The converse, of course, is not true. In the game on the right in Example 5.2 the mixed strategy $(1/2, 1/2, 0)$ is (strictly) dominated (by the third row), but its support does not contain any dominated pure strategy.

Whatever the type of strategy (to which dominance gets applied) happens to be, no rational player will ever use a strictly dominated strategy. Hence, eliminating all strictly dominated strategies for all players leaves a "smaller" game which can be perceived as a "solution" of the original game. Such a solution, of course, in general is a *set* of strategy combinations, rather than a single strategy combination. The prisoners' dilemma (Example 5.1) shows that this solution is selective.

To see existence, observe that strict dominance partially orders the set S_i of strategies for each player $i = 1, ..., n$.[2] If S_i is compact,[3] then it follows that the set of *undominated* strategies of player i is nonempty. Therefore, the product of those sets over $i = 1, ..., n$ is also nonempty, and the simple solution concept of eliminating all strictly dominated strategies for all players satisfies existence.

Iterated Dominance   That rational players will not use a strictly dominated strategy is an implication of common knowledge of rationality. But the latter assumption has stronger implications. In particular, consider the "smaller" game obtained by eliminating all strictly dominated strategies for all players. If it is commonly known that players will use only undominated strategies, then some player i may now find that in the "smaller" game *more* of her strategies are dominated; i.e., some of her strategies were originally undominated only because they were good against strategy combinations where some other players used dominated strategies. But if player i knows that her opponents will not use dominated strategies, then she can judge her own strategies against the smaller set of undominated strategy combinations among opponents. This may lead her to discard more of her strategies.

Since every player can reconstruct this reasoning, this process of elimination may actually continue. Round after round, players may discard strategies on grounds that other players will not use certain strategies. The conclusion is that common knowledge of rationality implies that a rational player will not use any strategy that gets discarded at any one of those rounds of elimination.

To make this formal, let $\mathcal{R}$ be the collection of all nonempty compact Borel product subsets[4] of S; i.e., $R \in \mathcal{R}$ if $R = R_1 \times ... \times R_n$ and each $R_i \neq \emptyset$ is a compact (Borel) subset of S_i, for all $i = 1, ..., n$. Define the mapping $\alpha_i : S \rightarrow S_i$ as the *upper contour set* of $s_i \in S_i$ at $s_{-i} \in S_{-i}$; i.e.,

$$\alpha_i(s) = \{s_i' \in S_i \, | u_i(s_{-i}, s_i') \geq u_i(s)\} \tag{5.3}$$

for all $s \in S$ and all $i = 1, ..., n$. Next, let $D_i(R)$ be the set of all of i's strategies that are undominated, given that opponents play in $R_{-i} = \times_{j \neq i} R_j$; i.e., $D_i(R)$

---

[2]   A *partial ordering* is an irreflexive and transitive relation. Since no strategy can dominate itself, dominance is irreflexive. If s_i^1 strictly dominates s_i^2 and s_i^2 strictly dominates s_i^3, then s_i^1 must strictly dominate s_i^3, viz. transitivity.

[3]   A subset of a topological space is *compact* if every open covering contains a finite subcovering. If this is too abstract, think of it as a subset of some finite-dimensional Euclidean space, in which case it is compact precisely if it is closed and bounded. In particular, it is compact if it has finitely many elements. Every sequence from a compact metric space contains a subsequence which converges to a point from the space.

[4]   *Borel sets* are sets belonging to the Borel σ-algebra on a metric space. A σ-*algebra* is a collection of subsets of the underlying space which contains the whole space and is closed under countable union and complementation. The *Borel* σ-algebra is the (unique) smallest (or coarsest) σ-algebra containing all open sets (generated by the metric on the space).

is the set of all $s_i \in S_i$ such that for all $s_i' \in S_i$ there is $s_{-i}' \in R_{-i}$ such that $u_i(s_{-i}', s_i) \geq u_i(s_{-i}', s_i')$. Using upper contour sets, this is equivalent to

$$D_i(R) = \cap_{s_i \in S_i} \cup_{s_{-i} \in R_{-i}} \alpha_i(s_{-i}, s_i) \qquad (5.4)$$

for all i. Finally, the mapping $D : \mathcal{R} \to \mathcal{R}$ is defined by $D(R) = D_1(R) \times ... \times D_n(R)$, for all $R \in \mathcal{R}$. In words, $D(R)$ is the set of undominated strategy combinations when players are restricted to play in R. That the image of $R \in \mathcal{R}$ under D is in $\mathcal{R}$ requires the assumption that the payoff function u is continuous and that S is a compact (sub)set (of a complete metric space). This will always be assumed. Let D^k, for $k = 1, 2, ...$, be defined recursively by $D^k(R) = D\left(D^{k-1}(R)\right)$, where $D^0(R) = R$, and $D^\infty(R) = \cap_{k=1}^{\infty} D^k(R)$, for all $R \in \mathcal{R}$.

Definition 5.2 *For a (finite or infinite) game $\Gamma = (S, u)$ with compact strategy space S and continuous payoff function u, a strategy for player i is **iteratively undominated** if it belongs to the i-th component of $D^\infty(S) = D_1^\infty(S) \times ... \times D_n^\infty(S)$.*

As pointed out, common knowledge of rationality implies that rational players will never use strategies that are not iteratively undominated. Hence, the set $D^\infty(S)$ constitutes potentially a solution concept. While in the prisoners' dilemma one round of elimination coincides with $D^\infty(S)$, in other games more predictive power is obtained from iteration.

Example 5.3 *Consider a Cournot duopoly where two firms wish to choose their output levels $q_i \in [0, 4]$, for $i = 1, 2$, so as to maximize profits, given that the market price p depends on their output choices according to*

$$p = p(q_1, q_2) = \max\{0, 8 - q_1 - q_2\}.$$

Each firm has a cost function $C_i(q_i) = 2q_i$ with constant marginal cost. Accordingly, profits are $\Pi_i(q_i, q_{3-i}) = q_i[p(q_1, q_2) - 2]$, for $i = 1, 2$.[5] Let $x_t = 2 - 2^{1-t}$ and $y_t = 2 + 2^{1-t}$, for all $t = 0, 1, 2, ...$, so $x_0 = 0$ and $y_0 = 4$. Suppose it is common knowledge among the firms that they will choose output levels only in the intervals $[x_t, y_t]$. Then, for any output level $q_i \in [x_t, 3 - y_t/2) = [2 - 2^{1-t}, 2 - 2^{-t})$ and all $q_{3-i} \in [x_t, y_t]$, one has

$$\Pi_i(q_i, q_{3-i}) = q_i(6 - q_1 - q_2) < \Pi_i\left(3 - \frac{y_t}{2}, q_{3-i}\right)$$

$$= \left(3 - \frac{y_t}{2}\right)\left[3 + \frac{y_t}{2} - q_{3-i}\right]$$

---

[5]   The special assumptions on the "inverse demand function" $p(.)$ and the cost functions $C_i(.)$ are, in fact, far more restrictive than is necessary. What is really needed is that cost functions are increasing and weakly convex, that the slope $p'(.)$ is negative, and that the cross partials $\partial^2\Pi_i/\partial q_i\partial q_{3-i}$ are negative, for $i = 1, 2$.

if and only if

$$0 < q_i^2 - 2\left(3 - \frac{q_{3-i}}{2}\right)q_i + \left(3 - \frac{y_t}{2}\right)\left[2\left(3 - \frac{q_{3-i}}{2}\right) - \left(3 - \frac{y_t}{2}\right)\right]$$

where the latter (quadratic) inequality follows from $q_i < 3 - y_t/2$, for $i = 1, 2$. Hence, all output levels q_i between x_t and $3 - y_t/2$ are strictly dominated (by $q_i' = 3 - y_t/2$) in the game where firms are constrained to choose output levels in $[x_t, y_t]$. Similarly, $q_i > 3 - x_t/2$ implies $\Pi_i(q_i, q_{3-i}) < \Pi_i(3 - x_t/2, q_{3-i})$ for all $q_{3-i} \in [x_t, y_t]$, for $i = 1, 2$. Hence, all output levels q_i between $3 - x_t/2$ and y_t are strictly dominated (by $q_i' = 3 - x_t/2$) in the game where firms choose from $[x_t, y_t]$. Eliminating these strictly dominated strategies yields $q_i \in [3 - y_t/2, 3 - x_t/2] = [2 - 2^{-t}, 2 + 2^{-t}] = [x_{t+1}, y_{t+1}]$, for $i = 1, 2$. Since x_t and y_t both converge to 2 as t goes to infinity, the set of iteratively undominated strategies yields $q_i = 2$, for $i = 1, 2$. In this example, therefore, the solution by iteratively undominated strategies is extremely selective. It picks a single strategy combination.

In this example, at each round of elimination *some* dominated strategies were discarded. But there was no argument that these were all dominated strategies. Yet, a unique solution was obtained. This is no coincidence. The third statement of the following result shows that it does not matter which (strictly) dominated strategies are eliminated at the various rounds of the iteration. (A proof is provided in the appendix to this section.)

Theorem 5.1 *For any (finite or infinite) game $\Gamma = (S, u)$ with compact strategy space S and continuous payoff function u,*

    (a) $D^\infty(S)$ is nonempty and compact;

    (b) $R = D^\infty(S)$ if and only if R is the largest (with respect to set inclusion) set in $\mathcal{R}$ that satisfies $R = D(R)$;

    (c) if $\{R^t\}_{t=0}^\infty$ is any sequence from $\mathcal{R}$ that satisfies $D(R^t) \subseteq R^{t+1} \subseteq R^t$, $R^0 = S$, and that $R^t = R^{t+1}$ implies $R^t = D(R^t)$, for all t, then $R^t = R^{t+1}$ if and only if $R^t = D^\infty(S)$.

The first statement (a) is an existence result. It ensures that iteratively undominated strategies qualify as a solution concept with respect to the existence requirement. Selectiveness is, again, established by the prisoners' dilemma (Example 5.1). Hence, iteratively undominated strategies satisfy the two basic requirements for a viable solution concept.

The second statement (b) says that the set $D^\infty(S)$ of iteratively undominated strategies is the largest "fixed point" under the mapping D.[6] It can be shown (see Samuelson [157], Theorem 2) that this is a sufficient condition for

---

    [6] A *fixed point* of a function mapping a set into itself is a point that (as a point in the domain) maps into itself (as a point in the range). Since D is a mapping from sets to sets, the terminology "fixed *point*" can be confusing, because a "point" in $\mathcal{R}$ is really a set (a subset of S).

it to be common knowledge that players will use only iteratively undominated strategies. Hence, the second statement says that the result of iterative elimination of strictly dominated strategies can be common knowledge among the players.

The third statement (c) says that the order of elimination is immaterial. Any iterative procedure that eliminates at each round some, but not necessarily all, strictly dominated strategies (and starts at S) will come to a halt only at $D^\infty(S)$. Hence, the technique used in Example 5.3 is justified.

Weak Dominance Neither of the two statements (b) and (c) survives extension to weakly dominated strategies. This is because it may not always be common knowledge that players will not use weakly dominated strategies. Indeed, the concepts of weak dominance and common knowledge are inherently contradictory. The former involves an implicit assumption that *all* of the opponents' strategy combinations are possible. The latter yields cases in which it is *known* that some strategies will not be played.

To see that the statement of Theorem 5.1(b) fails for weak dominance, we start with an example which shows that assuming common knowledge of admissibility may imply a contradiction.

Example 5.4 *Consider the two-player bimatrix game below, where player 1 controls rows and player 2, columns. The difficulty here is that s_1^1 weakly dominates s_1^2 for player 1. However, the two differ only if player 2 plays s_2^2 with positive probability, and s_2^2 is inferior if and only if player 1 chooses s_1^1. Admissibility then leads 1 to play s_1^1 if there is any chance that 2 plays s_2^2, but 2 is then led to play only s_2^1, at which point admissibility recommends s_1^1 and s_1^2 to 1, causing s_2^1 and s_2^2 to be recommendations for 2, and so on.*

	s_2^1	s_2^2
s_1^1	1 1	1 0
s_1^2	1 0	0 1

More precisely, suppose it is common knowledge that players will only use strategies in $R = R_1 \times R_2$ on account of admissibility. If $R_2 = \{s_2^1\}$, then $\{s_1^1, s_1^2\} \subseteq R_1$, in which case $\{s_2^1, s_2^2\} \subseteq R_2$, a contradiction. If $R_2 = \{s_2^2\}$, then $R_1 = \{s_1^1\}$, implying $R_2 = \{s_2^1\}$, a contradiction. Finally, if $\{s_2^1, s_2^2\} \subseteq R_2$, then $R_1 = \{s_1^1\}$, in which case $R_2 = \{s_2^1\}$, again a contradiction. Hence, there exists no fixed point under the mapping assigning admissible strategies. This argument can be extended to show that it cannot be common knowledge that players will not use weakly dominated strategies (see Samuelson [157], Theorem 3 and Example 8).

This example illustrates that there are games where it cannot be common knowledge that players do not use weakly dominated strategies. This entails the nonexistence of a fixed point under the mapping assigning admissible strategies.

To see that the independence from the order of elimination (Theorem 5.1(c)) fails for weak dominance, consider the following example.

Example 5.5 *Consider the two-player game given below. Eliminating the third strategy, which is weakly dominated, for both players initially yields a 2×2 game where the second strategy is weakly dominated for each player. Eliminating these strategies yields $\{s_1^1\} \times \{s_2^1\}$ as the final result of iterated admissibility.*

	s_2^1		s_2^2		s_2^3	
s_1^1	1		1		2	
		1		1		1
s_1^2	1		0		3	
		1		0		1
s_1^3	1		1		1	
		2		3		1

On the other hand, if s_1^3 is eliminated initially, then in the next round s_2^2 is weakly dominated. Eliminating s_2^2 yields a 2×2 game where player 1's first strategy s_1^1 is weakly dominated, yielding $\{s_1^2\} \times \{s_2^1, s_2^3\}$ as the final result of iterated admissibility. By symmetry, eliminating first s_2^3, then s_1^2, and finally s_2^1 yields $\{s_1^1, s_1^3\} \times \{s_2^2\}$ as the result of iterating admissibility. Hence, iterated elimination of weakly dominated strategies is sensitive to the order of elimination.

These difficulties with weak dominance can be traced to the fundamental conflict between common knowledge and admissibility. In Example 5.5, if it is common knowledge that players will not use weakly dominated strategies, then player 2 (resp. 1) is well justified in eliminating s_2^2 (resp. s_1^2) as the first strategy from her strategy set, because she can count on player 1 (resp. 2) not using s_1^3 (resp. s_2^3).

This yields paradoxes, such as when admissibility itself is common knowledge but players cannot deduce what this implies, because the outcome of common knowledge of admissibility is not known to the players.

Example 5.6 *To study what players know in Example 5.5, use a Bayesian game (the Harsanyi transformation—see Section 3.5). Let there be three states, ω_1, ω_2, and ω_3, which correspond to the possible reasoning processes of players such that each player knows her own reasoning process.*

State ω_1 corresponds to both players discarding simultaneously their third strategies, so at ω_1 player i uses only s_i^1 and s_i^2, for $i = 1, 2$. State ω_2 corresponds to player 1 discarding s_1^3 and player 2, counting on that, discarding s_2^2,

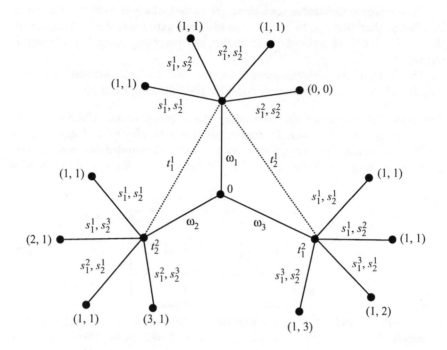

Figure 5.1: What admissibility implies remains unknown

so that at ω_2 player 1 uses only s_1^1 and s_1^2 and player 2 uses only s_2^1 and s_2^3. State ω_3 corresponds to player 2 discarding s_2^3 and player 1, understanding this, discarding s_1^2, so that at ω_3 player 1 has available s_1^1 and s_1^3 and player 2 uses s_2^1 and s_2^2.

Figure 5.1 illustrates the resulting Bayesian game, with the root in the middle (marked "0" for chance). At each immediate successor of the root both players have two choices available simultaneously. The first entry of each payoff vector is 1's payoff and the second is 2's.

Since players know their own reasoning, types are $t_1^1 = \{\omega_1, \omega_2\}$ and $t_1^2 = \{\omega_3\}$ for player 1 and $t_2^1 = \{\omega_1, \omega_3\}$ and $t_2^2 = \{\omega_2\}$ for player 2; i.e., type $t_1^1 = \{\omega_1, \omega_2\}$ of player 1 corresponds to 1 first discarding s_1^3 and type $t_1^2 = \{\omega_3\}$ for 1 corresponds to 1 first discarding s_1^2, and similarly for player 2. (In Figure 5.1 the dashed line indicates t_1^1 and the dash-dotted line, t_2^1.)

All three states correspond to some iteration of admissibility. Therefore, it can be shown (see [157], Lemma 2) that it is common knowledge at each state that players will not use weakly dominated strategies. However, type $t_1^1 = \{\omega_1, \omega_2\}$ of player 1 does not know the event that player 2 has discarded s_2^3 (viz. the event $\{\omega_1, \omega_3\}$), nor does type $t_2^1 = \{\omega_1, \omega_3\}$ of player 2 know that 1 *has discarded s_1^3 (viz. the event $\{\omega_1, \omega_2\}$).[7] Therefore, at state ω_1 the outcome of common knowledge of admissibility is not known to any one of the players.*

We conclude that there are games where admissibility is common knowledge, but what this implies may be unknown to the players.

The above examples are particularly disturbing, because they also apply to games that are commonly perceived as being easy to solve.

In particular, a game is *dominance solvable* if all players are indifferent between all strategy combinations that survive iterated elimination of weakly dominated strategies when at each stage *all* weakly dominated strategies are eliminated. The two games in Example 5.4 and 5.5 are both dominance solvable. Yet, in the first assuming common knowledge of admissibility leads to a contradiction, and in the second admissibility *is* common knowledge, but its implications are not known to the players.

The significance of dominance solvable games derives from extensive form games with perfect information. For almost all such games[8] weak dominance relations in the normal form reflect what Kuhn's algorithm (see Theorem 3.1 in Section 3.1) identifies in the extensive form. It can be shown (see [125]) that iterated elimination of weakly dominated strategies in the normal form of such a game generates the same outcome as Kuhn's algorithm does in the extensive form.[9]

Since in a generic extensive form game with perfect information Kuhn's algorithm gives a unique recommendation, the normal form of such a game is dominance solvable. (That perfect information in an extensive form game generates only weak, but not necessarily strict, dominance relations between strategies in its normal form is due to possibly unreached parts of the tree. Picking a superior choice at a certain move matters only if this move is indeed reached.)

Finally, one could ask what happens if the common knowledge assumption is slightly weakened. Indeed, in such cases some weakly dominated strategies may be eliminated, because if it is "not quite" certain that some strategy will not be used, then admissibility gets some bite. For instance, if there is small uncertainty about the payoff function, then one round of elimination of weakly dominated strategies followed by arbitrarily many rounds of eliminating strictly dominated strategies turns out to be the right solution (see Dekel and Fudenberg [43]). The same procedure emerges if rationality itself is only "almost" common knowledge (see Börgers [27]).

---

[7]   Recall that a decision maker *knows* an event *at a state* if this event is a superset of the decision maker's information set at this state (see Section 2.4).

[8]   The "generic" class of extensive form games with perfect information required here is defined by the property that no player is indifferent between any two distinct plays.

[9]   If not all weakly dominated strategies are removed at each stage, the resulting payoffs are still the same as those generated by Kuhn's algorithm, but the surviving strategy combinations need not induce plays consistent with Kuhn's algorithm.

Summary

- A solution concept assigns to every game a strategy combination or an outcome. Solutions should exist for all games and should select somewhat less than everything possible for some games.

- A strategy is strictly dominated if there is another strategy that always does better. It is weakly dominated if there is another strategy that is always at least as good as the first and sometimes better.

- Common knowledge of rationality implies that no player will use a strategy that gets removed at some stage of iterated elimination of strictly dominated strategies.

- The set of iteratively undominated strategies is the largest fixed point under the mapping assigning undominated strategies. It is reached by any order of elimination of strictly dominated strategies which starts from the full set of all strategies. Neither of these two statements holds for weakly dominated strategies.

Appendix

Proof of Theorem 5.1 The key is the monotonicity of the mapping D defined in (5.4). If $R, R' \in \mathcal{R}$ are such that $R \subseteq R'$, then $s_i \in D_i(R)$ implies that for all $s_i' \in S_i$ there is some $s_{-i}' \in R_{-i} \subseteq R_{-i}'$ such that $u_i(s_{-i}', s_i) \geq u_i(s_{-i}', s_i')$, hence, $s_i \in D_i(R')$. Since this holds for all $i = 1, ..., n$, $D(R) \subseteq D(R')$ if $R \subseteq R'$.

(a) By continuity of u and compactness of R, the set $D(R)$ is nonempty and compact for all $R \in \mathcal{R}$ and, hence, in $\mathcal{R}$. Since D is monotone, $D^{k+1}(S) \subseteq D^k(S)$ for all $k = 0, 1, 2, ...$ follows from $D(S) \subseteq S$, so $D^k(S)$ is nonempty and compact by induction, for all $k = 1, 2, ...$ Since each $D^k(S)$ is nonempty and compact, so is the intersection $D^\infty(S) = \cap_{k=0}^\infty D^k(S)$ by the Cantor intersection theorem.

(b) If $R = D^\infty(S)$, then for each k we have $R \subseteq D^{k-1}(S)$ and so $D(R) \subseteq D^k(S)$, whence $D(R) \subseteq R$.

Suppose there is $s \in R = D^\infty(S)$ with $s \notin D(R)$. Then for some i there is $s_i' \in S_i$ such that $u_i(s_{-i}', s_i) < u_i(s_{-i}', s_i')$ for all $s_{-i} \in R_{-i}$. Since $s \in D^k(S)$, there is $s_{-i}^k \in D_{-i}^{k-1}(S)$ such that $u_i(s_{-i}^k, s_i) \geq u_i(s_{-i}^k, s_i')$ for all k. By compactness, the sequence $\{s_{-i}^k\}_{k=1}^\infty$ (contains a subsequence which) converges to s_{-i}^o. Since at most $k - 1$ points of this (sub)sequence are outside $D_{-i}^k(S)$ and $D_{-i}^k(S)$ is compact, we have $s_{-i}^o \in D_{-i}^k(S)$, for all k, and so $s_{-i}^o \in R_{-i}$. But then continuity of u_i implies that $u_i(s_{-i}^o, s_i) \geq u_i(s_{-i}^o, s_i')$, in contradiction to the hypothesis. Therefore, $s \in R$ implies $s \in D(R)$ and we conclude $R \subseteq D(R)$. Combined with the earlier observation, this yields $R = D(R)$.

Let $R = D^\infty(S)$. If there is $R' \in \mathcal{R}$ with $R \subset R'$ and $R' = D(R')$, then $R' \subseteq S$ implies $R' = D^k(R') \subseteq D^k(S)$, for all $k = 1, 2, ...$, such that $R' \subseteq D^\infty(S)$ contradicts $D^\infty(S) = R \subset R'$.

Conversely, if R is the largest set in $\mathcal{R}$ that satisfies $R = D(R)$, then $R \subseteq S$ implies $R = D^k(R) \subseteq D^k(S)$, for all $k = 1, 2, ...$, hence, $R \subseteq D^\infty(S)$. If $R \subset D^\infty(S)$, then $D^\infty(S) = D(D^\infty(S))$ contradicts the hypothesis that R is the largest fixed point under D. Therefore, $R = D^\infty(S)$.

(c) Let $\{R^t\}_{t=0}^\infty$ be any sequence from R that satisfies $D(R^t) \subseteq R^{t+1} \subseteq R^t$, $R^0 = S$, and that $R^{t+1} = R^t$ implies $D(R^t) = R^t$, for all t. Monotonicity of D implies $D^2(R^t) \subseteq D(R^{t+1})$, for all $t = 1, 2, ...$, such that

$$D^{t+1}(S) \subseteq ... \subseteq D^2(R^{t-1}) \subseteq D(R^t) \subseteq R^{t+1} \subseteq R^t,$$

for all t. If $R^t = R^{t+1}$, then by hypothesis $R^t = D(R^t)$ and by monotonicity $D^\infty(S) \subseteq D^{t+1}(S) \subseteq R^t$. Hence, by (b) $R^t = D^\infty(S)$. Conversely, if $R^t = D^\infty(S)$, then $D(R^t) = R^t$ implies $R^t = R^{t+1}$. This completes the proof of Theorem 5.1.

Exercises

Exercise 5.1 Let strategy spaces S_i for n players $i = 1, ..., n$ be given by $S_i = [0, \bar{s}]$ for some common $\bar{s} > 0$. Each player names a number $s_i \in S_i$ and tries to choose this number as close as possible to a fraction α of the average across all numbers named, $0 < \alpha < 1$. Hence, if $f : R \to R$ is a function which is strictly monotone decreasing (resp. increasing) on the positive (resp. negative) reals, then payoff functions can be taken $u_i(s_1, ..., s_n) = f\left(s_i - \frac{\alpha}{n}\sum_{j=1}^n s_j\right)$ for all $i = 1, ..., n$. Determine the iteratively undominated strategies. (Hint: If it is common knowledge that $s_i \leq \alpha^k \bar{s}$, for some $k = 0, 1, 2, ...$, for all $i = 1, ..., n$, then $s_i > \alpha^{k+1}\bar{s}$ implies $0 \leq \alpha^{k+1}\bar{s} - \frac{\alpha}{n}\sum_{j=1}^n s_j < s_i - \frac{\alpha}{n}\sum_{j=1}^n s_j.$)

Exercise 5.2 Find two sequences of iterated elimination of weakly dominated strategies (where not necessarily all dominated strategies are eliminated at each stage) which give different final results in the game below. (Player 1 controls rows and 2 columns; the upper left entry is 1's payoff and the lower right, 2's.)

	s_2^1	s_2^2	s_2^3
s_1^1	-1 / 1	1 / -1	1 / 3
s_1^2	1 / -1	-1 / 1	1 / 2

Exercise 5.3 Consider a private-value second-price sealed-bid auction. An object is allocated to the highest bidder at a price equal to the second highest bid. Each of the n (risk neutral) bidders $i = 1, ..., n$ has a (maximal) willingness-to-pay v_i which is an independent draw from a common distribution; i.e., the v_i's are independently identically distributed. Participants in the auction make their bids privately and independently (simultaneously). Show that the strategy to bid precisely the willingness-to-pay v_i weakly dominates any other strategy, for any $i = 1, ..., n$.

Exercise 5.4 Each of two players $i = 1, 2$ announces a non-negative integer s_i not exceeding 100. If $s_1 + s_2 \leq 100$, then each player receives payoff s_i, for $i = 1, 2$. If $s_1 + s_2 > 100$ and $s_i < s_{3-i}$, then player i receives s_i and player $3 - i$ receives $100 - s_i$, for $i = 1, 2$. If $s_1 + s_2 > 100$ and $s_1 = s_2$, then each player receives 50. Show that this game is dominance solvable and find the set of surviving outcomes.

Exercise 5.5 Find the set of iteratively undominated strategies for the mixed extension of the following two-player game (player 1 controls rows and 2, columns):

	s_2^1		s_2^2		s_2^3	
s_1^1	0		0		1	
		3		0		1
s_1^2	1		1		0	
		0		3		1

Notes on the Literature: Iterated elimination of dominated strategies was first discussed by Luce and Raiffa ([105], pp. 106-109). Iterative procedures for eliminating strictly dominated strategies, similar to the one presented here, are studied by Gilboa, Kalai, and Zemel [66] and by Dufwenberg and Stegeman [53], with analogous results.[10] The discussion of weak dominance and common knowledge follows Samuelson ([157]; Examples 5.4 and 5.5 are also from this source). The notion of dominance solvability given above is due to Moulin [123], and so is its relation to Kuhn's algorithm [125]. Example 5.3 and Exercise 5.1 are also inspired by Moulin ([124] and [125], p. 72, respectively). An experimental investigation of Exercise 5.1 by Nagel [129] reveals that human subjects in experiments usually do not go through all stages of

---

[10]   The difference concerns which strategies are allowed to dominate. Here it is any strategy $s_i' \in S_i$; Dufwenberg and Stegeman use $s_i' \in R_i$; and Gilboa, Kalai, and Zemel use $s_i' \in D_i(R)$. Dufwenberg and Stegeman, however, prove a lemma to the effect that these differences do not matter.

iterative elimination of strictly dominated strategies. The implications of weakening the assumptions of common knowledge of the game and rationality are studied by Dekel and Fudenberg [43] and Börgers [27].

5.2 Rationalizability

Best Replies Dominance is concerned with what players will *not* choose, because they have unambiguously better alternatives at their disposal. Therefore, that players will use only iteratively undominated strategies is an implication of common knowledge of rationality. But the latter also carries implications about what players *will* choose.

In particular, if a rational player *does* use a certain strategy, then this strategy must be *optimal* (payoff maximizing) against at least some belief about what opponents do. Since this follows from rationality, each player will expect her opponents to choose only strategies that are optimal against some belief. Further, each player will expect every other player to reason this way and, hence, will think that every other player believes that every other player's strategy is optimal against some belief; and so on.

Hence, that players choose only strategies that are optimal against some belief constrains also what a player may expect her opponents to do, i.e. the beliefs she may hold about her opponents' choices. Since this applies iteratively, it ultimately leads to an infinite regress. So, the questions arises of what characterizes the set of all strategies that a rational player could use under common knowledge of rationality.

To formalize this idea of "rationalizable" strategies, start by defining the pure and mixed "best reply" correspondences. Fix a game $\Gamma = (S, u)$ with n players $i = 1, ..., n$. The *pure* best reply correspondence of player i, $\beta_i : \Theta \to S_i$, assigns to every belief $\sigma \in \Theta$ about opponents' behavior the set of all pure strategies that maximize the payoff function of i against this belief; i.e.,

$$\begin{aligned}
\beta_i(\sigma) &= \{s_i \in S_i \,|\, U_i(\sigma_{-i}, s_i) \geq U_i(\sigma_{-i}, s_i'),\text{ for all } s_i' \in S_i\} \quad (5.5)\\
&= \arg\max_{s_i \in S_i} U_i(\sigma_{-i}, s_i) = \bigcap_{s_i \in S_i} \alpha_i(\sigma_{-i}, s_i),
\end{aligned}$$

where α_i is as in (5.3). The *mixed* best reply correspondence of player i, $\tilde{\beta}_i : \Theta \to \Delta_i$, assigns to every belief $\sigma \in \Theta$ about what opponents do the set of all mixed strategies that maximize i's payoff against this belief; i.e.,

$$\begin{aligned}
\tilde{\beta}_i(\sigma) &= \{\sigma_i' \in \Delta_i \,|\, U_i(\sigma_{-i}, \sigma_i') \geq U_i(\sigma_{-i}, \sigma_i''),\text{ for all } \sigma_i'' \in \Delta_i\} \quad (5.6)\\
&= \arg\max_{\sigma_i' \in \Delta_i} U_i(\sigma_{-i}, \sigma_i') = \bigcap_{\sigma_i' \in \Delta_i} \alpha_i(\sigma_{-i}, \sigma_i').
\end{aligned}$$

Note that both these correspondences are functionally independent of the i-th components σ_i of the belief $\sigma = (\sigma_1, ..., \sigma_n) \in \Theta$, which is accounted for only to simplify notation. The associated product mappings $\beta = \beta_1 \times ... \times \beta_n : \Theta \to S$

and $\tilde{\beta} = \tilde{\beta}_1 \times ... \times \tilde{\beta}_n : \Theta \to \Theta$ are denoted by dropping the subscript for the player identity.

The new element concerning rationalizable strategies—compared with iterated dominance—is the presence of *beliefs* (and sets of beliefs). To relate (sets of) pure strategy combinations to sets of beliefs, proceed as follows. Let $R = R_1 \times ... \times R_n \in \mathcal{R}$ be a (nonempty compact Borel) product subset of S. For each $i = 1, ..., n$, associate with $R_i \subseteq S_i$ the set $\Delta_i(R_i)$ of probability distributions with support in R_i; i.e.,

$$\Delta_i(R_i) = \{\sigma_i \in \Delta_i \mid \text{supp}(\sigma_i) \subseteq R_i\},$$

and with the product set R associate *the face spanned by R,*

$$\Theta(R) = \Delta_1(R_1) \times ... \times \Delta_n(R_n). \tag{5.7}$$

The face $\Theta(R)$ spanned by R is the set of all beliefs with support in R that satisfy the independence property that the joint distribution is the product of the marginals. That beliefs have this product structure follows directly from the non-cooperative approach (see Section 3.3). Since the game is with complete rules, it is common knowledge that players make their decisions independently. The largest face (with respect to set inclusion) is $\Theta = \times_{i=1}^n \Delta_i = \Theta(S)$.

The pure and mixed best reply correspondence are closely related, owing to the linearity of payoff functions under the expected utility hypothesis. Linearity implies that, if two strategies are equally good, then so is any convex combination. So, the next result effectively shows—in the notation just developed—that $\tilde{\beta}_i(\sigma) = \Delta_i(\beta_i(\sigma))$, for all $\sigma \in \Theta$, for each player i, or $\tilde{\beta}(\sigma) = \Theta(\beta(\sigma))$. (A proof is given in the appendix to this section.)

Lemma 5.2 *For all $\bar{\sigma} \in \Theta$ and all $i = 1, ..., n$, $\sigma_i \in \tilde{\beta}_i(\bar{\sigma})$ if and only if* $\text{supp}(\sigma_i) \subseteq \beta_i(\bar{\sigma})$.

In words, a mixed strategy is a (mixed) best reply against some belief if and only if every pure strategy in its support is a (pure) best reply against the same belief.

If the strategy space S is compact (more precisely, a compact subset of a complete metric space) and the payoff function u is continuous, then the pure best reply correspondence is nonempty- and compact-valued and (by Berge's maximum theorem[11]) upper hemi-continuous. These properties transfer to the mixed best reply correspondence,[12] which, moreover, is convex-valued by Lemma 5.2.

---

[11]   The maximum theorem states that, if a continuous function is maximized on a constraint set given by a continuous correspondence, then the maximum is continuous and the maximizers form an upper hemi-continuous correspondence in the parameters (see Berge [17], p. 116). On hemi-continuity notions for correspondences, see the appendix to the next section (Definition 5.8).

[12]   To make this precise for infinite S, the space Θ is endowed with the topology of weak convergence. A sequence $\{\sigma^t\}_{t=1}^{\infty}$ from Θ converges weakly to $\sigma \in \Theta$ if $\int f \, d\sigma^t \to_{t \to \infty} \int f \, d\sigma$ for all continuous real-valued functions f on S.

Rationalizable Strategies   With this in hand, an iterative process on best replies can be defined which identifies rationalizable strategies. Define the mapping $B : \mathcal{R} \to \mathcal{R}$ by the composition of the pure best reply correspondence β (defined in (5.5)) and the assignment of faces $\Theta(\,.\,)$ from (5.7); i.e.,

$$B_i(R) \;=\; \beta_i\left(\Theta(R)\right) \equiv \cup_{\sigma \in \Theta(R)} \beta_i(\sigma), \text{ for } i = 1, ..., n, \text{ and} \qquad (5.8)$$
$$B(R) \;=\; B_1(R) \times ... \times B_n(R),$$

for all sets $R \in \mathcal{R}$. In words, $B(R)$ is the set of all (pure) best replies against beliefs with support in R. That the image of $R \in \mathcal{R}$ under B is in $\mathcal{R}$ requires that S is compact and u is continuous. For each $k = 1, 2, ...$ let $B^k(R) = B\left(B^{k-1}(R)\right)$ be defined recursively, and $B^\infty(R) = \cap_{k=1}^\infty B^k(R)$, for all $R \in \mathcal{R}$.

Definition 5.3 *For a (finite or infinite) game* $\Gamma = (S, u)$ *with compact strategy space S and continuous payoff function u, a strategy of player i is **rationalizable** if it belongs to the i-th component of* $B^\infty(S) = B_1^\infty(S) \times ... \times B_n^\infty(S)$.

As with iterated dominance, rationalizable strategies are defined by an iterative procedure which eliminates at each stage strategies that are not best replies to any belief supported in strategy combinations that have not yet been eliminated.

Since mixed strategies here enter only as beliefs, one could ask whether mixed strategies can also be rationalizable. There are two responses. One is that Γ could already be the mixed extension of a finite game (with $S = \Theta$ and $u = U$) and Definition 5.3 applies directly. This requires appropriately defined beliefs over mixed strategies.

The other is to define rationalizable mixed strategies as those that are best replies to beliefs with support in the rationalizable pure strategies. Then, a *mixed* strategy of player i is *rationalizable* if it belongs to the i-th component of

$$\tilde{\beta}\left(\Theta\left(B^\infty(S)\right)\right) = \tilde{\beta}_1\left(\Theta\left(B^\infty(S)\right)\right) \times ... \times \tilde{\beta}_n\left(\Theta\left(B^\infty(S)\right)\right). \qquad (5.9)$$

By Lemma 5.2 this does not add any new beliefs, nor does it eliminate further pure best replies; therefore, it is a valid definition of rationalizable mixed strategies.

Since rationalizable strategies are defined by an iterative procedure, similar to iterated dominance, a statement analogous to Theorem 5.1 applies.

Theorem 5.2 *For any (finite or infinite) game $\Gamma = (S, u)$ with compact strategy space S and continuous payoff function u,*

  (a) $B^\infty(S)$ *is nonempty and compact;*

  (b) $R = B^\infty(S)$ *if and only if R is the largest set (with respect to set inclusion) in $\mathcal{R}$ that satisfies $R = B(R)$;*

  (c) *if $\{R^t\}_{t=0}^\infty$ is any sequence from $\mathcal{R}$ that satisfies $B(R^t) \subseteq R^{t+1} \subseteq R^t$, $R^0 = S$, and that $R^t = R^{t+1}$ implies $R^t = B(R^t)$, for all t, then $R^t = R^{t+1}$ if and only if $R^t = B^\infty(S)$.*

The first statement (a), again, ensures existence. Combined with selectiveness (from the prisoners' dilemma, Example 5.1), it turns rationalizable strategies into a viable solution concept. The second statement (b) says that the set of rationalizable strategies is the largest "fixed point" under the composition of the (pure) best reply correspondence with the assignment of faces (5.7). The third statement (c) points out that the order of elimination of non-best replies is immaterial. (In Exercise 5.6 below you are asked to prove Theorem 5.2.)

Theorem 5.2(b) has a further interpretation. If $R = B^\infty(S)$, then $R = B(R)$ and each strategy of player i in R_i is justified by being a best reply against some belief with support in R_{-i}. Formally, $R_i \subseteq B_i(R)$ for all i. Moreover, if player i thinks that each of her opponents j will choose a strategy in R_j, then i will optimally also play in R_i. Formally, $B_i(R) \subseteq R_i$ for all i.

CURB Sets The latter defines *Closure Under Rational Behavior*, CURB for short. It is the property that a set of strategy combinations contains all best replies against beliefs with support in this set. Intuitively, if it is commonly known that, for some reason, players will only use strategies in R, then every rational player i will optimally choose a strategy from R_i, for all $i = 1, ..., n$. In fact, choosing a strategy outside R_i entails a loss for player i (because it cannot be a best reply) if all other players use strategies in R_{-i}.

Definition 5.4 *For a (finite or infinite) game $\Gamma = (S, u)$ with compact strategy space S and continuous payoff function u, a nonempty product subset $R \in \mathcal{R}$ of S is **closed under rational behavior**, or CURB, if $\beta(\Theta(R)) \subseteq R$. It is a **tight** CURB set if it is CURB and $R \subseteq \beta(\Theta(R))$, i.e., $\beta(\Theta(R)) = R$.*

Theorem 5.2(b) says that the *maximal* (or largest, with respect to set inclusion) tight CURB set is precisely the set of rationalizable strategies. By definition, a tight CURB set is a fixed point of the mapping B. So, the set of rationalizable strategies is the largest fixed point of B.

But a game has, in general, many more CURB sets. For instance, the whole strategy space S is CURB, because $\beta(\Theta) \subseteq S$. In particular, it may have more tight CURB sets. It can be shown ([14], Propositions 1 and 2) that every game $\Gamma = (S, u)$ (with compact strategy space S and continuous payoff function u) possesses at least one *minimal* (with respect to set inclusion) CURB set and that every minimal CURB set is tight.

The CURB concept thus represents a "localized" version of rationalizability. The existence of minimal CURB sets shows that information beyond common knowledge of rationality (some further coordination of expectations) can refine solutions beyond rationalizable strategies. Hence, rationalizability is "tight CURB without further information beyond common knowledge of rationality". In this sense, rationalizability is about as much as common knowledge of rationality alone implies.

But how much is that? It is clear that the set of rationalizable strategies will be a subset of the set of iteratively undominated strategies. After all, a strictly dominated strategy cannot be a best reply. Whether it will be a *proper* subset is less clear. The next result identifies a particular class of games for which rationalizable and iteratively undominated strategies coincide. (A proof is provided in the appendix to this section.)

Proposition 5.1 *If* Γ *is the mixed extension of a finite two-player game, then a strategy is rationalizable if and only if it is iteratively undominated.*

Hence, for finite two-player games the set of rationalizable mixed strategies is precisely the set of iteratively undominated mixed strategies. Note that this implies that iterative elimination of *weakly* dominated strategies may eliminate rationalizable strategies.

For more than two players, however, the set of rationalizable strategies may well be a proper subset of iteratively undominated strategies. The next example illustrates this.

Example 5.7 *Consider the three-player game given below. Player 2 controls rows, player 3 controls columns, and player 1 chooses a matrix. The numbers are player 1's payoffs.*

	s_3^1	s_3^2
s_2^1	0	2
s_2^2	0	3

s_1^1

	s_3^1	s_3^2
s_2^1	3	0
s_2^2	2	0

s_1^2

	s_3^1	s_3^2
s_2^1	2	0
s_2^2	0	2

s_1^3

The claim is that player 1's strategy s_1^3 *(the matrix on the right) is iteratively undominated, but not rationalizable. Denote* $\sigma_2(s_2^1) = x$ *and* $\sigma_3(s_3^1) = y$, *with* $0 \le x, y \le 1$ *(where* $\sigma_2(s_2^2) = 1 - x$ *and* $\sigma_3(s_3^2) = 1 - y$*). If* s_1^3 *were strictly dominated, then there would be some* $\sigma_1' \in \Delta_1$ *such that*

$$U_1(\sigma_{-1}, \sigma_1') - U_1(\sigma_{-1}, s_1^3) = \sigma_1'(s_1^1)\,[1 + x - y - 3xy]$$
$$+ \sigma_1'(s_1^2)\,[2x + 4y - 2 - 3xy] > 0$$

for all $\sigma_{-1} \in \Theta_{-1} = \Delta_2 \times \Delta_3$. *In particular, this inequality would hold for* $\sigma_{-1} = (s_2^1, s_3^1)$ *and for* $\sigma_{-1} = (s_2^2, s_3^2)$; *i.e.,*

$$U_1(s_2^1, s_3^1, \sigma_1') - U_1(s_2^1, s_3^1, s_1^3) = \sigma_1'(s_1^2) - 2\sigma_1'(s_1^1) > 0,$$
$$U_1(s_2^2, s_3^2, \sigma_1') - U_1(s_2^2, s_3^2, s_1^3) = \sigma_1'(s_1^1) - 2\sigma_1'(s_1^2) > 0.$$

The latter two inequalities would imply $\sigma_1'(s_1^1) > 2\sigma_1'(s_1^2) > 4\sigma_1'(s_1^1)$, *which is obviously impossible. Therefore,* s_1^3 *is not strictly dominated. (Nor is any* *other strategy strictly dominated.) But s_1^3 is never a best reply. If it were, then there would be some $\sigma_{-1} \in \Theta_{-1}$ (resp. some $x, y \in [0,1]$) such that*

$$U_1(\sigma_{-1}, s_1^3) - U_1(\sigma_{-1}, s_1^1) = y - x - 1 + 3xy \geq 0 \text{ and}$$
$$U_1(\sigma_{-1}, s_1^3) - U_1(\sigma_{-1}, s_1^2) = 2 - 2x - 4y + 3xy \geq 0.$$

The first of these inequalities is equivalent to $y \geq (1+x)/(1+3x)$ and the second is equivalent to $1 - x \geq 2y(1 - 3x/4)$. Since $(1+x)/(1+3x) \geq 1/2$, this, combined with the second inequality, yields $1-x \geq 1-3x/4$, which implies $x = 0$. It follows (from the first inequality) that $y = 1$. Substituting this into the second inequality yields the contradiction $1 \geq 2$. Hence, s_1^3 is never a best reply and, therefore, is not rationalizable.

Some authors prefer to define rationalizability not by best replies against beliefs that respect the independence property (that they correspond to mixed strategy combinations), but by best replies against beliefs that are *arbitrary* probability distributions over the opponents' strategy combinations. This is sometimes referred to as *correlated rationalizability*. Under such a definition rationalizability (for the mixed extension of finite games) becomes always equivalent to iteratively undominated strategies, because the proof of Proposition 5.1 (see appendix) then extends to n-player games. But this, of course, takes the bite out of rationalizability, since it ignores that in a non-cooperative game it is common knowledge that players decide independently. The current definition makes use of this information and, therefore, yields sharper predictions.

Still, rationalizable strategies are a rather coarse solution concept—first, because it is a set-valued concept, and second, because the set of rationalizable strategies may be large. The next example illustrates this. Recall that in Example 5.3, with two firms, iteratively undominated strategies, and, therefore, rationalizable strategies, give a unique prediction. This fails when there are more than two firms.

Example 5.8 *Consider again the Cournot market (Example 5.3), but now with $n \geq 3$ firms. Let $Q_i = \sum_{j \neq i} q_j$ denote the aggregate quantity supplied by all other firms, except firm i. The profit function of firm i is*

$$\Pi_i(q_i, Q_i) = q_i [\max\{0, 8 - Q_i - q_i\} - 2]$$

and the (pure) best reply correspondence (in this context also called the "reaction function") of firm i is given by

$$\beta_i(Q_i) = \max\{0, 3 - Q_i/2\}.$$

Since for all $0 \leq Q_i \leq 8$ the profit function is linear in Q_i, forming beliefs about the opponents' choices boils down to taking an expectation (viz. is a number between 0 and $n - 1$ times the maximum quantity that an individual firm can choose). Now iterate B. Initially, $q_i \in [0, 4]$ for all $i = 1, ..., n$. Hence, $0 \leq \beta_i(Q_i) \leq 3$; i.e., $B\left(\times_{i=1}^n [0, 4]\right) = \times_{i=1}^n [0, 3]$. With $q_i \in [0, 3]$ for all $i = 1, ..., n$, one obtains again $0 \leq \beta_i(Q_i) \leq 3$ for all i, or $B\left(\times_{i=1}^n [0, 3]\right) = \times_{i=1}^n [0, 3]$ if $n \geq 3$. Hence, all strategies between inactivity, $q_i = 0$, and the monopoly output, $q_i = 3$, are rationalizable (and, henceforth, iteratively undominated).

Fixed Sets   Rationalizability and the CURB concept portray players essentially as choosing *pure* strategies (except when the game is the mixed extension of some other game). Mixed strategies represent only beliefs. But, if players use only pure strategies, why would they consider beliefs that correspond to a (nondegenerate) randomization by some opponent? The usual justification is that a player may be uncertain as to which of their rationalizable strategies her opponents will use. (If this "uncertainty" is not based on a lottery, but comes from the strategic dimension of the problem, let us refer to it as "strategic uncertainty".) But, conceptually, this is already taken care of by considering best replies against *all* rationalizable strategy combinations among opponents, i.e. by taking the union in (5.8). Hence, rationalizability uses the players' strategic uncertainty twice: first by forming beliefs, and second by accounting for *all* best replies against rationalizable strategies of the opponents.

If strategic uncertainty is modeled purely by considering best replies against all "justifiable" possibilities, a different concept arises.

Definition 5.5   *For a (finite or infinite) game $\Gamma = (S, u)$ with compact strategy space S and continuous payoff function u, a nonempty subset $R \in \mathcal{R}$ of S is a **fixed set under the best reply correspondence** if*

$$R = \beta(R) \equiv \cup_{s \in R} \beta(s) \equiv \times_{i=1}^n \cup_{s_{-i} \in R_{-i}} \cap_{s_i \in S_i} \alpha_i(s_{-i}, s_i), \qquad (5.10)$$

where α_i is defined in (5.3) for $i = 1, ..., n$.

Again, every strategy combination in R is justified as a best reply against some strategy combination in R; formally, $R \subseteq \beta(R)$. Second, if R_{-i} records what others may do, then each player i will optimally also play in R_i; formally, $\beta(R) \subseteq R$. The prisoners' dilemma (Example 5.1) again shows that this solution is selective. Existence is established in the next result.

Theorem 5.3   *For every game $\Gamma = (S, u)$ with compact strategy space S and continuous payoff function u, there is a nonempty compact set $R \in \mathcal{R}$ such that $R = \beta(R)$.*

The proof uses again an iteration argument[13] based on the monotonicity of (the extension of) the best reply correspondence (to $\mathcal{R}$). The latter follows directly from taking the union in the extension of β to $\mathcal{R}$.

Like with the previous concepts, the game in Definition 5.5 may be the mixed extension of a finite game. In this case, β is the mixed best reply correspondence $\tilde{\beta}$ and S is replaced by Θ. Moreover, it can be shown that the solution concept of fixed sets under the best reply correspondence also applies to games where the preferences of players satisfy milder assumptions than the von Neumann-Morgenstern axioms (see [150]). This is so, because the existence theorem for fixed sets does not draw upon the linearity of the payoff function in probabilities.

Fixed sets under the best reply correspondence differ conceptually from rationalizability not only because they can do without the concept of beliefs (the structure of which depends heavily on the expected utility hypothesis), but also because "localized" solutions are allowed for. That is, not only is the largest fixed set identified by Definition 5.5, but a game may have several and smaller fixed sets. Smaller ones may be of interest if there is some coordination of players' expectations beyond the common knowledge of rationality.

Summary

- Common knowledge of rationality implies that players will exclusively use rationalizable strategies.

- A set of pure strategies is closed under rational behavior (CURB) if it contains all best replies against beliefs with support in the set.

- The set of rationalizable strategies is the largest fixed point of the composition of the (pure) best reply correspondence with the assignment of faces, or the largest tight CURB set.

- Alternatively, rationalizable strategies are obtained from iteratively eliminating strategies that are never best replies against beliefs with support in strategy combinations that have not yet been eliminated.

- Every rationalizable strategy is iteratively undominated. For finite two-player games the set of iteratively undominated mixed strategies coincides with the set of rationalizable mixed strategies. For more than two players, the inclusion may be strict.

---

[13] The statement of Theorem 5.3 follows immediately from Theorem 8 of Berge ([17], p. 113). The only difference from the proof of Theorem 5.1(b) is that the maximum theorem has to be invoked to establish the upper hemi-continuity of β (see [150], Theorem 2).

Appendix

Proof of Lemma 5.2 If there is $s_i \in \text{supp}(\sigma_i)$ with $s_i \notin \beta_i(\bar{\sigma})$, then there is $s_i' \in S_i$ such that $U_i(\bar{\sigma}_{-i}, s_i') > U_i(\bar{\sigma}_{-i}, s_i)$. Choose $\sigma_i' \in \Delta_i$ such that $\sigma_i'(s_i') = \sigma_i(s_i') + \sigma_i(s_i) > 0$, $\sigma_i'(s_i) = 0$, and $\sigma_i'(s_i'') = \sigma_i(s_i'')$, for all $s_i'' \in S_i \setminus \{s_i, s_i'\}$. Then $U_i(\bar{\sigma}_{-i}, \sigma_i') > U_i(\bar{\sigma}_{-i}, \sigma_i)$, which implies $\sigma_i \notin \beta_i(\bar{\sigma})$.

Conversely, if $s_i \in \beta_i(\bar{\sigma})$, for all $s_i \in \text{supp}(\sigma_i)$, then $U_i(\bar{\sigma}_{-i}, s_i) \geq U_i(\bar{\sigma}_{-i}, s_i')$, for all $s_i' \in S_i$, for all $s_i \in \text{supp}(\sigma_i)$, implies $U_i(\bar{\sigma}_{-i}, \sigma_i) \geq U_i(\bar{\sigma}_{-i}, s_i')$, for all $s_i' \in S_i$, and, therefore, $U_i(\bar{\sigma}_{-i}, \sigma_i) \geq U_i(\bar{\sigma}_{-i}, \sigma_i')$, for all $\sigma_i' \in \Delta_i$, by linearity of U_i in σ_i, yielding $\sigma_i \in \beta_i(\bar{\sigma})$. This completes the proof of Lemma 5.2.

Proof of Proposition 5.1[14] Let Γ be the mixed extension of a finite two-player game. Notation, at this point, becomes very confusing, because here $s_i \in S_i$ denotes a mixed strategy of player i $(= 1, 2)$, while outside this proof $s_i \in S_i$ denotes a pure strategy. Similarly, outside this proof $\sigma_i \in \Delta_i$ denotes a mixed strategy of player i, while here it denotes a belief over mixed strategies of player i. And here β_i denotes the mixed best reply correspondence, while outside this proof β_i denotes the pure best reply correspondence.

Hence, here S_i is a finite-dimensional simplex, and $\sigma_i \in \Delta_i$ denotes a probability measure on the Borel sets of S_i, $i = 1, 2$. To distinguish what used to be pure strategies in the game of which Γ is the mixed extension, denote by $z_i \in S_i$ the vertices (vectors with precisely one positive entry) of the simplex S_i, for $i = 1, 2$.

It can be shown that the mean of a subset of S_i (the support of σ_i) with respect to σ_i lies in the convex hull (the smallest convex superset) of this subset ([140], Lemma 1), and that the expected utility associated with a belief σ_i can be calculated using this mean, viz. a strategy in the convex hull of the support of σ_i ([140], Lemma 2). Therefore, $\Delta_i(R_i)$ can be replaced by the convex hull of R_i, for all compact subsets R_i of the strategy simplex S_i, for $i = 1, 2$.

To prove Proposition 5.1, proceed inductively from the viewpoint of player $i = 1$. (Player 2 is analogous.) Clearly, $D^0(S) = B^0(S) = S$. Now assume $D^k(S) = B^k(S)$ for some integer $k > 0$. Every strategy $s_2 \in D_2^k(S)$ must be a convex combination of finitely many vertices $z_2 \in D_2^k(S)$, because if s_2 can be represented only as a convex combination of vertices that include (at least) one vertex outside $D_2^k(S)$, then s_2 would be strictly dominated (against $D^{k-1}(S)$) by Lemma 5.1 and would not, therefore, be in $D_2^k(S)$.

Hence, consider the set $\mathcal{V}_1$ of (finite-dimensional) vectors

$$v_1(s_1) = \left(u_1(z_2, s_1) \right)_{z_2 \in D_2^k(S)}$$

for all $s_1 \in S_1$. This set is convex, because S_1 is convex.

---

[14]   Pearce [140] gives a different proof, based on the existence of a value for two-player zero-sum games.

Pick any $s_1 \in D_1^{k+1}(S)$. If there is $s_1' \in S_1$ such that[15] $v_1(s_1') \gg v_1(s_1)$, then $u_1(z_2, s_1') > u_1(z_2, s_1)$, for all $z_2 \in D_2^k(S)$, implies that $u_1(s_2, s_1') > u_1(s_2, s_1)$, for all $s_2 \in D_2^k(S)$, because every s_2 is a convex combination of vertices in $D_2^k(S)$. Hence, s_1 will be strictly dominated (by s_1') against $D^k(S)$ and cannot be in $D_1^{k+1}(S)$. Therefore, $s_1 \in D_1^{k+1}(S)$ implies that there is no $s_1' \in S_1$ such that $v_1(s_1') \gg v_1(s_1)$. Hence, all $s_1 \in D_1^{k+1}(S)$ are associated with *boundary points* of the convex set $\mathcal{V}_1$.

It follows from the separating hyperplane theorem[16] that, for every $s_1 \in D_1^{k+1}(S)$, there exists a non-negative vector

$$q = (q(z_2))_{z_2 \in D_2^k(S)} \geq 0$$

(which can be chosen so that its components sum to 1) such that

$$q \cdot (v_1(s_1) - v_1) \geq 0, \text{ for all } v_1 \in \mathcal{V}_1.$$

Since q is in the convex hull of $D_2^k(S) = B_2^k(S)$ by construction, this says that each $s_1 \in D_1^{k+1}(S)$ represents a best reply against some belief with support in $B_2^k(S)$. Therefore, $D_1^{k+1}(S) \subseteq B_1^{k+1}(S)$. Combined with the obvious inclusion $B_1^{k+1}(S) \subseteq D_1^{k+1}(S)$ this yields $D_1^{k+1}(S) = B_1^{k+1}(S)$.

Since a symmetric argument applies to $i = 2$ and the last equality holds for all k, we conclude that $D_i^\infty(S) = B_i^\infty(S)$, for $i = 1, 2$. This completes the proof of Proposition 5.1.

Exercises

Exercise 5.6 Prove Theorem 5.2. (Hint: First show that B, as defined in (5.8), is monotone, and then apply arguments as in the proof of Theorem 5.1. In one version of the proof you may want to invoke Berge's maximum theorem.)

Exercise 5.7 Consider the three-player game below, where player 2 controls rows, players 3 columns, and player 1 chooses matrices. Find out whether s_1^4 can be a (pure) best reply against some mixed strategy combination of players

---

[15]   The symbol $x \gg y$ for two real vectors $x, y \in \mathbb{R}^m$ means that every component of x exceeds the corresponding component of y; i.e., $x_j > y_j$, for all $j = 1, ..., m$.

[16]   Consider a closed convex subset $\mathcal{V}$ of some finite-dimensional Euclidean space $\mathbb{R}^m$ and a point $\overline{x} \in \mathbb{R}^m$ which is not in $\mathcal{V}$. The separating hyperplane theorem states that there is $y \in \mathbb{R}^m \setminus \{0\}$ such that $y \cdot \overline{x} < y \cdot x$ for all $x \in \mathcal{V}$. (The inner product of two vectors $x, y \in \mathbb{R}^m$ is $y \cdot x = \sum_{j=1}^m x_j y_j$.) In words, a convex set and a point outside of it can be "separated" by a hyperplane (see Exercise 5.28 below).

2 and 3. Discuss whether or not s_1^4 is strictly dominated?

	s_3^1	s_3^2
s_2^1	3	0
s_2^2	0	0

s_1^1

	s_3^1	s_3^2
s_2^1	0	3
s_2^2	3	0

s_1^2

	s_3^1	s_3^2
s_2^1	0	0
s_2^2	0	3

s_1^3

	s_3^1	s_3^2
s_2^1	2	0
s_2^2	0	2

s_1^4

The numbers given are player 1's payoffs.

Exercise 5.8 Find the set of rationalizable strategies in the two-player game given below. (Player 1 controls rows and obtains the upper left payoff; player 2 controls columns and obtains the lower right payoff.)

	s_2^1	s_2^2	s_2^3	s_2^4
s_1^1	2 9	4 7	9 2	2 3
s_1^2	7 4	5 5	7 4	2 3
s_1^3	9 2	4 7	2 9	2 3
s_1^4	2 2	2 0	2 2	12 1

Exercise 5.9 Find the minimal CURB sets of the two-player game below. (The conventions on row and column players and on payoffs are as before.)

	s_2^1	s_2^2	s_2^3
s_1^1	1 1	0 0	0 0
s_1^2	0 0	2 3	3 2
s_1^3	0 0	3 2	2 3

Exercise 5.10 Show that if $R \in \mathcal{R}$ is a singleton, i.e. if $R = \{s\}$ for some $s \in S$, then $D_i(R) = \beta_i(\Theta(R)) = B_i(R)$, for all $i = 1, ..., n$, where D_i is defined in (5.4), β_i is defined in (5.5), $\Theta(R)$ is defined in (5.7), and B_i is defined in (5.8).

Notes on the Literature: Rationalizable strategies have been introduced by Bernheim [18] and Pearce [140]. Correlated rationalizability was introduced by Brandenburger and Dekel [30], though the concept had already been alluded to by Pearce [140]. The CURB concept is due to Basu and Weibull [14]. Statements (a) and (b) of Theorem 5.2 follow Bernheim ([18], Propositions 3.1 and 3.2; see also Pearce [140], Proposition 2); Theorem 5.2(c) is new. Proposition 5.1 is from Pearce ([140], Lemma 3). Example 5.7 is based on an example by Ritzberger and Weibull ([152], p. 1393) and Example 5.8 on one by Bernheim ([18], pp. 1024). Fixed sets under the best reply correspondence have been introduced by Ritzberger [150]. Exercise 5.9 is from Basu and Weibull [14].

5.3   Nash Equilibrium

Definition   For many games all strategies are rationalizable. Still, in given applications there may often be reasons to expect more coordination among the players. The CURB concept and fixed sets under the best reply correspondence do account for this. Yet, these concepts still do not entirely eliminate the strategic uncertainty of players. Rather, they account for it by formulating set-valued solution concepts. These do not unambiguously tell the player what she should do when she actually plays the game. To obtain a solution that completely eliminates the strategic uncertainty, a heroic assumption is required.

Suppose players can, for some reason, figure out correctly what others will do. Maybe players are smarter than the analyst, who plainly deduces implications of rationality. Or they know from experience and history. Or they have been instructed by an umpire, who is equipped with an authoritative game theory book. Or they have discussed matters before. This assumption says nothing about how the coordination of expectations is brought about. It simply assumes it. What can we say about such a solution?

First of all, if rational players have these capabilities, then *all* of them can figure out the solution, and so this particular solution would be commonly known among all the players. Thus, the problem boils down to investigating the implications of *common knowledge of the solution*. This is a considerably stronger hypothesis than common knowledge of rationality. It stipulates that, for some reason, there is an established way in which the game will be played.

If a solution is commonly known among rational players, two implications arise. First, all rational players will optimize (use best replies) given their expectations. Second, expectations will be supported on best replies against the solution. That is, if some element in the solution is not a best reply against the solution, then players should attribute zero probability to it.

To formalize this, consider an arbitrary game $\Gamma = (S, u)$. If a strategy combination $s \in S$ is a solution that players correctly anticipate, then all players will use best replies against s, since all expect the others to choose according to s. Formally, $s \in \beta(s)$. This is a rather strong conclusion—from a strong hypothesis.

Definition 5.6 *For a (finite or infinite) game $\Gamma = (S, u)$, a pure strategy combination $s \in S$ is a (pure) **Nash equilibrium** if $s \in \beta(S)$. A mixed strategy combination $\sigma \in \Theta$ is a Nash equilibrium of the mixed extension, or a mixed Nash equilibrium, if $\sigma \in \tilde{\beta}(\sigma)$.*

Any pure strategy $s_i \in S_i$ used in a mixed Nash equilibrium $\sigma \in \Theta$ (i.e., $\sigma_i(s_i) > 0$) is a best reply against σ and, therefore, is *rationalizable* and *iteratively undominated*. Hence, when for a game Γ the set $E(\Gamma) \subseteq \Theta$ denotes the *set of all* (pure and mixed) Nash equilibria (viewing S as the subset of vertices of Θ), it follows that $E(\Gamma) \subseteq B^\infty(\Theta) \subseteq D^\infty(\Theta)$. Neither of those inclusions must be strict (as the prisoners' dilemma, Example 5.1, illustrates).

The above definition of a Nash equilibrium (or, an *equilibrium*) stresses that an equilibrium is a fixed point of the best reply correspondence.[17] There are many ways to rephrase this. For pure equilibria it is equivalent to requiring that

$$u_i(s) \geq u_i(s_{-i}, s_i'), \text{ for all } s_i' \in S_i \text{ and all } i = 1, ..., n, \qquad (5.11)$$

which, intuitively, says that an (pure) equilibrium $s \in S$ is a strategy combination such that no player has an incentive to deviate from it, if she expects no other player to deviate. This is, of course, a necessary condition for a (single) strategy combination to qualify as a commonly known solution.

If (5.11) were not true, the solution would be "self-defeating", because some player would find it optimal to violate what it recommends. Since the others can perfectly predict this, they too may have an incentive to deviate. Thus, nobody should expect the solution to be played.

Alternatively, the latter aspect can be stressed by requiring that the solution $s \in S$ must not be destabilized by becoming common knowledge. Formally, $\{s\} \subseteq \beta(s)$ provides an equivalent definition of an equilibrium. Note that this simply says that s is justified, because all players behave optimally (and expect others to do so). This inclusion will be referred to as the *Nash property*.

Finally, for the mixed extension a Nash equilibrium may equivalently be defined as a mixed strategy combination $\sigma \in \Theta$ such that

$$\text{if } U_i(\sigma_{-i}, s_i) < \max_{s_i' \in S_i} U_i(\sigma_{-i}, s_i'), \text{ then } \sigma_i(s_i) = 0, \text{ for all } s_i \in S_i \qquad (5.12)$$

and all $i = 1, ..., n$. That is, every strategy that i is expected to use (by the others) must be a best reply (for i) against the solution.

Here, the mixed equilibrium $\sigma \in \Theta$ can be thought of as a *consistent common* belief about what to expect from the opponents. In particular, it need not imply that players actually employ random devices. They may still choose pure strategies. Rather, $\sigma \in \Theta$ describes a consistent way to reason about the opponents (who may have several pure best replies against σ).

In this interpretation the focus of the analysis becomes not the strategy choices of players, but the assessments (or "beliefs") of the players about the choices of others. The basic consistency condition which then arises is that a

---

[17] A "fixed point" for a correspondence (which maps a set into itself) is a generalization of the notion of a fixed point for a function. It is a point which is contained in its own image under the correspondence (see the appendix to this section).

player reasoning through the conclusions that others would draw from their assessments should not be led to revise her own assessment.

Existence   Indeed, it is important to consider the mixed extension, because otherwise the existence of Nash equilibria cannot be guaranteed, as the following example illustrates.

Example 5.9 *(Matching pennies) By going through all four possibilities, it can be verified that the 2×2 game below does not have an equilibrium in pure strategies. (As usual, player 1 controls rows and 2 columns, and the upper left entry is player 1's payoff and the lower right, player 2's.)*

	s_2^1		s_2^2	
s_1^1	1		0	
		0		2
s_1^2	0		2	
		1		0

This example is informative, because at each pure strategy combination one of the players wants to deviate. So, what should a given player expect from her opponent? She must form assessments or "beliefs" about her opponent's behavior, i.e. probability distributions over his strategy set, that are consistent with his reasoning about herself. And this is precisely what a mixed strategy equilibrium captures.

Hence, the existence requirement on solutions has to refer to the mixed extension of a game. It turns out that, at least for finite games, this is enough to ensure existence.

Theorem 5.4 *(Nash's theorem) For every finite game $\Gamma = (S, u)$, there is a Nash equilibrium of its mixed extension (Θ, U).*

There are various ways to prove it. One, which is the archetypal existence proof, employs Kakutani's fixed point theorem. The latter applies to a correspondence, mapping a nonempty compact convex[18] subset of some finite-dimensional Euclidean space to itself, which has nonempty compact convex values and is upper hemi-continuous. (See Theorem 5.10 in the appendix to this section.) For such a mapping, it guarantees the existence of a fixed point.

Now, the correspondence $\tilde{\beta} : \Theta \to \Theta$ maps the nonempty compact convex subset Θ of a finite-dimensional Euclidean space to itself. Since U is continuous, by the maximum theorem, $\tilde{\beta}$ is nonempty compact valued and upper

---

[18]   A set is *convex* if, for any two of its elements, x and y, the convex combination $\lambda x + (1 - \lambda)y$ also belongs to the set, for all $0 \leq \lambda \leq 1$. It is *strictly convex* if, for any two distinct x, y from it, the convex combination $\lambda x + (1 - \lambda)y$ belongs to its interior, for all $0 < \lambda < 1$.

hemi-continuous. By Lemma 5.2 $\tilde{\beta}_i$ is convex valued and so is $\tilde{\beta} = \tilde{\beta}_1 \times \dots \times \tilde{\beta}_n$. Hence, all conditions of Kakutani's fixed point theorem are fulfilled.

Nash himself has devised a second proof which uses a more fundamental result, known as Brouwer's fixed point theorem. (See Theorem 5.7 in the appendix.) It works as follows. Map each vector $\sigma \in \Theta$ into another vector in Θ by[19]

$$\sigma \mapsto \left(\left(\frac{\sigma_i(s_i) + \max\left\{0, U_i(\sigma_{-i}, s_i) - U_i(\sigma)\right\}}{1 + \sum_{s_i' \in S_i} \max\left\{0, U_i(\sigma_{-i}, s_i') - U_i(\sigma)\right\}} \right)_{s_i \in S_i} \right)_{i=1,\dots,n}$$

Clearly, the components of the image are non-negative and sum to 1 for each i. Hence, this "Nash mapping", by construction, defines a continuous function from Θ to itself. Next, it will be shown that the fixed points of the Nash mapping coincide with the Nash equilibria of the game.

Suppose $\sigma \in \Theta$ is a Nash equilibrium. Then $U_i(\sigma) \geq U_i(\sigma_{-i}, s_i)$ for all $s_i \in S_i$ and all i. Thus, $\max\left\{0, U_i(\sigma_{-i}, s_i) - U_i(\sigma)\right\} = 0$ for all $s_i \in S_i$ and all i. Hence, the image of σ under the Nash mapping is σ itself; i.e., it is a fixed point.

Conversely, suppose that the Nash mapping takes $\sigma \in \Theta$ into itself. Then, multiplying through and cancelling terms yields

$$\sigma_i(s_i) \sum_{s_i' \in S_i} \max\left\{0, U_i(\sigma_{-i}, s_i') - U_i(\sigma)\right\} = \max\left\{0, U_i(\sigma_{-i}, s_i) - U_i(\sigma)\right\}$$

for all $s_i \in S_i$ and all i. If there were some $s_i \in S_i$ that does better than σ_i, then $U_i(\sigma_{-i}, s_i) > U_i(\sigma)$ would imply

$$\sigma_i(s_i) = \frac{U_i(\sigma_{-i}, s_i) - U_i(\sigma)}{\sum_{s_i' \in S_i} \max\left\{0, U_i(\sigma_{-i}, s_i') - U_i(\sigma)\right\}} > 0.$$

Summing over all such better strategies for player i would yield

$$\sum_{U_i(\sigma_{-i}, s_i) > U_i(\sigma)} \sigma_i(s_i) = 1.$$

But this would be a contradiction, because a convex combination of numbers that all exceed a common lower bound cannot equal this lower bound. Hence, $U_i(\sigma_{-i}, s_i) \leq U_i(\sigma)$ for all $s_i \in S_i$ and all i—which is equivalent to a Nash equilibrium of the mixed extension. Thus, $\sigma \in \Theta$ is a fixed point of the Nash mapping if and only if it is a Nash equilibrium.

Since the Nash mapping is a continuous function from a compact convex subset of a finite-dimensional Euclidean space to itself, by Brouwer's theorem it has a fixed point. Thus, Nash's theorem is proved.

---

[19]   Instead of the maximum function, $\max\left\{0, x\right\}$, any continuous function that is zero for non-positive reals and positive for positive reals can be used in the proof. For instance, the function f defined by $f(x) = 0$, for all $x \leq 0$, and $f(x) = \exp\left\{-\frac{1}{x}\right\}$, for all $x > 0$, is an alternative that is smooth even at $x = 0$.

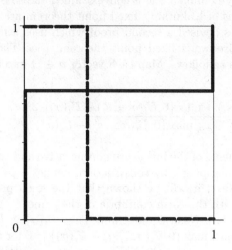

Figure 5.2: Matching pennies

Example 5.10 *Returning to Example 5.9 and writing $\sigma_1\left(s_1^1\right) = x = 1 - \sigma_1\left(s_1^2\right)$, and $\sigma_2\left(s_2^1\right) = y = 1 - \sigma_2\left(s_2^2\right)$ with $0 \le x, y \le 1$, the best replies (taking values x and y, respectively) are given by*

$$\tilde{\beta}_1(x,y) = \begin{cases} 1 & \text{if } y > 2/3 \\ [0,1] & \text{if } y = 2/3 \\ 0 & \text{if } y < 2/3 \end{cases} \qquad \tilde{\beta}_2(x,y) = \begin{cases} 1 & \text{if } x < 1/3 \\ [0,1] & \text{if } x = 1/3 \\ 0 & \text{if } x > 1/3 \end{cases}$$

In Figure 5.2, $x = \sigma_1\left(s_1^1\right)$ is on the horizontal and $y = \sigma_2\left(s_2^1\right)$ is on the vertical axis. The thick solid graph is $\tilde{\beta}_1$ and the thick dashed one is $\tilde{\beta}_2$. Where the two graphs intersect, the Nash equilibrium is located. From the previous discussion, it is clear that there is no pure strategy equilibrium, so x and y cannot equal 0 or 1. Hence, x must equal 1/3 and y must equal 2/3. If player 1 expects that player 2 will play her first pure strategy with probability 2/3, then she has no reason to prefer one of her strategies over the other. Thus, player 2 is justified in expecting player 1 to play her first pure strategy with probability 1/3 (which is as good as any other); in which case player 2 has no reason to prefer one strategy over the other, and player 1's belief is justified.

In this example, if players do indeed "toss coins", they both choose the probabilities of the coin toss so as to keep the opponent indifferent between her strategies. Hence, both are indifferent between all the possibilities they have. In particular, they have no further reason for playing the (mixed) Nash equilibrium strategies.

This illustrates that Nash equilibrium does not imply that "players have an incentive to play the solution". The latter would be true only if $\tilde{\beta}(\sigma) \subseteq \{\sigma\}$, i.e. if σ were a (singleton) CURB set. However, Nash equilibrium does not require this, but only the Nash property (viz. the reverse inclusion).

Yet, there exist games that have Nash equilibria where each player would incur a loss if she were to deviate.

Example 5.11 *Return to the version of Example 5.3 in Example 5.8. The pure best reply correspondence of firm i $(= 1, ..., n$ with $n \geq 3)$ is given by $\beta_i(Q_i) = \max\{0, 3 - Q_i/2\}$, where $Q_i = \sum_{j \neq i} q_j$ is the aggregate quantity supplied by all other firms. Hence, firm i will optimally want to produce positive output if and only if $Q_i < 6$. Let there be $k \leq n$ active firms in equilibrium. Then for each active firm $q_i = 6/(k+1) < 6$, so $k = n$ and in equilibrium each firm produces $q^* = 6/(n+1)$ units of output. Given that all $j \neq i$ choose $q_j^* = 6/(n+1)$, the pure best reply correspondence of firm i evaluates to $\beta_i(6(n-1)/(n+1)) = 6/(n+1)$, so q_i^* is the unique best reply to q_{-i}^*. Every deviation from q_i^* would entail a strictly smaller profit for firm i, given that all other firms $j \neq i$ choose q_j^*. Therefore, $\beta(q^*) \subseteq \{q^*\}$ verifies the CURB property.*

Indeed, the (unique) equilibrium identified in this example has the attractive feature of giving players an incentive to actually "stick to the solution". Clearly, if this is so, then the equilibrium must be in pure strategies. Otherwise, at least one player has multiple best replies (by Lemma 5.2).

Definition 5.7 *A Nash equilibrium is a* **strict equilibrium** *if it is a singleton CURB set.*

A strict equilibrium satisfies both the Nash property, $\{\sigma\} \subseteq \tilde{\beta}(\sigma)$, and the CURB property, $\tilde{\beta}(\sigma) \subseteq \{\sigma\}$. Hence, $\sigma \in \Theta$ is a strict equilibrium if and only if $\{\sigma\} = \tilde{\beta}(\sigma)$. (And, therefore, it is in pure strategies, $\sigma \in S \subseteq \Theta$.) Consequently, a strict equilibrium is a (singleton) fixed set under the best reply correspondence (Definition 5.5), both in the original game (where players choose among pure strategies) and in its mixed extension. And every singleton fixed set under the best reply correspondence must be a strict equilibrium.

Though strict equilibrium sounds like the most compelling equilibrium notion, it fails the existence requirement, even in very simple finite games (as Example 5.9 illustrates). This is so, because a solution that consists of a single strategy combination may often have to resolve the players' strategic uncertainty by randomized beliefs.

Another issue raised by Example 5.11 is that it does not fall under the Nash theorem (Theorem 5.4), because strategy spaces are infinite. Still, it belongs to the class of games with continuous payoff functions on a compact strategy space. For this class the Nash theorem needs to be generalized to the following.

Theorem 5.5 *(Glicksberg's theorem) For any game* $\Gamma = (S, u)$ *with contin-uous payoff function* u *on a strategy space* S, *which is a nonempty compact subset of a complete metric space, there exists a Nash equilibrium for its mixed extension.*

Here mixed strategies are (Borel) probability measures over pure strate-gies. The space of mixed strategies gets endowed with the topology of weak convergence.[20] The set of probability measures endowed with this topology is compact. Hence, strategy spaces can be approximated by a sequence of finite grids. By Nash's theorem, each (game defined by the finite) grid has a mixed strategy equilibrium. It can then be shown that, since the space of probability measures is (weakly) compact, the sequence of these approximating equilibria has a cluster point. Exploiting the continuity of the payoff function, it can be verified that the cluster point is a Nash equilibrium of the infinite game.

There are examples of (infinite) games with discontinuous payoff func-tion which do not even have a mixed Nash equilibrium (see e.g. [169]). Yet, the hypothesis of a continuous payoff function can be slightly relaxed. If the discontinuities are "well behaved" and the payoff function satisfies two as-sumptions expressed in terms of "semi-continuity", then mixed equilibria still exist (see [40]).

These equilibria and those identified by Glicksberg's theorem allow for nondegenerate randomizations. To obtain existence of pure strategy Nash equilibria, considerably stronger assumptions are needed.

Let S_i be a convex compact subset of some finite-dimensional Euclidean space (more generally, a complete metric space), for all $i = 1, ..., n$. A (payoff) function $u_i : S \to \mathbb{R}$ is (strictly) *quasi-concave* in $s_i \in S_i$ if the upper contour set $\alpha_i(s)$ (defined in (5.3)) of $s_i \in S_i$ at $s_{-i} \in S_{-i}$ is a (strictly) convex set, for all $s \in S$. If u_i is quasi-concave, then for any two maximizers $s_i', s_i'' \in \beta_i(s) \subseteq S_i$ their convex combination $\lambda s_i' + (1 - \lambda)s_i'' \in S_i$ must belong to $\alpha_i(s_{-i}, s_i') = \alpha_i(s_{-i}, s_i'')$, and, therefore, be itself a maximizer, for any $0 \le \lambda \le 1$. (If it is strictly quasi-concave, then any convex combination $\lambda s_i' + (1 - \lambda)s_i''$ with $0 < \lambda < 1$ would do strictly better if s_i' and s_i'' were distinct, implying $s_i' = s_i''$, i.e. that the maximizer is unique.)

Hence, if u_i is quasi-concave (resp. strictly quasi-concave) in $s_i \in S_i$, then the pure best reply correspondence β_i is convex valued (resp. single valued, i.e. a function). An application of the maximum theorem to the continuous payoff function u_i yields non-emptyness and upper hemi-continuity (u.h.c.) of β_i. Thus, β_i is an u.h.c. correspondence with nonempty convex compact values (resp. a continuous function, if u_i is strictly quasi-concave). These properties are inherited by the product mapping $\beta = \beta_1 \times ... \times \beta_n : S \to S$. Now apply Kakutani's (resp. Brouwer's) fixed point theorem to obtain the following.

[20]   A sequence of probability measures converges weakly to a limiting measure if the means with respect to the elements of the sequence converge to the mean with respect to the limiting measure, for any real-valued continuous function.

Theorem 5.6 *Let* $\Gamma = (S, u)$ *be a game where each strategy space* S_i *is a nonempty convex compact subset of some finite-dimensional Euclidean space, for all* $i = 1, ..., n$, *and* $u : S \to \mathbb{R}^n$ *is continuous. If, moreover,* u_i *is quasi-concave (resp. strictly quasi-concave) in* $s_i \in S_i$ *for all* i, *then* Γ *has a Nash equilibrium (resp. a strict equilibrium) in pure strategies.*

This sounds like good news for the general existence of pure strategy equilibria. But the hypothesis, quasi-concavity, is very restrictive. It happens to be true in Example 5.11. But in general this is too much to ask for. In economics examples abound that violate quasi-concavity of the payoff function. In fact, a combinatorial argument establishes that a large (but finite) game with randomly drawn payoff function has no pure strategy equilibrium with probability close to $1/e \approx .37$ (see [52]).

Most of the games that appear in applications also have *multiple* Nash equilibria. Uniqueness, as in the examples above, is really the exception rather than the rule. Since Nash equilibrium is driven by the coordination of expectations, there are usually many ways to coordinate. And this gives rise to many Nash equilibria. (See the exercises.)

What may be less obvious is that in some Nash equilibria players may use *weakly dominated* strategies. For finite games there is always at least one equilibrium where all strategies used (with positive probability) are admissible (see Proposition 6.1(c) in Section 6.2). But there may be other equilibria where players use weakly dominated strategies. For infinite games the situation is worse.

Example 5.12 *Consider the two-player game with (pure) strategy sets* $S_i = [0, 1/2]$ *and payoff function*

$$u_i(s_{3-i}, s_i) = \min\left\{ s_i, \frac{s_{3-i}(1 - s_i)}{2 - s_{3-i}} \right\}$$

for $i = 1, 2$. *Note that* $s_i \leq s_{3-i}(1 - s_i) / (2 - s_{3-i})$ *if and only if* $s_i \leq s_{3-i}/2$ *for* $i = 1, 2$.

By Theorem 5.5 the game has an equilibrium, because the payoff function is continuous and S *is compact. Let* s_i^* *be the infimum over the set of all* $s \in [0, 1/2]$ *such that player* i's *mixed equilibrium strategy assigns mass 1 to the interval* $[0, s]$, *for* $i = 1, 2$. *For all* s_2 *with* $0 \leq s_2 \leq s_2^*$, *player 1's payoff function* $u_1(s_2, s_1)$ *is strictly decreasing in any* s_1 *that satisfies* $s_2^*/2 \leq s_1 \leq 1/2$. *Therefore, player 1's best replies must assign mass 1 to the interval* $[0, s_2^*/2]$. *Hence,* $s_1^* \leq s_2^*/2$. *By symmetry, player 2's best replies must put mass 1 on the interval* $[0, s_1^*/4]$ *so that* $s_2^* \leq s_2^*/4$. *The latter implies* $s_2^* = 0$ *and, consequently,* $s_1^* = 0$. *Thus,* $s_1^* = s_2^* = 0$ *constitutes the unique Nash equilibrium (which is in pure strategies).*

On the other hand, $u_i(s) \geq 0$ *for all* $s = (s_1, s_2) \in S$, $u_i(s_{3-i}, 0) = 0$, *and* $u_i(s) > 0$ *for all* $s = (s_1, s_2) \gg 0$ *establishes that* $s_i = 0$ *is a weakly dominated strategy. Hence, the unique equilibrium is inadmissible.*

This example illustrates that for some games the unique Nash equilibrium may prescribe that players use exclusively weakly dominated strategies. Again, weak dominance implicitly considers all strategy combinations among opponents possible. But Nash equilibrium is driven by common knowledge of the solution. Thus, certain strategy combinations are known to be excluded—and admissibility loses its bite.

The next two sections concern applications of Nash equilibrium to specific classes of games.

Summary

- Common knowledge of the solution yields Nash equilibrium, i.e., a strategy combination such that no player can gain by deviating, given that nobody else deviates.

- Any strategy used in a Nash equilibrium is rationalizable and, henceforth, iteratively undominated.

- Nash's theorem asserts the existence of an equilibrium for (the mixed extension of) any finite game.

- A strict equilibrium is a Nash equilibrium which constitutes the unique best reply against itself.

- In general, for a given game, there are multiple Nash equilibria. Moreover, equilibrium strategies may be weakly dominated.

Appendix: Fixed Points

Since fixed point theory is crucial for the existence of solutions, some of its basic ingredients are discussed in this appendix. The most useful statement of one of its fundamental results is as follows.

Theorem 5.7 *(Brouwer's fixed point theorem) Let $X \subseteq \mathbb{R}^m$ be nonempty, compact, and convex and $f : X \to X$ a continuous function. Then f has a fixed point; i.e., there is $x \in X$ such that $x = f(x)$.*

There are at least two ways to prove this. One is via simplicial subdivisions of X and Sperner's lemma (see e.g. [28], Chapters 4-6). Another, used here for a sketch of the proof, employs methods from differential topology (see e.g. [121] or [77]). Hence, a few basic notions from this branch of mathematics are introduced now.

A function $f : X \to \mathbb{R}^m$, with $X \subseteq \mathbb{R}^l$, is *smooth* if it is infinitely often continuously differentiable in a neighborhood of each point in X. (A neighborhood of $x \in X$ is an open set containing x.) A smooth map $f : X \to Y$, with $Y \subseteq \mathbb{R}^m$, is a *diffeomorphism* if it is one-to-one (injective) and onto (surjective) and its inverse $f^{-1} : Y \to X$ is also smooth. If such a map exists, X and Y are *diffeomorphic*.

From the topologist's viewpoint, all diffeomorphic sets are alike, since they all count as copies of the same abstract space. Let X be a subset of some ambient Euclidean space $\mathbb{R}^k$. Then X is an *m-dimensional manifold* without boundary if it is locally diffeomorphic to $\mathbb{R}^m$, i.e. if each point $x \in X$ has a neighborhood in X that is diffeomorphic to an open set of $\mathbb{R}^m$. A manifold (locally) "looks like" a Euclidean space, at least after some (mild) stretching and "ironing" has been applied.[21] Therefore, differentiation can be defined on manifolds.

Now let $H^m = \{x \in \mathbb{R}^m \,|\, x_m \geq 0\}$ be the closed half-space where each point has a non-negative last coordinate. The *boundary* ∂H^m of H^m is the hyperplane $\mathbb{R}^{m-1} \times \{0\} \subseteq \mathbb{R}^m$. The set $X \subseteq \mathbb{R}^k$ is an *m-dimensional manifold with boundary* if each $x \in X$ has a neighborhood the intersection of which with X is diffeomorphic to a relatively open subset of H^m (viz. the intersection of an open set in $\mathbb{R}^m$ with H^m). The *boundary* ∂X is the set of all points in X that correspond to points of ∂H^m under such a diffeomorphism. (Beware: the boundary ∂X is, in general, not equal to the topological boundary of X.)

Let $f : X \to Y$ be a smooth function from an l-dimensional manifold to an m-dimensional manifold, with $l \geq m$. Denote by $D_x f(\overline{x})$ the Jacobian (matrix of first-order partial derivatives) evaluated at $\overline{x} \in X$. An element $\overline{x} \in X$ is a *critical point* for f if the rank of $D_x f(\overline{x})$ is (strictly) less than m. Otherwise, it is a *regular point*. The image of all critical points in X under f is the set of *critical values*. All points in its complement (in Y) are *regular values*. That is, $y \in Y$ is a regular value if, for all $\overline{x}$ in the preimage $f^{-1}(y) = \{x \in X \,|\, f(x) = y\}$, the Jacobian $D_x f(\overline{x})$ has full rank ($= m$). (Beware: $y \in Y$ qualifies as a regular value if it is not a value of f at all.)

A fundamental result in differential topology states that almost all elements of the range Y are regular values. A variant of it is as follows. (A proof can be found e.g. in [121], §3.)

Theorem 5.8 *(Sard's theorem) The closure[22] of the set of regular values for a smooth map $f : X \to Y$ between manifolds covers Y.*

(Note that for the case $l < m$ this simply says that $f(X) \equiv \cup_{x \in X} f(x)$ is of "negligible size" in Y.) To exploit the power of Sard's theorem for the present

---

[21] In principle, the "stretching and ironing" could be more severe; i.e., the bijection (one-to-one and onto mapping) may only be continuous with continuous inverse, rather than smooth. In this case it would be a "homeomorphism" and the sets would be "homeomorphic".

[22] The closure of a set Y is the smallest closed set containing Y.

purpose, we need to combine it with another result, vaguely reminiscent of the implicit function theorem. (A proof can be found e.g. in [121], pp. 11 and 13, or in [77], pp. 21 and 60.)

Theorem 5.9 *(Preimage theorem) Let $f : X \to Y$ be a smooth map between an l-dimensional manifold (with boundary) and an m-dimensional manifold, where $l \geq$ (resp. $>$) m. If $y \in Y$ is a regular value (resp. both for f and for the restriction $f \,|\partial X$ of f to the boundary of X), then the preimage $f^{-1}(y) \subseteq X$ is an $(l-m)$-dimensional manifold (resp. with boundary, where the boundary $\partial f^{-1}(y)$ equals $f^{-1}(y) \cap \partial X$).*

In particular, if X and Y are manifolds with equal dimensions, $l = m$, then $f^{-1}(y)$ consists of isolated points (zero-dimensional manifolds) for any regular value $y \in Y$. Moreover, if X has boundary and $Y = \partial X$, then the preimage $f^{-1}(y)$ of a regular value $y \in Y$ is a one-dimensional manifold, because the dimension of ∂X is $l - 1$.

With these two ingredients at hand, the proof of a "smooth version" of Brouwer's fixed point theorem works as follows. We first establish that for a compact manifold X with boundary there is no smooth function $f : X \to \partial X$ such that $f(x) = x$ for all $x \in \partial X$.

Suppose there is such a function. Let $y \in \partial X$ be a regular value for f (which exists by Sard's theorem). Since y is certainly a regular value for the identity map $f \,|\partial X$, by the preimage theorem, $f^{-1}(y)$ is a one-dimensional manifold with boundary consisting of the single point $f^{-1}(y) \cap \partial X = \{y\}$. But $f^{-1}(y)$ is also compact. A classification of all compact 1-manifolds (see [121], pp. 55) shows they are finite disjoint unions of circles and arcs. Since, among those only arcs have boundary points (each of them always two), the number of boundary points of $f^{-1}(y)$ must be even. This is a contradiction, establishing the desired conclusion.

In particular, the unit disc $D^l = \{x \in \mathbb{R}^l \mid \|x\| \leq 1\}$ is a compact manifold with boundary $\partial D^l = \{x \in \mathbb{R}^l \mid \|x\| = 1\}$ equal to the unit sphere. (The Euclidean norm is defined by $\|x\| = \sqrt{x \cdot x}$.) Therefore, it can now be shown that every smooth map $f : D^l \to D^l$ has a fixed point. If this would not be true, one could define $g(x)$ for each $x \in D^l$ by

$$g(x) = x + \left(\sqrt{1 - x \cdot x + \left(\frac{x \cdot x - x \cdot f(x)}{\|x - f(x)\|} \right)^2} - \frac{x \cdot x - x \cdot f(x)}{\|x - f(x)\|} \right) \frac{x - f(x)}{\|x - f(x)\|}.$$

(The function g maps each point $x \in D^l$ to the point $g(x)$ at the intersection of ∂D^l with the ray starting at $f(x)$ and passing through x, see Figure 5.3.) Clearly, g would be smooth by construction if there is no x with $x = f(x)$. Moreover, because

$$\left(\frac{x - f(x)}{\|x - f(x)\|} \right) \cdot \left(\frac{x - f(x)}{\|x - f(x)\|} \right) = 1,$$

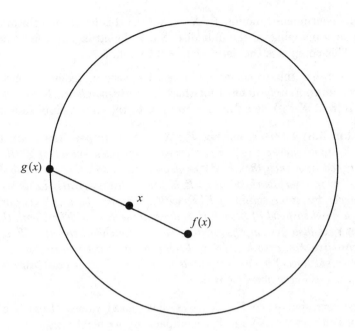

Figure 5.3: Mapping the disc to its boundary

it would follow that $\|g(x)\| = 1$. So, g would map D^l into ∂D^l. But, $\|x\| = 1$ (i.e., $x \in \partial D^l$) would imply $g(x) = x$, yielding a contradiction to the above. Hence, f must have a fixed point.

(Intuitively, if f did not have a fixed point on the unit disc, then the disc could smoothly be mapped to its boundary (the unit sphere) by g. But mapping the whole disc onto its boundary, keeping boundary points fixed, requires "ripping" the interior of the disc; i.e., there must be points "close together" in the interior of the disc which under g are shoved "far apart". Thus, the mapping g cannot be continuous, in contradiction to its construction.)

To extend this to merely continuous functions f on D^l, the Weierstrass approximation theorem is invoked (see [49], p. 133). This ensures that a continuous function f on a compact subset of a Euclidean space can be arbitrarily closely approximated by a polynomial (hence, smooth) function. Choosing the smooth approximation such that it (a) maps D^l into D^l and (b) it has a fixed point only if f has a fixed point, establishes Brouwer's fixed point theorem for $X = D^l$ (see [121], pp. 14).

To go from $X = D^l$ to an arbitrary compact convex subset X of some Euclidean space, one shows that such a compact convex set X can be mapped onto a unit disc (of appropriate dimension) by a continuous one-to-one and onto mapping with continuous inverse (by a "homeomorphism"). The composition of a continuous function $f : X \to X$ with this homeomorphism induces a continuous mapping on the unit disc. Since the latter has a fixed point, so does f. This completes the derivation of Theorem 5.7.

To generalize this to correspondences, its basic ingredient—continuity—has to be appropriately defined for these. A *correspondence* F (or set-valued function) from X to Y is a function from X to the collection of subsets of Y.

Definition 5.8 *A correspondence $F : X \to Y$ is **upper hemi-continuous** at $x \in X$ if, whenever $F(x)$ is contained in an open subset of Y, there is a neighborhood of x such that $F(x')$ is also contained in this open subset of Y, for all x' in the neighborhood of x. It is **lower hemi-continuous** at $x \in X$ if, whenever the intersection of $F(x)$ with an open subset of Y is nonempty, there is a neighborhood of x such that the intersection of $F(x')$ with this open subset of Y is also nonempty, for all x' in the neighborhood of x. F is **upper hemi-continuous**, or u.h.c. (resp. **lower hemi-continuous**, or l.h.c.) if it is upper (resp. lower) hemi-continuous at all $x \in X$. It is **continuous** if it is both upper and lower hemi-continuous.*

If the correspondence $F : X \to Y$ has closed values ($F(x)$ is a closed subset of Y for all $x \in X$) and Y is compact, upper hemi-continuity at $x \in X$ is equivalent to the property

$$\text{if} \quad x_k \to_{k \to \infty} x, \ y_k \to_{k \to \infty} y, \text{ and} \tag{5.13}$$
$$y_k \in F(x_k) \text{ for all } k = 1, 2, ..., \text{ then } y \in F(x),$$

defined in terms of sequences. This simply says that F has a closed *graph* $\mathcal{G}(F) = \{(x, y) \in X \times Y \,|\, y \in F(x)\}$. More intuitively, the images under F do not "implode" if F is u.h.c. Likewise, if F is compact-valued, then lower hemi-continuity at $x \in X$ is equivalent to

$$\text{if } x_k \to_{k \to \infty} x \text{ and } y \in F(x), \text{ then there is } \{y_k\}_{k=1}^{\infty} \tag{5.14}$$
$$\text{such that } y_k \in F(x_k) \text{ for all } k = 1, 2, ... \text{ and } y_k \to_{k \to \infty} y,$$

in terms of sequences. Intuitively, the images under F do not "explode" if F is l.h.c.

A *fixed point* for a correspondence F mapping a set X to itself is a point $x \in X$ such that $x \in F(x)$. The most widely used (in economics) fixed point theorem uses upper hemi-continuity combined with convexity.

Theorem 5.10 *(Kakutani's fixed point theorem) Let $X \subseteq \mathbb{R}^m$ be compact and convex and $F : X \to X$ an u.h.c correspondence with nonempty convex compact values. Then there is $x \in X$ such that $x \in F(x)$.*

One way to prove this utilizes an approximation result, known as *von Neumann's approximation lemma* (see e.g. [28], p. 68). It states that, for any u.h.c correspondence F with nonempty compact convex values, from a compact domain to a convex range, and any $\varepsilon > 0$, there is a continuous function the graph of which is ε-close to the graph $\mathcal{G}(F)$ of the correspondence (where the distance between two compact sets is Hausdorff distance). Since each of these approximating functions has a fixed point by Brouwer's theorem, compactness and convex-valuedness can be exploited to show that a sequence of fixed points of approximating functions converges to a point which constitutes a fixed point for the correspondence (for details see e.g. [28], p. 72). Thus, Kakutani's theorem emerges as a consequence of Brouwer's theorem.

Exercises

Exercise 5.11 A two-player game is symmetric if $S_1 = S_2$ and $u_1(s_2, s_1) = u_2(s_1, s_2)$ for all $(s_1, s_2) \in S$. Consider symmetric 2×2 games given by the bimatrix representation below.

	s_2^1	s_2^2
s_1^1	a a	0 0
s_1^2	0 0	b b

Each such game can be identified with a point $(a, b) \in R^2$ in the plane. Derive the set of all Nash equilibria for all such generic (i.e., $a \neq b$, $a \neq 0$, and $b \neq 0$) games. Which of these games admit asymmetric equilibria? (Hint: It is helpful to draw pictures of Θ, which here happens to be the unit square.)

Exercise 5.12 (The Santa Fé bar problem) There is a bar in Santa Fé called "El Farol" which has 60 seats. Every night 100 people would like to come to this bar. But all of them would rather stay home when the bar is so crowded that they can't get a seat. How many people will turn up at the bar in (any one pure Nash) equilibrium at any given night? (Hint: It may be illuminating to start on this problem by considering a bar with two seats which three people would like to attend.)

Exercise 5.13 There are $n \geq 2$ consumers who have an identical constant marginal valuation θ for a public good, where $0 < \theta < 1$. The public good is produced by a linear technology converting 1 unit of money contribution into 1 unit of public good. The following auction is used to finance its production. All players are (simultaneously) asked to contribute voluntarily to the production of the public good, but those who make the highest contributions will receive a prize worth $v \geq 1$ to them. The utility function of player $i = 1, ..., n$ is, therefore, given by

$$u_i(s_1, ..., s_n) = \begin{cases} v + \theta \sum_{j=1}^n s_j - s_i & \text{if} \qquad s_i \geq s_j, \text{ for all } j = 1, ..., n \\ \theta \sum_{j=1}^n s_j - s_i & \text{otherwise} \end{cases}$$

where $s_i \in S_i = R_+$ denotes the contribution of player i. What are the possible (pure strategy) Nash equilibria of the corresponding game?

Exercise 5.14 (Bank run) A large number n of investors $i = 1, ..., n$ make a joint investment (of 1 unit each) which yields a return $r > 1$ (per unit invested) if it matures. For the project to mature requires that at least L, $1 < L < n$, investors do not withdraw their money early. Yet, each investor may experience an interim liquidity shock which forces her to withdraw. More formally, an investor may be one of two types. Type 1 of investor i occurs with a small probability θ, $0 < \theta < 1$, making i's payoff function $u_{i1}(s) = 1 - s_i$, where $s_i \in \{0, 1\}$. This type will always withdraw ($s_i = 0$). With probability $1 - \theta$, investor i is of type 2 and has payoff function

$$u_{i2}(s) = \begin{cases} rs_i & \text{if } \sum_{j=1}^{n} s_j \geq L \\ 1 - s_i & \text{if } \sum_{j=1}^{n} s_j < L \end{cases}$$

where $s_i \in S_i = \{0, 1\}$ and $s = (s_1, ..., s_n) \in S = \{0, 1\}^n$. Each investor learns her own type, but not the types of other investors, and decides whether to withdraw ($s_i = 0$) or not ($s_i = 1$) conditional on her type. Assume that

$$\frac{r}{1+r} > \sum_{k=1}^{L-1} \binom{n}{k} \theta^k (1 - \theta)^{n-k}$$

and show that there are at least two pure Nash equilibria. In one equilibrium only investors of type 1 choose $s_i = 0$, and in the other all investors choose $s_i = 0$. (The latter is the bank run equilibrium.)

Exercise 5.15 (Bertrand duopoly) Two suppliers of mineral water can pump it from the ground at (identical) constant marginal cost $c \geq 0$. They compete by making price offers, $p_i \in R_+ = S_i$ for $i = 1, 2$. The market is represented by a downward sloping continuous demand function $D(p)$, where $D(0) > 0$ and $\lim_{p \to \infty} D(p) = 0$. Consumers always buy from the supplier who asks the lower price. If both ask the same price, they share the market demand equally. Hence, payoff functions are

$$u_i(p_{3-i}, p_i) = \begin{cases} (p_i - c)D(p_i) & \text{if} & p_i < p_{3-i} \\ \frac{1}{2}(p_i - c)D(p_i) & \text{if} & p_i = p_{3-i} \\ 0 & \text{otherwise} \end{cases}$$

for $i = 1, 2$. Show by construction that a pure Nash equilibrium exists and that in equilibrium $p_1 = p_2 = c$ holds.

Exercise 5.16 Consider the following two-player extensive form game with perfect information. Player 1 can make two offers on how to split a pie of size 5. She can either offer $(3, 2)$, where the first entry is her payoff and the second is player 2's, or she may offer $(4, 1)$. If she offers $(3, 2)$, the game terminates with these payoffs. If she offers $(4, 1)$, player 2 can either accept, which terminates the game and implements these payoffs, or reject. If 2 rejects, it becomes again the turn of player 1, who can now choose between (terminal nodes yielding payoffs) $(2, 2)$ and $(2, 0)$. Derive the set of all (pure and mixed) Nash equilibria for this game. Which of those use a weakly dominated strategy?

Exercise 5.17 A number $n \geq 2$ of bidders $i = 1, ..., n$ participate in a first-price sealed bid auction. Each bidder i has a valuation v_i for the (indivisible) object drawn independently from a uniform distribution on the unit interval $[0, 1]$. The valuations are private information. The object is allocated to the highest bidder at the price she bids. (If several bids are highest, some tie breaking is applied.) Bidders make their offers simultaneously in sealed envelopes. Bidder i has payoff $v_i - b_i$ if she wins the auction with bid $b_i \in [0, 1]$; otherwise her payoff is zero. Pure strategies are functions from valuations in $[0, 1]$ to bids in $[0, 1]$. Find a pure strategy Nash equilibrium for this game of "incomplete information". (Hint: If the other bidders uses the bid function $b(v_j) = (n - 1)v_j/n$, for $j \neq i$, then the probability of i winning with a bid b_i is

$$\Pr\left(\frac{n-1}{n}v_j \leq b_i, \forall j \neq i\right) = \Pr\left(v_j \leq \frac{n}{n-1}b_i, \forall j \neq i\right) = \left(\frac{n}{n-1}b_i\right)^{n-1}$$

viz. the probability that $n - 1$ independent draws from a uniform distribution will not exceed $b^{-1}(b_i)$.)

Exercise 5.18 Each of $n \geq 2$ agents, $i = 1, ..., n$, can make contributions $s_i \in [0, w] = S_i$ to the production of a public good, for some $w > 0$. Their payoff functions are given by

$$u_i(s) = n \min\{s_1, ..., s_n\} - s_i$$

for all $s = (s_1, ..., s_n) \in S = [0, w]^n$ and all $i = 1, ..., n$. Find all pure strategy Nash equilibria for this game.

Exercise 5.19 (Rock-paper-scissors) Find the equilibria of the two-player game below (which is a version of a childrens' game).

	s_2^1	s_2^2	s_2^3
s_1^1	0 0	0 1	1 0
s_1^2	1 0	0 0	0 1
s_1^3	0 1	1 0	0 0

Notes on the Literature: That common knowledge of the solution yields Nash equilibrium is one way to interpret results from a literature on Bayesian games; see e.g. Armbruster and Böge [5], Böge and Eisele [26], Mertens and Zamir [116], and Tan and Werlang [174], among others. The idea behind Nash equilibrium goes back at least to Cournot [33]. Theorem 5.4 is due to Nash ([131] and [132]). Theorem 5.5 is due to Glicksberg [69]. An extension to discontinuous games is by Dasgupta and Maskin [40]. Theorem 5.6 has been proved by Glicksberg [69], Debreu [41], and Fan [56]. Example 5.12 is taken from Simon and Stinchcombe ([168], p. 1428). The material in the appendix draws on Milnor [121] and Border [28]. Exercise 5.14 is based on Diamond and Dybvig [47].

5.4   Zero-Sum Games

Strictly Competitive Games   In general, very little can be said about the set of Nash equilibria. Yet, in a limited class of games we can say something about the qualitative character of equilibria. The class consists of games between two players, whose preferences are diametrically opposed, because what one wins the other loses.

Definition 5.9 *A (finite or infinite) game is **strictly competitive** if it is a two-player game* $\Gamma = (S_1 \times S_2, u_1 \times u_2)$ *satisfying*

$$u_1(s) \leq u_1(s') \text{ if and only if } u_2(s) \geq u_2(s') \text{ for all } s, s' \in S. \qquad (5.15)$$

A strictly competitive game occurs, for instance, if the two players' payoffs sum to the same constant for all strategy combinations. By the expected utility hypothesis the constant is arbitrary, so it can be normalized to zero. Such (two-player) games where

$$u_1(s) + u_2(s) = 0 \text{ for all } s \in S, \qquad (5.16)$$

are called *zero-sum* games. Clearly, (5.16) implies (5.15) by $u_1 = -u_2$, so every zero-sum game is strictly competitive. The converse is a priori false. It takes payoff transformations (which do not affect the best replies) to convert a strictly competitive game into a zero-sum game (see Exercise 5.24 below).

Example 5.13 *Consider the 2×2 game below for some ε with $0 \leq \varepsilon \leq 1$. For $\varepsilon = 0$ this is precisely the game from Example 5.9 which is not strictly competitive, because $u_2\left(s_1^1, s_2^1\right) = 0 \leq u_2\left(s_1^2, s_2^2\right) = 0$ and $u_1\left(s_2^1, s_1^1\right) = 1 < u_1\left(s_2^2, s_1^2\right) = 2$, in violation of (5.15). For $0 < \varepsilon < 1$ it is strictly competitive, but not zero-sum. Only for $\varepsilon = 1$ is this game zero-sum. (Subtract 1 from all of player 1's payoffs, to normalize to zero.)*

	s_2^1	s_2^2
s_1^1	1 / 0	$-\varepsilon$ / 2
s_1^2	0 / 1	2 / $-\varepsilon$

As before, the game Γ in Definition 5.9 may be the mixed extension of a finite game, so strategies may be mixed ($S = \Theta$ and $u = U$). It is easy to show that the mixed extension of a (finite) two-player game is zero-sum *if and only if* the original game is zero-sum. (In Exercise 5.25 you are asked to prove this claim.)

But this is not so for merely strictly competitive games. The mixed extension of a strictly competitive game *need not* be strictly competitive (see Exercise 5.23 below). This is because of possible risk considerations. Both players may prefer one (pure) strategy combination to a (nondegenerate) mixture between two other strategy combinations.

Example 5.14 *The mixed extension of the (strictly competitive) game from Example 5.13 with $0 < \varepsilon < 1$ is not strictly competitive. To see this, denote $\sigma_1\left(s_1^1\right) = 1 - \sigma_1\left(s_1^2\right) = x$ and $\sigma_2\left(s_2^1\right) = 1 - \sigma_2\left(s_2^2\right) = y$. Then payoffs in the mixed extension are*

$$\begin{aligned} U_1(\sigma) &= x\left[(3+\varepsilon)\,y - (2+\varepsilon)\right] + 2\,(1-y) \quad and \\ U_2(\sigma) &= y\left[1+\varepsilon - (3+\varepsilon)\,x\right] + (2+\varepsilon)\,x - \varepsilon. \end{aligned}$$

Now let $\sigma, \sigma' \in \Theta$ be such that $U_1(\sigma) \leq U_1(\sigma')$, which is equivalent to

$$(3+\varepsilon)\,xy - (2+\varepsilon)\,x - 2y \leq (3+\varepsilon)\,x'y' - (2+\varepsilon)\,x' - 2y',$$

and $y < y'$. The latter implies $-2y' + (1-\varepsilon)\,y < -(1+\varepsilon)\,y'$ which, combined with the previous inequality, yields

$$(3+\varepsilon)\,xy - (2+\varepsilon)\,x - (1+\varepsilon)\,y < (3+\varepsilon)\,x'y' - (2+\varepsilon)\,x' - (1+\varepsilon)\,y'$$

(by adding $(1-\varepsilon)\,y$ to both sides of the earlier inequality), which is precisely $U_2(\sigma) < U_2(\sigma')$. Hence both players prefer σ' to σ.

A distinguishing property of a Nash equilibrium for a strictly competitive game is as follows. If player 1 assumes (as she must in a Nash equilibrium) that player 2 behaves optimally, given $s_1 \in S_1$, she has to conclude that by maximizing $u_2(s_1, s_2)$ player 2 will choose $s_2 \in S_2$ so as to *minimize $u_1(s_2, s_1)$*, because of (5.15); likewise for player 2. By maximizing $u_1(s_2, s_2)$ given $s_2 \in S_2$ player 1 will minimize $u_2(s_1, s_2)$. That is, each player has to choose her own strategy on the assumption that her opponent will try to hurt her as much as possible.

More formally, by (5.15) a strategy is a best reply for player 1 (resp. 2) if and only if it minimizes u_2 (resp. u_1) over S_1 (resp. S_2); i.e.,

$$s_i' \in \beta_i\,(\tilde{s}) \text{ if and only if} \tag{5.17}$$
$$u_{3-i}\,(s_i', \tilde{s}_{3-i}) \leq u_{3-i}\,(s_i, \tilde{s}_{3-i}) \text{ for all } s_i \in S_i,$$

for all $\tilde{s} \in S_1 \times S_2$ and $i = 1, 2$.

Hence, an equilibrium strategy must maximize the minimal payoff. This follows directly from the assumption of a strictly competitive game and does not require any coordination of expectations. Call a strategy $\bar{s}_i \in S_i$ for player $i \ (= 1, 2)$ in a strictly competitive game a *maxminimizer* if

$$\min_{s_{3-i} \in S_{3-i}} u_i(s_{3-i}, \bar{s}_i) \geq \min_{s_{3-i} \in S_{3-i}} u_i(s_{3-i}, s_i) \text{ for all } s_i \in S_i. \tag{5.18}$$

The value v_i of the payoff function at a maxmimimizer,

$$v_i = \max_{s_i \in S_i} \min_{s_{3-i} \in S_{3-i}} u_i(s_{3-i}, s_i), \tag{5.19}$$

is called player i's *security level*, for $i = 1, 2$. It is the highest payoff that player i can guarantee herself. The security level v_i cannot exceed the lowest payoff to which the opponent can hold player i down to; i.e.,

$$v_i \leq \min_{s_{3-i} \in S_{3-i}} \max_{s_i \in S_i} u_i(s_{3-i}, s_i), \tag{5.20}$$

because $u_i(s_{3-i}, s_i) \leq \max_{s_i'} u_i(s_{3-i}, s_i')$, for all $(s_{3-i}, s_i) \in S$, implies that

$$\min_{s_{3-i}} u_i(s_{3-i}, s_i) \leq \min_{s_{3-i}} \max_{s_i'} u_i(s_{3-i}, s_i'),$$

for all $s_i \in S_i$, and, since the right hand side of the latter is a constant, it implies (5.20).

The Minimax Theorem The unique feature of strictly competitive games (which have a Nash equilibrium) is that a pair of strategies constitutes a Nash equilibrium if and only if each player's strategy is a maxminimizer. This is striking, because it implicitly prescribes a "decision theory" that is considerably more "cautious" than what expected utility suggests. It recommends to players to do "worst case planning", i.e. to always assume that the opponent will do the least preferred thing. Since this makes her (maxminimizing) strategy independent of the opponent's choice, beliefs are redundant. The key result is as follows.

Theorem 5.11 *(Minimax theorem) For any (finite or infinite) strictly competitive game* $\Gamma = (S_1 \times S_2, u_1 \times u_2)$, *the following three conditions are equivalent:*

(a) There exists a Nash equilibrium of Γ.

(b) $\max_{s_1 \in S_1} \min_{s_2 \in S_2} u_1(s_2, s_1) = \min_{s_2 \in S_2} \max_{s_1 \in S_1} u_1(s_2, s_1)$.

(c) There exist a real number $v_1 \in \mathbb{R}$ *and a pair of strategies* $(\bar{s}_1, \bar{s}_2) \in S_1 \times S_2$ *such that* $u_1(s_2, \bar{s}_1) \geq v_1 \geq u_1(\bar{s}_2, s_1)$ *for all* $(s_1, s_2) \in S_1 \times S_2$.

The proof shows that (a) implies (b), that (b) implies (c), and that (c) implies (a). To see that (a) implies (b), suppose there is a Nash equilibrium $(s_1^*, s_2^*) \in S_1 \times S_2$ for Γ. Then,

$$\min_{s_2 \in S_2} \max_{s_1 \in S_1} u_1(s_2, s_1) \leq \max_{s_1 \in S_1} u_1(s_2^*, s_1) = u_1(s^*)$$

$$= \min_{s_2 \in S_2} u_1(s_2, s_1^*) \leq \max_{s_1 \in S_1} \min_{s_2 \in S_2} u_1(s_2, s_1).$$

The first inequality follows from the definition of a minimum, the two equalities from the Nash property and (5.17), and the final inequality follows from the definition of a maximum. Combined with (5.20), this yields (b).

If (b) is true, let $v_1 = \min_{s_2} \max_{s_1} u_1(s_2, s_1) = \max_{s_1} \min_{s_2} u_1(s_2, s_1)$ and $\bar{s}_i \in S_i$ be a maxminimizer for player $i = 1, 2$. Then, for all $(s_1, s_2) \in S_1 \times S_2$,

$$u_1(s_2, \bar{s}_1) \geq \min_{s_2' \in S_2} u_1(s_2', \bar{s}_1) = v_1 = \max_{s_1' \in S_1} u_1(\bar{s}_2, s_1') \geq u_1(\bar{s}_2, s_1),$$

where the first inequality follows from the definition of a minimum, the two equalities follow from the definition of a maxminimizer and the hypothesis, and the final inequality follows from the definition of a maximum. This verifies (c).

Finally, if (c) holds true, then, from $u_1(s_2, \bar{s}_1) \geq v_1 \geq u_1(\bar{s}_2, s_1)$ and (5.15), $u_2(\bar{s}_1, s_2) \leq u_2(s_1, \bar{s}_2)$, for all $(s_1, s_2) \in S_1 \times S_2$. Choose $s_i^* = \bar{s}_i$, for $i = 1, 2$. This yields $v_1 = u_1(s_2^*, s_1^*)$ and $u_2(\bar{s}_1, \bar{s}_2) = u_2(s_1^*, s_2^*) \geq u_2(s_1^*, s_2)$ for all $s_2 \in S_2$, so that $(s_1^*, s_2^*) \in S_1 \times S_2$ constitutes a Nash equilibrium. This completes the proof of Theorem 5.11.

In words, it says that, for any strictly competitive game, existence of a Nash equilibrium is equivalent to the property that the security level of a player coincides with the lowest payoff to which the opponent can hold her. And this is equivalent to the existence of a unique number v_1 such that any maxminimizer for player 1 guarantees her *at least* v_1 and any maxminimizer for player 2 guarantees that player 1 obtains *at most* v_1. Hence, player 1 obtains v_1 (and player 2 obtains v_2) in all equilibria. If the game is zero-sum, then $v_1 = -v_2$ by (5.16).

This number v_1 is called the *value* of the game Γ. Hence, any strictly competitive game that has an equilibrium also has a value (and vice versa). The value measures, in a sense, what the game is "worth" to a player, because it is the payoff which she will obtain in any one equilibrium. This is peculiar to this class of games, because in general games the players' (equilibrium) payoffs vary considerably with the equilibria that are played (see Example 5.15 below).

The existence of a value also implies a "rectangularity" property of the set of Nash equilibria. If $s, s' \in S_1 \times S_2$ are both equilibria, then by the Nash property $u_1(s_2', s_1) \leq u_1(s')$ and $u_2(s_1', s_2) \leq u_2(s')$, hence $u_1(s') \leq u_1(s_2, s_1')$ from (5.15), and $u_1(s_2, s_1') \leq u_1(s)$ and $u_2(s_1, s_2') \leq u_2(s)$, hence $u_1(s) \leq u_1(s_2', s_1)$ from (5.15), imply

$$u_1(s_2', s_1) \leq u_1(s') \leq u_1(s) \leq u_1(s_2', s_1) \text{ and}$$
$$u_1(s_2, s_1') \leq u_1(s) \leq u_1(s') \leq u_1(s_2, s_1'),$$

so all these payoffs are equal to v_1. Applying the same argument for player 2, this demonstrates the following.

Proposition 5.2 *For any (finite or infinite) strictly competitive game, if* $s, s' \in S_1 \times S_2$ *are both equilibria, then* (s_1, s'_2) *and* (s'_1, s_2) *are also equilibria.*

Hence, in these games there is no need to coordinate expectations. Any player simply picks a maxminimizer. Whatever maxminimizer the opponent picks, the resulting strategy combination will constitute a Nash equilibrium.

The latter also implies that it never pays a player to communicate with the opponent prior to a playing of the game. This could only make the opponent better off and, therefore, hurt. If the game were extended by a pre-play communication stage, where players could inform each other about what they will do, it would be a weakly dominated strategy to reveal one's choice beforehand.

Neither of these two properties holds outside the realm of strictly competitive games. For instance, in the mixed extension of the game from Example 5.13 with $0 < \varepsilon < 1$, player 1's unique maxminimizer is $\bar{\sigma}_1 \left(s_1^1 \right) = 2/(3 + \varepsilon)$ while her (unique) Nash equilibrium strategy is $\sigma_1^* \left(s_1^1 \right) = (1 + \varepsilon)/(3 + \varepsilon)$ which is smaller than $\bar{\sigma}_1 \left(s_1^1 \right)$. (In Exercise 5.26 below you are asked to verify this.) If player 2 did indeed expect her opponent to use a maxminimizer, i.e. to choose $\bar{\sigma}_1$, then she would be best off playing s_2^2 with certainty. But player 1 would foresee this and respond with s_1^1 for sure; and so on. Thus a "solution" by maxminimizers would be "self defeating", because once it becomes common knowledge it is destabilized by deviations. (This argument also holds for $\varepsilon = 0$.)

For the game from Example 5.13 the security levels coincide with the Nash equilibrium payoffs, for all $0 \leq \varepsilon \leq 1$. So, one may feel that it might be "safer" to play a maxminimizer, since it would yield the same payoff as the Nash equilibrium. But, if $\varepsilon < 1$, this cannot be a self enforcing "solution theory", because if players believed in this "theory" player 2 would want to deviate, inducing player 1 to deviate, setting off an infinite regress which would invalidate the players' beliefs. Hence, unless $\varepsilon = 1$, expectations about the opponent's choice do make a difference.

To see that, for games that are not strictly competitive, it might pay to communicate, consider the following example.

Example 5.15 *The* 2×2 *game below is a "coordination game" with three Nash equilibria for the mixed extension.*

	s_2^1	s_2^2
s_1^1	3 1	−1 −1
s_1^2	−1 −1	1 3

Both players using their first and both using their second (pure) strategies constitute strict equilibria. Furthermore, there is a mixed equilibrium where

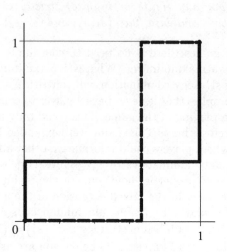

Figure 5.4: Coordination game

player 1 *uses her first strategy with probability* 2/3 *and player* 2 *uses her first strategy with probability* 1/3. *(See Figure 5.4, where* $\sigma_1\left(s_1^1\right)$ *is on the horizontal and* $\sigma_2\left(s_2^1\right)$ *is on the vertical axis. The thick solid graph is* $\tilde{\beta}_1$ *and the broken one is* $\tilde{\beta}_2$. *The intersection points constitute Nash equilibria.) The corresponding payoff vectors are* $(3,1)$, $(1,3)$, *and* $(1/3,1/3)$, *respectively. (The combination of maxminimizers does not constitute an equilibrium, because the unique maxminimizer for player* 1 *is* $\bar{\sigma}_1\left(s_1^1\right) = 1/3$ *and for player* 2 *it is* $\bar{\sigma}_2\left(s_2^1\right) = 2/3$.) *If player* 1 *could credibly communicate that she will use her first pure strategy for sure, then she would get her best possible payoff. Hence, pre-play communication may pay.*

Moreover, suppose players could agree to flip a fair coin, observable to both, and to play $\left(s_1^1, s_2^1\right)$ *if heads come up, and* $\left(s_1^2, s_2^2\right)$ *if tails come up. Then, expected payoffs (from this "correlated strategy") would be* $(2,2)$. *Yet, the payoff vector* $(2,2)$ *is infeasible without pre-play communication: Since*

$$U_1(\sigma) = \sigma_1\left(s_1^1\right)\left[6\sigma_2\left(s_2^1\right) - 2\right] + 1 - 2\sigma_2\left(s_2^1\right) \ \ and$$
$$U_2(\sigma) = \sigma_2\left(s_2^1\right)\left[6\sigma_1\left(s_1^1\right) - 4\right] + 3 - 4\sigma_1\left(s_1^1\right),$$

an assumption $U_1(\sigma) = U_2(\sigma) = 2$ *would imply* $\sigma_1\left(s_1^1\right) + \sigma_2\left(s_2^1\right) = 1$ *(by subtracting the two payoffs), which (when substituted back into the payoff function) would imply* $1/2 = \sigma_1\left(s_1^1\right)\left[1 - \sigma_1\left(s_1^1\right)\right]$, *which is a contradiction because* $\sigma_1\left(s_1^1\right)\left[1 - \sigma_1\left(s_1^1\right)\right] \leq 1/4$.

Hence, while strictly competitive games have very attractive properties, these are peculiar to this type of game. When players' preferences are not diametrically opposed, or when there are more than two players, then there is much less structure on the set of Nash equilibria.

Summary

- A two-player game is strictly competitive if whatever player 1 prefers is worse for player 2. It is zero-sum if the payoffs of the two players always sum to the same constant.

- The mixed extension of a two-player game is zero-sum if and only if the original game is zero-sum. But the mixed extension of a strictly competitive game need not be strictly competitive.

- In a strictly competitive game, a strategy is a Nash equilibrium strategy if and only if it maximizes the minimal payoff, such that all equilibria yield the security level.

- Hence, the playing of Nash equilibria does not require the coordination of expectations, nor can communication be profitable. Yet, these properties are peculiar to strictly competitive games.

Exercises

Exercise 5.20 Player 1 has to hide a treasure in one of $K > 1$ places. Player 2 can access one of those and collect the treasure if it is there. The treasure is worth one unit to either player. Hence, $S_i = \{1, ..., K\}$ for $i = 1, 2$, and payoffs in the mixed extension of the induced game are

$$U_1(\sigma) = \sum_{k=1}^{K} \sigma_1(k)\left(1 - \sigma_2(k)\right) \text{ and } U_2(\sigma) = \sum_{k=1}^{K} \sigma_1(k)\sigma_2(k),$$

where $\sigma_i(k)$ stands for the probability of choosing location $k \in \{1, ..., K\}$, for $i = 1, 2$. The sum of these payoffs is 1. To normalize the payoff sum to zero, subtract, say, the constant $1/2$ from both payoff functions. Then, find the value of this game.

Exercise 5.21 Army 1 has a single plane which it can target at one of three possible locations. Army 2 has one anti-aircraft system which it can place at one of the three locations. The values of the three targets are $w_1 > w_2 > w_3 > 0$. Army 1 can destroy a target only if it is undefended and attacked. Army 1 wishes to maximize the expected value of damage and army 2 wishes to minimize it. Formulate this as a zero-sum game and find its mixed Nash equilibria.

Exercise 5.22 Show that the 2×2 game below, with $0 < \delta \leq 1$ and $0 < \beta < 1$, has a strictly competitive mixed extension.

	s_2^1	s_2^2
s_1^1	$-\delta\beta$ $\quad$ β	-1 $\quad$ $\frac{1}{\delta}$
s_1^2	$-\delta$ $\quad$ 1	0 $\quad$ 0

Exercise 5.23 Consider the two-player game given below. Show that it is strictly competitive, but its mixed extension is not.

	s_2^1	s_2^2
s_1^1	2 $\quad$ -1	2 $\quad$ -1
s_1^2	3 $\quad$ -3	0 $\quad$ 0

(Hint: Compare the strategy combinations $\sigma^1 \in \Theta$ and $\sigma^2 \in \Theta$, where $\sigma_1^1\left(s_1^1\right) = 0$, $\sigma_2^1\left(s_2^1\right) = 1/2$, and $\sigma_1^2\left(s_1^1\right) = 1$, $0 \leq \sigma_2^2\left(s_2^1\right) \leq 1$.)

Exercise 5.24 Reconsider the game from Exercise 5.22. Though it is strictly competitive (and has a strictly competitive mixed extension), it is not zero-sum. To see how it can be turned zero-sum, prove the following general result. For an arbitrary finite n-player game $\Gamma = (S, u)$, player i's preference ordering over her strategies $s_i \in S_i$, given a fixed strategy combination $\bar{s}_{-i} \in S_{-i}$ for the other players, is unaffected by adding one and the same constant to all payoffs for i that are associated with $\bar{s}_{-i} \in S_{-i}$. It follows that the best reply correspondence $\beta_i : \Theta \rightarrow S_i$ for player i is unaffected when the same constant is added to all payoffs for i associated with a fixed $\bar{s}_{-i} \in S_{-i}$. Also show that this holds for the mixed best reply correspondence $\tilde{\beta}_i : \Theta \rightarrow \Delta_i$ as well. (Invoke Lemma 5.2.)

Exercise 5.25 Show that the mixed extension of a finite two-player game $\Gamma = (S_1 \times S_2, u_1 \times u_2)$ is zero-sum if and only if Γ is zero-sum. (Hint: To show the "only if" part, use the fact that pure strategy combinations are special cases of mixed strategy combinations. For the "if" part, write out explicitly the extensions of payoff functions to mixed strategies.)

Exercise 5.26 Solve explicitly for the Nash equilibrium and the maxminimizers in Example 5.13. Also determine the security levels for the two players and the Nash equilibrium payoffs.

Exercise 5.27 Find all Nash equilibria for the two-player game below. Does this game have a value?

	s_2^1	s_2^2	s_2^3
s_1^1	1 2	2 1	1 2
s_1^2	0 3	3 0	−1 4

Exercise 5.28 Prove the following mathematical result: If X and Y are two nonempty compact convex subsets of some finite-dimensional Euclidean space, then there are $\bar{x} \in X$ and $\bar{y} \in Y$ and a real number v such that $\bar{x} \cdot y \geq v \geq x \cdot \bar{y}$ for all $(x,y) \in X \times Y$. (This is a variant of the separating hyperplane theorem. Hint: Combine Theorems 5.6 and 5.11.)

Notes on the Literature: Historically, the analysis of two-player zero-sum games has been key for the development of game theory, because of its attractive properties (see von Neumann and Morgenstern [134], Chapters 12-17). The first minimax theorem was proved by von Neumann [133]. Generalizations were established by Sion [170] and Fan [57]. The version presented here is taken from Luce and Raiffa ([105], p. 388).

5.5   Correlated Equilibrium

Correlating Devices   Example 5.15 is suggestive. There, players can achieve an efficient symmetric outcome by tossing a coin which both can observe. In practice this is often done. Think of two people who have to coordinate on where to meet for dinner but have different tastes when it comes to cuisine. They may well toss a coin to decide whether to meet at the first person's favorite place or the second person's.

More generally, in some situations all players might gain if they could implement a (possibly random) coordination device, even if they merely coordinate on the occurrence of "sunspots". Nash equilibrium allows for that only insofar as this coordination device is part of the rules of the game. Yet, often one may want to model the strategic interaction as sparsely as possible, thereby ignoring the potential existence of a communication system. To bring it back into the picture, possible coordination devices can be made part of the solution concept.

This is largely the idea of *correlated equilibrium*. It is derived from considering the Nash equilibria of an expanded game with such a (random) "correlating device" available and then characterizing these equilibria.

To see how this works, identify a *correlating device* with a triple $Q = (\Omega, \{T_i\}_{i=1}^{n}, \mu)$. Here Ω is a (finite) state space; each T_i is player i's information partition[23] (of Ω) for $i = 1, ..., n$; and μ is a probability measure on the state space. In Example 5.15 the state space Ω is (the set consisting of) "heads" and "tails", and the probability measure μ is the probabilities of the coin flip; i.e., this is what the players have to implement. The information partition T_i determines what player i learns about the outcome of the random device. In Example 5.15 each T_i is the finest possible partition of the state space, so both players get the best possible information. But, in general, some players might not directly observe the state, but see only a superset thereof. (The latter is her "type" $t_i \in T_i$ in the terminology of Bayesian games—see Section 3.5.)

Now, given any game $\Gamma = (S, u)$ and some correlating device Q, define an enlarged game $\Gamma(Q)$ by making pure strategies functions from the state space Ω to S_i which are measurable with respect to the information partition T_i for all $i = 1, ..., n$.[24] The idea is that player i will learn that the possible state is an element of her type $t_i \in T_i$ and that she can condition her choice of strategy (from S_i) on this information. This conditioning is expressed by making pure strategies in $\Gamma(Q)$ (T_i-measurable) functions $f_i : \Omega \to S_i$.

---

[23]   Recall that a *partition* is a collection of pairwise disjoint ("mutually exclusive") subsets which jointly cover the underlying space ("jointly exhaustive").

[24]   Recall that a function is *measurable* with respect to a partition of its domain if it is constant on any element of the partition.

Mixed strategies can be defined as usual. But if the state space is large enough, they will be redundant. For instance, if player i wants to play $s_i \in S_i$ and $s_i' \in S_i$ with equal probabilities when observing information set $t_i \in T_i$, we can replace each $\omega \in t_i$ by two equally likely states, ω^1 and ω^2, and t_i by two information sets, t_i^1 and t_i^2, where t_i^1 contains all ω^1 and t_i^2 all ω^2 (and the other players' information sets at ω contain all ω^1 and ω^2). Then player i can use the strategy "play s_i if t_i^1 and play s_i' if t_i^2". This will be equivalent to the original mixed strategy.

The expanded game $\Gamma(Q)$ captures that players have the "correlating device" Q at their disposal and can condition their strategy choices on the signals communicated to them through this device, *knowing* that other players receive a (possibly) correlated signal. The latter allows them to predict (at least up to some error) what other players will do, enabling a coordinated action.

A Nash equilibrium of $\Gamma(Q)$ is an n-tuple $f = (f_1, ..., f_n)$ of functions $f_i : \Omega \to S_i$ that are measurable with respect to T_i such that

$$\sum_{\omega \in \Omega} \mu(\omega) u_i \left(f_{-i}(\omega), \hat{f}_i(\omega) \right) \leq \sum_{\omega \in \Omega} \mu(\omega) u_i \left(f(\omega) \right) \qquad (5.21)$$

for all T_i-measurable functions $\hat{f}_i$ and all $i = 1, ..., n$.[25]

Such an equilibrium is called a *correlated equilibrium relative to Q* for the original game Γ. It is often also called an "objective correlated equilibrium" (relative to Q for Γ), because players agree on the prior beliefs μ. In a "subjective correlated equilibrium" players are allowed to hold different prior beliefs μ_i.

Condition (5.21) requires that f_i maximizes player i's "ex ante" payoff, before she learns her type t_i. This implies that f_i maximizes player i's *conditional* payoff *given* t_i (or the "interim" payoff) for each type $t_i \in T_i$ that occurs with positive probability. That is, (5.21) is equivalent to

$$\sum_{\omega \in t_i} \mu\left(\omega \,|\, t_i\right) u_i \left(f_{-i}(\omega), \hat{f}_i(\omega) \right) \leq \sum_{\omega \in t_i} \mu\left(\omega \,|\, t_i\right) u_i \left(f(\omega) \right) \qquad (5.22)$$

for all $t_i \in T_i$ with $\mu(t_i) \equiv \sum_{\omega \in t_i} \mu(\omega) > 0$, all T_i-measurable functions $\hat{f}_i$, and all $i = 1, ..., n$. Here, $\mu\left(\omega \,|\, t_i\right) = \mu(\omega)/\mu(t_i)$ for all $\omega \in t_i$ and $\mu\left(\omega \,|\, t_i\right) = 0$ for all $\omega \notin t_i$ denotes player i's posterior beliefs about the states.

Correlated Equilibrium The drawback of such a construction is that it depends on the correlating device Q. Yet, there is an infinite number of possible correlating devices with all sorts of type partitions. On the other hand, any joint distribution over strategy combinations (in S) that forms a correlated equilibrium relative to some correlating device Q can be attained by

---

[25] If μ is a probability measure on the finite set Ω, $\mu(\omega) = \mu(\{\omega\})$ denotes the probability assigned to the singleton set $\{\omega\} \subseteq \Omega$.

a "universal device" which directly recommends a strategy to each player. Rather than sending players arbitrary signals (types), this "direct mechanism" tells each player what she should do. Learning the recommendation, the player can still deviate; i.e., the recommendation is not binding. Formally, let $\Delta(S) = \left\{ \varphi : S \to \mathbb{R}_+ \left| \sum_{s \in S} \varphi(s) = 1 \right. \right\}$ be the set of all probability distributions on strategy combinations $s \in S$.[26]

Proposition 5.3 *For any (finite) game* $\Gamma = (S, u)$, *a probability distribution* $\varphi \in \Delta(S)$ *over pure strategy combinations is generated by a correlated equilibrium relative to some correlating device* Q *(see (5.21)) if and only if*

$$\sum_{s \in S} \varphi(s) u_i \left(s_{-i}, g_i(s_i) \right) \leq \sum_{s \in S} \varphi(s) u_i(s) \qquad (5.23)$$

for all functions $g_i : S_i \to S_i$ *and all* $i = 1, ..., n$.

To see the "if" part, let $\Omega = S$ and define the t_i's by the sets of all strategy combinations with fixed strategy for player i; i.e., $t_i = \{ s' \in S \, | s'_i = s_i \}$ for all $s_i \in S_i$ and all $i = 1, ..., n$. Then (5.23) simply says that the identity map $(f_i(s_i) = s_i)$ is optimal, i.e. that it is always optimal to follow the recommendation, whence (5.21).

For the "only if" part, let f be a correlated equilibrium relative to some $Q = (\Omega, \{T_i\}_{i=1}^n, \mu)$. The preimage $f_i^{-1}(s_i) = \{\omega \in \Omega \, | f_i(\omega) = s_i \}$ is the set of all states where i chooses s_i, for all $s_i \in S_i$ and all i. Let $\varphi \in \Delta(S)$ with $\varphi(s) = \mu \left(\cap_{i=1}^n f_i^{-1}(s_i) \right)$, for all $s = (s_1, ..., s_n) \in S$, be the induced distribution on S. By construction, for every $t_i \in T_i$ there is some $s_i \in f_i(\Omega) \equiv \cup_{\omega \in \Omega} f_i(\omega)$ such that $t_i \subseteq f_i^{-1}(s_i)$, for all i. Hence, $\left\{ f_i^{-1}(s_i) \right\}_{s_i \in S_i}$ represents coarser information than T_i. Conditional upon observing a type to which f_i assigns s_i, i.e. $\omega \in f_i^{-1}(s_i)$, player i's posterior belief about the other players' strategy combinations is

$$\varphi \left(s_{-i} \, | s_i \right) = \frac{\mu \left(\cap_{j=1}^n f_j^{-1}(s_j) \right)}{\mu \left(f_i^{-1}(s_i) \right)}$$

if this is defined, and zero otherwise, by Bayes' rule. Since player i cannot gain by deviating at any $t_i \subseteq f_i^{-1}(s_i)$ by (5.22), she cannot gain by deviating from s_i when her finer information partition T_i is replaced by $\left\{ f_i^{-1}(s_i) \right\}_{s_i \in S_i}$. Since (5.23) aggregates (5.22) over all $f_i^{-1}(s_i)$'s, condition (5.23) must hold true. This completes the proof of Proposition 5.3.

Proposition 5.3 characterizes the correlated equilibrium outcomes relative to *all* correlating devices Q or, alternatively, the set of all probability distributions on S that can be generated by Nash equilibria of expanded games $\Gamma(Q)$ for *some* Q. Hence, it is independent of the particular correlating device. This

---

[26]   The set $\Delta(S)$ is a superset of Θ. On how the set of mixed strategy combinations Θ is embedded in $\Delta(S)$ see Section 3.3, in particular Figure 3.5.

can now be used to "hide" the dependence on Q in a solution concept for the original game Γ.

Since the values of functions on a finite domain can be enumerated by finite lists, the following definition is equivalent to (5.23).

Definition 5.10 *A **correlated equilibrium** for a (finite) game $\Gamma = (S, u)$ is a probability distribution $\varphi \in \Delta(S)$ such that*

$$\sum_{s_{-i} \in S_{-i}} \varphi \left(s_{-i}, s_i\right) u_i \left(s_{-i}, s_i'\right) \leq \sum_{s_{-i} \in S_{-i}} \varphi \left(s_{-i}, s_i\right) u_i(s_{-i}, s_i) \qquad (5.24)$$

for all $s_i, s_i' \in S_i$ and all $i = 1, ..., n$.

Different correlated equilibria may require different correlating devices Q and not all correlated equilibria may be feasible under a fixed Q. For, if $T_i = \Omega$ for all i, i.e. if no player learns anything about the state (has a single type), then the only equilibria of $\Gamma(Q)$ are the Nash equilibria of Γ.

The latter observation provides a convenient existence proof for correlated equilibrium. Since every Nash equilibrium for Γ is a correlated equilibrium for Γ, every game that has a Nash equilibrium also has a correlated equilibrium.

In fact, the set of correlated equilibria has a simpler structure than the set $E(\Gamma)$ of Nash equilibria. If $\varphi^1, \varphi^2 \in \Delta(S)$ are both correlated equilibria, then for any λ with $0 \leq \lambda \leq 1$ we have $\lambda \varphi^1 + (1 - \lambda) \varphi^2 \in \Delta(S)$, because $\Delta(S)$ is convex, and

$$\sum_{s \in S} \left(\lambda \varphi^1(s) + (1 - \lambda) \varphi^2(s)\right) u_i \left(s_{-i}, s_i'\right) \leq \sum_{s \in S} \left(\lambda \varphi^1(s) + (1 - \lambda) \varphi^2(s)\right) u_i(s)$$

for all $s_i' \in S_i$ and all i, from (5.24). Hence, $\lambda \varphi^1 + (1 - \lambda) \varphi^2$ is also a correlated equilibrium.

Therefore, the set of correlated equilibria is convex. Moreover, it is completely characterized by finitely many linear inequalities: those that correspond to (5.24), the non-negativity constraints $\varphi(s) \geq 0$ for all $s \in S$, and $\sum_{s \in S} \varphi(s) = 1$.

Since all Nash equilibria are correlated equilibria, it follows that the convex hull (in $\Delta(S)$) of the set of Nash equilibria is contained in the set of correlated equilibria. This inclusion may be proper, as the following example illustrates.

Example 5.16 *Consider the three-player game below. Player 1 chooses rows, player 2 columns, and player 3 matrices; the upper left entry is 1's payoff, the lower right 3's, and the middle player 2's. The unique Nash equilibrium is $\left(s_1^2, s_2^1, s_3^1\right)$ with payoffs $(1, 1, 1)$. In this game player 3 can gain in a correlated equilibrium by limiting her information.*

If the state space has two equally likely elements, $\Omega = \{\omega_1, \omega_2\}$ with $\mu(\omega_1) = \mu(\omega_2) = 1/2$, types are $T_1 = T_2 = \{\{\omega_1\}, \{\omega_2\}\}$, but $T_3 = \Omega$, so players 1 and 2 perfectly observe the state while player 3 does not learn anything, then the following is a correlated equilibrium relative to this device: $f_1(\omega_1) = s_1^1$ and $f_1(\omega_2) = s_1^2$, $f_2(\omega_1) = s_2^1$ and $f_2(\omega_2) = s_2^2$, and $f_3(\omega) = s_3^2$ for all $\omega \in \Omega$.

Player 3 then faces the distribution $\varphi_{-3}\left(s_1^1, s_2^1\right) = 1/2$ and $\varphi_{-3}\left(s_1^2, s_2^2\right) = 1/2$, making s_3^2 the unique best reply for player 3 which gives her a payoff of 2. This correlated equilibrium is outside the convex hull of the (unique) Nash equilibrium.

Game s_3^1:

	s_2^1	s_2^2
s_1^1	0 / 1 / 3	0 / 0 / 0
s_1^2	1 / 1 / 1	1 / 0 / 0

Game s_3^2:

	s_2^1	s_2^2
s_1^1	2 / 2 / 2	0 / 0 / 0
s_1^2	2 / 2 / 0	2 / 2 / 2

Game s_3^3:

	s_2^1	s_2^2
s_1^1	0 / 1 / 0	0 / 0 / 0
s_1^2	1 / 1 / 0	1 / 0 / 3

If player 3 also could observe the state, she would choose s_3^1 conditional upon ω_1 and s_3^3 conditional upon ω_2, which would induce players 1 and 2 also to deviate. Hence, if all players observed the state, the only correlated equilibrium would be the Nash equilibrium with payoffs $(1,1,1)$.

Thus, in the correlated equilibrium above, where player 3 cannot condition on the state, she is better off. This contrasts with the intuition that more information should be beneficial.

One may ask why the focus is on correlating devices rather than on pre-play communication. After all, the notion of correlated equilibrium might be particularly appropriate with pre-play communication, for then players might be able to implement a mechanism for obtaining (private) signals.

And indeed, it can be shown (see Barany [12] and Forges [60]) that, for games with at least four players, a system of direct unmediated costless communication between *pairs* of players can simulate any correlating device, provided the communication between any pair of players is unobservable by others.

In this sense, correlated equilibrium also captures what may happen with pre-play "cheap talk".

Summary

- A correlated equilibrium for a game is a Nash equilibrium of an enlarged game expanded by a correlating device from which players can receive private signals on which they may condition their strategy choices.

- A correlated equilibrium is a probability distribution on pure strategy combinations such that no player has an incentive to deviate conditional on observing her part of the recommended strategy combination.

- The convex hull of the set of Nash equilibria is contained in the set of correlated equilibria, but this inclusion may be proper.

- Correlated equilibria may serve as models of games with pre-play "cheap-talk" communication.

Exercises

Exercise 5.29 Consider the "Battle of the Sexes" game below. Show that the following is a correlated equilibrium relative to $Q = \left(\Omega, \{T_i\}_{i=1,2}, \mu\right)$ with $\Omega = \{\omega_1, \omega_2, \omega_3\}$ and $T_1 = \{\{\omega_1\}, \{\omega_2, \omega_3\}\}$, $T_2 = \{\{\omega_1, \omega_2\}, \{\omega_3\}\}$, and $\mu(\omega_1) = \mu(\omega_2) = \mu(\omega_3) = 1/3$: For player 1 we have $f_1(\omega_1) = s_1^2$ and $f_1(\omega_2) = f_1(\omega_3) = s_1^1$, and for player 2 we have $f_2(\omega_1) = f_2(\omega_2) = s_2^1$ and $f_2(\omega_3) = s_2^2$.

	s_2^1		s_2^2	
s_1^1	6		2	
		6		7
s_1^2	7		0	
		2		0

Also show that this correlated equilibrium yields a payoff vector outside the convex hull of Nash equilibrium payoffs.

Exercise 5.30 In the game from Exercise 5.29, find the correlated equilibrium that maximizes the sum of payoffs for the two players. (Hint: Solving this problem boils down to solving a linear programming problem.)

Exercise 5.31 In the game from Exercise 5.29, find all correlated equilibria and all correlated equilibrium payoffs. Compare the latter with the Nash equilibrium payoffs.

Exercise 5.32 Show that in the two-player game below it is a correlated equilibrium to assign probability 1/6 to each off-diagonal (and zero to the diagonal) strategy combinations. What are the expected payoffs to the players from this correlated equilibrium? Construct a correlating device that supports this correlated equilibrium. Also determine the (unique) Nash equilibrium for this game and the Nash equilibrium payoffs.

	s_2^1	s_2^2	s_2^3
s_1^1	0 0	0 1	1 0
s_1^2	1 0	0 0	0 1
s_1^3	0 1	1 0	0 0

Notes on the Literature: Correlated equilibrium has been introduced by Aumann [7]; see also Aumann [9]. Example 5.16 and Exercise 5.29 are a slight modifications of examples by Aumann [7]. A fuller treatment of correlated equilibrium with a discussion of its relation to the literature on "mechanism design" is given by Myerson ([128], Chapter 6). Pairwise communication systems have been studied by Barany [12] and Forges [60].

6

Refined Nash Equilibria

Refinement concepts are designed to select from among multiple Nash equilibria of a game. They take Nash equilibrium as a *necessary* condition for a solution to be self-enforcing and aim at adding extra necessary conditions. This is to be distinguished from "equilibrium selection", as advocated by Harsanyi and Selten [82]. The latter starts from the presumption that a theory of rational behavior should assign a *single* strategy combination to every game as "its" solution. This is as if rationality in itself implied *sufficient* conditions for a self-enforcing solution, and the task of game theory were to discover this point-valued solution for every game. So, while refinements see problems with some of the Nash equilibria, equilibrium selection theory sees a problem with multiplicity itself.

If multiplicity of equilibria in itself is not seen as an obstacle, the agenda is more modest. It boils down to eliminating Nash equilibria with "undesirable" properties, leaving still (possibly) several solutions. And this is what equilibrium refinements are about. There is a vast literature on refinements of Nash equilibrium. No attempt is made here to review it. (For an excellent survey see van Damme [37].) This chapter concentrates on a few key ideas, organized around methodological approaches.

6.1   Backwards Induction

Subgame Perfection   Very broadly, "backwards induction" refers to the idea that any solution to a given "large" problem should induce solutions to all its "small" subproblems. So, any "large" problem can be solved by first solving its "small" subproblems, then replacing the subproblems by their solutions, and finally solving the so reduced original problem. In an extensive form representation of a game, this means that a solution to the overall game entails solutions to all subgames.

Example 6.1 *(Durable goods monopoly) Consider a profit maximizing monopolist who supplies an indivisible durable good over several periods. Each consumer buys at most one unit. Since the good is durable, a consumer who has bought will never buy again. The monopolist sets the prices for all periods, but may revise them every period. Each consumer is characterized by a valuation v which summarizes her discounted expected benefits from possessing the good. Both the monopolist and consumers discount future payoffs at a common rate δ with $0 < \delta < 1$. Hence, if a consumer with valuation v buys in period t at the price p_t, her utility is $u_v = \delta^t (v - p_t)$, and it is zero if she never buys the good. Assume, for simplicity, that consumers' valuations v are uniformly distributed on the unit interval $[0, 1]$ (henceforth we refer to "consumer $v \in [0, 1]$"), that the monopolist's production cost is zero, and that there are only two periods, $t = 0, 1$.*

This defines an extensive form game between the monopolist and a continuum of consumers. The monopolist starts by setting a price p_0 which all consumers observe, where $0 \le p_0 \le 1$. So, every p_0 defines the root of a subgame at which all consumers decide simultaneously whether to buy at p_0 or to hold out. Every player observes these decisions. Hence, at the beginning of period $t = 1$ a subgame starts with a new price offer p_1 by the monopolist. Again all consumers see p_1 and decide simultaneously whether or not to buy, $0 \le p_1 \le 1$.

One Nash equilibrium is as follows. The monopolist initially sets $p_0 = 2/(4 - \delta)$ and then asks $p_1 = p_0/2 = 1/(4 - \delta)$ if all consumers with $v \ge p_0$ have bought in period $t = 0$ and $p_1 = 1$ if some consumers with $v \ge p_0$ failed to do so. Consumers $v \ge p_0$ buy in the first period $t = 0$, and consumers $v < p_0$ buy in period $t = 1$ if and only if $v \ge p_1$. This is an equilibrium, because, given the consumers' strategies, the monopolist faces in $t = 1$ the profit function $\pi_1(p_1) = p_1 (p_0 - p_1)$, so setting $p_1 = p_0/2$ is optimal. Given this, in $t = 0$ the monopolist faces the profit function

$$\pi_0\left(p_0\right) = p_0 \int_{p_0}^1 dv + \delta\pi_1\left(\frac{p_0}{2}\right) = p_0\left[1 - p_0\right] + \frac{\delta p_0}{4}$$

which attains its maximum at $p_0 = 2/(4 - \delta)$. Given the monopolist's strategy, any consumer $v \ge p_0$ will optimally buy in $t = 0$, because otherwise she will never buy (since the monopolist would then set $p_1 = 1$). Consumers $v < p_0$ will certainly not buy in $t = 0$, so buying in $t = 1$ if and only if $p_1 \le v$ is optimal. Hence, this is a Nash equilibrium where the monopolist earns $\pi_0 = 1/(4 - 2\delta) \to_{\delta \to 1} 1/2$, which is about twice as much as the static (one-period) monopoly profit.

But, consider a consumer v satisfying

$$p_0 = \frac{2}{4 - \delta} \le v < \frac{2 - \delta}{(1 - \delta)(4 - \delta)}.$$

She buys in $t = 0$ because the monopolist threatens to punish her for holding out by $p_1 = 1$. Yet, if she did indeed hold out, her unilateral deviation would not change market demand in period $t = 1$, so that upon reaching $t = 1$ the monopolist is faced with the choice between maximizing $\pi_1(p_1)$ or punishing by $p_1 = 1$. If the monopolist behaved optimally, she would not punish, but would set again $p_1 = p_0/2$. But then

$$\delta\left(v - \frac{p_0}{2}\right) > v - p_0 \text{ if and only if } \frac{2 - \delta}{2(1 - \delta)}p_0 > v$$

is equivalent to how v has been chosen. Hence, if consumer v expects the monopolist to respond optimally to her deviation, she gains by deviating.

The equilibrium in the example prescribes suboptimal behavior in unreached subgames. The monopolist "threatens" to ask an excessive price if some consumer (with $v \geq p_0$) fails to buy in the initial period. If, however, the monopolist were (counterfactually) called upon to move after some consumer with valuation $v \geq p_0$ has deviated (to holding out), she would *not* want to set an excessive price. Hence, this involves a threat which would not be carried through when tested for, a "non-credible threat".

Suboptimal behavior in a subgame and the subgame remaining unreached in equilibrium are two sides of the same coin. Because players trust that the subgame will not be reached, equilibrium does not constrain behavior within the subgame. Yet, the subgame will not be reached, because the expected irrational behavior in the event of it being reached induces players to prevent the subgame from being reached.

Though this is in accordance with the Nash equilibrium notion, it is at variance with "backwards induction". According to the latter, a solution of an extensive form game should *induce a solution* in every subgame. This requirement is captured by the following definition.

Definition 6.1 *A Nash equilibrium for an extensive form game is **subgame perfect** if it induces a Nash equilibrium in every subgame.*[1]

Recall that Kuhn's algorithm (see Section 3.1) solves for a Nash equilibrium of a perfect information extensive form game by breaking it down into (single player) decision problems. Without the perfect information assumption, this cannot be done. But what *can* be done in general is to decompose an extensive form game into all its subgames, and those into *their* subgames. Then the "small" subgames can be solved, and a solution for the overall game can be "glued" from the subgame solutions.

---

[1] In its original formulation by Selten, the definition of subgame perfection has a second part which applies to games with plays of infinite length. For such games the solution is required to be the limit of solutions for finite approximations of the infinite game, a property called "asymptotic insensitivity" by Selten [162].

The idea of "gluing" a solution from solutions of subgames, in fact, gives rise to an existence proof for subgame perfect Nash equilibrium. Let $G = (F, v)$ be a finite extensive form game and x a node (move) at which the subgame G_x rises. (Every nontrivial choice must be either contained in or disjoint from the set of successors of x.) Let $b \in B$ be a behavior strategy combination for G (see Section 3.3). Write b_x for the part of b that concerns choices in G_x. Denote by $G_{-x}(b_x)$ the "truncation at x" which is the game G where the subgame G_x has been replaced by a terminal node at which the payoffs that accrue to players are those that would accrue to them had they behaved according to b in G_x.

Lemma 6.1 *(Kuhn's lemma) For any finite game G in extensive form, if b_x is a Nash equilibrium for the subgame G_x and b_{-x} is a Nash equilibrium for the truncation $G_{-x}(b_x)$, then (b_{-x}, b_x) is a Nash equilibrium of G.*

The proof is driven by finiteness and the von Neumann-Morgenstern axioms which are responsible for additive separability of payoffs from subgames.

Let $\pi_b(w)$ denote the realization probabilities of plays $w \in W$ given a behavior strategy combination $b \in B$ (see (3.14) in Section 3.3). The probability that a move $x \in N$ will be reached, given $b \in B$, is $\pi_b(x) = \sum_{w \in x} \pi_b(w)$. By Bayes' rule the conditional probability of a play $w \in x$ (passing through a move $x \in N$), given that move $x \in N$ has been reached, under b is $\pi_b(w \,|\, x) = \pi_b(w)/\pi_b(x)$ whenever $\pi_b(x) > 0$. Hence, the conditional payoff for player i given that x has been reached (under b) is

$$U_i(b \,|\, x) = \sum_{w \in x} \pi_b(w \,|\, x)\, v_i(w) = \sum_{w \in x} \frac{\pi_b(w)}{\pi_b(x)}\, v_i(w). \qquad (6.1)$$

Owing to the expected utility hypothesis, therefore, if $x \in N$ is the root of a subgame, player i's payoff can be written as

$$U_i(b) = \sum_{w \in W \setminus x} \pi_b(w) v_i(w) + \pi_b(x) U_i(b \,|\, x); \qquad (6.2)$$

i.e., the payoff from a subgame G_x is additively separable. And for the conditional payoff $U_i(b \,|\, x)$, only the part b_x of b that concerns choices in G_x matters, because every nontrivial choice must be either contained in or disjoint from G_x.

Now let b_x be an equilibrium for the subgame G_x and b_{-x} an equilibrium of $G_{-x}(b_x)$. Then $U_i(b \,|\, x) \geq U_i(b_{-i}, b_i' \,|\, x)$ for all $b_i' \in B_i$ and all $i = 1, ..., n$. Because the game is finite, the function

$$\sum_{w \in W \setminus x} \pi_b(w) v_i(w) + \pi_b(x) \max_{b_i' \in B_i} U_i(b_{-i}, b_i' \,|\, x) \qquad (6.3)$$

is continuous and (the ith component of) b_{-x} maximizes it (by the choice of b_{-x}), for all players i. Therefore, $(b_{-x}, b_x) \in B$ constitutes an equilibrium of the overall game G. This completes the proof of Kuhn's lemma.

The assumption of finiteness is crucial for Kuhn's lemma. Without it the function in (6.3) may not have a maximum (even if v is continuous), because the opponents' strategies may not vary continuously in player i's strategy for $G_{-x}(b_x)$. (This point is illustrated below in the discussion of "ultimatum bargaining".)

If an extensive form game is finite, then all its subgames are finite games. For finite games a Nash equilibrium exists by Nash's theorem. Hence, Kuhn's lemma immediately implies that subgame perfect Nash equilibrium exists for all finite extensive form games.

Theorem 6.1 *For every finite extensive form game there is a subgame perfect Nash equilibrium.*

How much bite this concept has depends on the game. In perfect information games Kuhn's algorithm identifies the subgame perfect equilibria; and generically there will be only one. In Bayesian games (see Section 3.5) all Nash equilibria are subgame perfect, because these games do not have any proper subgames.

The "durable monopoly" game from Example 6.1, however, has a very rich subgame structure. While it is not a finite game, subgame perfect equilibrium exists for this game, too.

Example 6.2 *Example 6.1 allows for the following subgame perfect Nash equilibrium. The monopolist sets initially $p_0 = (2 - \delta)^2 / (8 - 6\delta)$, and in every subgame of period $t = 1$ she sets a price p_1 that maximizes her profit function $\pi_1(p_1)$. Consumers $v > 2p_0/(2 - \delta)$ buy in the initial period $t = 0$, and consumers $v \le 2p_0/(2 - \delta)$ buy in $t = 1$ if and only if $v \ge p_1$.*

Given the consumers' strategies, the monopolist faces in every subgame of the second period the profit function

$$\pi_1(p_1) = p_1 \left(\frac{2p_0}{2 - \delta} - p_1 \right)$$

which is maximized at $p_1 = p_0/(2-\delta)$. Substituting this into the initial period's profit function, the monopolist maximizes

$$\pi_0(p_0) = p_0 \left(1 - \frac{2p_0}{2 - \delta} \right) + \frac{\delta p_0^2}{(2 - \delta)^2}$$

in $t = 0$, yielding $p_0 = 2\left(1 - \frac{\delta}{2}\right)^2 / (4 - 3\delta)$. Given this strategy for the monopolist, consumers $v > 2p_0/(2 - \delta)$ are better off buying in $t = 0$, because

$$v - p_0 \ge \delta \left(v - \frac{p_0}{2 - \delta} \right) \quad \text{if and only if} \quad v \ge \frac{2p_0}{2 - \delta},$$

and consumers $v \leq 2p_0/(2-\delta)$ are better off holding out, for the same reason. Hence, this is a subgame perfect equilibrium where the monopolist earns

$$\pi_0 = \frac{\left(1 - \frac{\delta}{2}\right)^2}{4 - 3\delta} < \frac{1}{4}$$

which is less than the static monopoly profit.

That the monopolist earns less in this equilibrium than in a one-shot monopoly is due to intertemporal price competition. The monopolist's agents of period $t = 1$ try to steal away customers from the market in the initial period by cutting prices. Since consumers rationally anticipate this, they would hold out unless first-period prices are lowered.

Generic perfect information extensive form games are about the only class of games for which uniqueness of subgame perfect Nash equilibrium can be established. In general, an extensive form game has multiple subgame perfect equilibria.

This is so, because a rich subgame structure means that a lot of public information becomes available in the course of the game. And the players can *condition* their behavior on this rich flow of information. The more subgames there are, the more complicated the possible strategies become. This is also true for the "durable monopoly" game from Example 6.1.

Example 6.3 *On top of the equilibrium from Example 6.2, the durable goods monopoly game has the following subgame perfect equilibrium, provided $\delta \in (0,1)$ is large enough. The monopolist initially asks $p_0 = p_0^*$ (to be determined later); in any subgame of $t = 1$ preceded by some p_0 with $\varepsilon < p_0 \leq (1/2) - \varepsilon$, for small $\varepsilon > 0$, she sets the lowest price that maximizes her profit function $\pi_1(p_1)$ if all consumers $p_0 \leq v < 1/2$ and $f(p_0) \leq v \leq 1$ have bought in $t = 0$ and she sets the highest price that maximizes $\pi_1(p_1)$ if some consumer with such a valuation failed to buy at p_0, where*

$$f(p_0) = \frac{3}{2} - p_0 - \sqrt{1 - 2p_0}.$$

In subgames of $t = 1$ preceded by $p_0 \leq \varepsilon$ or $(1/2) - \varepsilon < p_0 \leq 1$, the monopolist sets $p_1 = p_0/(2 - \delta)$. Consumers confronted with an initial price $\varepsilon < p_0 \leq (1/2) - \varepsilon$ act as follows. If $p_0 \leq v < 1/2$ or $f(p_0) \leq v \leq 1$ they buy in $t = 0$. Consumers $0 \leq v < p_0$ and $1/2 \leq v < f(p_0)$ hold out and buy in $t = 1$ if and only if $v \geq p_1$. If $p_0 \leq \varepsilon$ or $(1/2) - \varepsilon < p_0 \leq 1$, then all consumers $2p_0/(2 - \delta) < v \leq 1$ buy in $t = 0$ and all consumers $0 \leq v \leq 2p_0/(2 - \delta)$ hold out and buy in $t = 1$ if and only if $v \geq p_1$.

To verify subgame perfection, consider a subgame of period $t = 1$ preceded by an initial price $\varepsilon < p_0 \leq (1/2) - \varepsilon$. Given the consumers' strategies, the *monopolist's profit function is*

$$\pi_1(p_1) = \begin{cases} p_1\left(1 - p_1 - \sqrt{1-2p_0}\right) & \text{if} \quad 0 \le p_1 \le p_0 \\ p_1\left(1 - p_0 - \sqrt{1-2p_0}\right) & \text{if} \quad p_0 < p_1 \le \tfrac{1}{2} \\ p_1\left(f(p_0) - p_1\right) & \text{if} \quad \tfrac{1}{2} < p_1 \le f(p_0) \\ 0 & \text{if} \quad f(p_0) < p_1 \le 1 \end{cases}$$

This is a continuous function, the first part of which is a downward opening parabola with peak at $p_1 = g(p_0) < p_0$, where

$$g(p_0) = \frac{1}{2}\left(1 - \sqrt{1-2p_0}\right).$$

The second part is linearly increasing until $p_1 = 1/2$, and the third part is a downward opening parabola with peak at $p_1 = f(p_0)/2 < 1/2$, so in its relevant range it is downward sloping. Hence, the profit function has two maxima,

$$\pi_1(g(p_0)) = \pi_1\left(\frac{1}{2}\right) = \frac{1}{2}\left(1 - p_0 - \sqrt{1-2p_0}\right).$$

Therefore, after $\varepsilon < p_0 \le (1/2) - \varepsilon$ the monopolist will set $p_1 = g(p_0)$ if all consumers with $p_0 \le v < 1/2$ and $f(p_0) \le v \le 1$ have bought in $t = 0$, and she will set $p_1 = 1/2$ otherwise. Given this, all consumers $p_0 \le v < 1/2$ are better off buying in $t = 0$, because otherwise they will not buy at all. Consumers $1/2 \le v < f(p_0)$ are better off holding out, because $v - p_0 < \delta\left(v - g(p_0)\right)$ holds if

$$p_0 - \frac{\delta}{2}\left(1 - \sqrt{1-2p_0}\right) \ge (1-\delta)f(p_0)$$

which is certainly true at $\delta = 1$ (because $p_0 \ge \varepsilon > 0$), but (by continuity) also for large enough $\delta < 1$. Consumers $f(p_0) \le v \le 1$ prefer to buy in $t = 0$, because otherwise they will face $p_1 = 1/2$ and $v - p_0 > \delta\left(v - (1/2)\right)$ holds for all δ large enough by $p_0 \le (1/2) - \varepsilon < 1/2$. (Consumers $v < p_0$ always hold out.)

If the game begins with some $p_0 \le \varepsilon$ or $(1/2) - \varepsilon < p_0 \le 1$ players behave like in Example 6.2. Thus, their behavior in those subgames constitutes equilibrium behavior.

At the root, finally, the monopolist chooses p_0 so as to maximize

$$\pi_0(p_0) = \begin{cases} p_0\sqrt{1-2p_0} + \frac{\delta}{2}\left(1 - p_0 - \sqrt{1-2p_0}\right) & \text{if} \qquad \varepsilon < p_0 \le \tfrac{1}{2} - \varepsilon \\ p_0\left(1 - \frac{2p_0}{2-\delta}\right) + \frac{\delta p_0^2}{(2-\delta)^2} & \text{otherwise} \end{cases}$$

Provided ε is small enough, the first part has an interior maximum at

$$p_0 = p_0^* \equiv \frac{1}{3} + \frac{5\delta}{36} - \frac{\delta}{36}\sqrt{\delta^2 + 12(1-\delta)} \xrightarrow{\delta \to 1} \frac{4}{9} < \frac{1}{2}$$

where the profit function takes a value that converges to 7/27 as δ → 1.
Since the second part cannot exceed $(2 - \delta)^2 / (16 - 12\delta) < 1/4$, the unique
maximum of the profit function occurs at $p_0 = p_0^*$ which, for δ large enough and
ε small enough, lies between ε and $(1/2) - \varepsilon$. This completes the demonstration
that a subgame perfect equilibrium has been constructed.

The pattern that this equilibrium generates looks rather fancy. Since

$$0 < g(p_0^*) < p_0^* < \frac{1}{2} < f(p_0^*) < 1,$$

there are five different (nonempty) intervals of consumers. Consumers with
low valuation ($v < g(p_0^*)$) never buy, those with somewhat higher valuation
($g(p_0^*) \leq v < p_0^*$) buy tomorrow (t = 1), those with valuation just below
average ($p_0^* \leq v < 1/2$) buy today (t = 0), those with valuation just above
average ($1/2 \leq v < f(p_0^*)$) buy tomorrow (t = 1), and consumers with high
valuation ($f(p_0^*) \leq v \leq 1$) buy today (t = 0). Yet, for large δ, the monopolist
earns more in this equilibrium than she does in the static monopoly.

The strength of subgame perfection is that it is an instance of "backwards induction" in a case where this principle is most compelling. A subgame is itself a game and the event of it being reached entails the subgame being common knowledge among the players. So, to extend the equilibrium requirement to all subgames seems compelling.

On the other hand, there are simple games where experiments indicate systematic violations of subgame perfection. Suppose one player has the power to divide a pie and the other, upon observing the proposed split, can only accept or reject. If an offer is accepted it is implemented, if it is rejected no player receives anything. In this "ultimatum bargaining game" the unique subgame perfect equilibrium prescribes that the proposer asks for the whole pie and the responder accepts this. Yet, experimental evidence reveals a significant proportion of rejections of "unfair" offers, where the proposer takes the lion's share (see [76]).

This simple game also illustrates another point, raised in connection with Kuhn's lemma. In finite games it is always possible to solve subgames first and then solve the overall game with the solutions to the subgames substituted in. Thus, for finite games *every* equilibrium of a subgame is part of *some* (subgame perfect) equilibrium of the overall game. In infinite games, like "ultimatum bargaining" with a perfectly divisible pie, this is not true. Confronted with an offer of zero for herself, it is a best reply for the responder to reject, so this is an equilibrium of the subgame initialized by the proposer asking for all the pie. Yet, this is not part of any equilibrium of the overall game. For, given that the responder rejects zero, but accepts any positive offer (by subgame perfection), the proposer has no best reply (since a strictly increasing function on an open interval does not have a maximum).

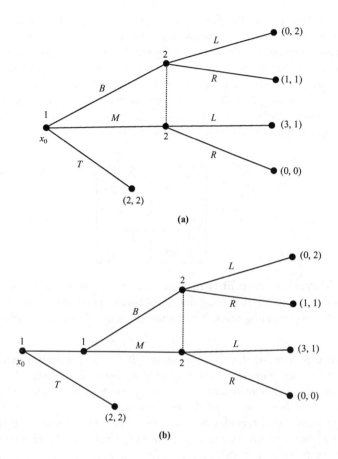

Figure 6.1: Subgames are sensitive to Thompson transformations

Moreover, in more complicated games there is a conceptual problem with subgame perfection. Suppose that reaching a subgame (off the equilibrium play) constitutes evidence that a player, who has a move in the subgame, has behaved "irrationally". Why should this player be assumed to act rationally *in* the subgame (see Exercises 6.1 and 6.2 below)?

Another drawback is that the presence of subgames is not invariant with respect to inessential (Thompson) transformations of the extensive form (see Section 4.2). The following example illustrates this.

Example 6.4 *Consider the extensive form game in Figure 6.1(a). (Moves are labeled with the identity of the player who decides there, and choices are identified by labeling the corresponding edges. Payoffs at terminal nodes are*

given by vectors where the first entry refers to player 1 and the second to
2.) Applying the inverse of "Coalescing information sets" to player 1's initial
information set produces the game in Figure 6.1(b). In the first game all Nash
equilibria are subgame perfect, because there is no proper subgame. In the
last game only a single equilibrium, (M, L), is subgame perfect. (The Nash
equilibria—the same for both games—are as follows: (M, L) and T together
with any mixture between L and R for which the probability of L does not
exceed 2/3; see the normal form of the game below.)

	L	R
T	2 2	2 2
M	3 1	0 0
B	0 2	1 1

Perfect Bayesian Equilibrium Moreover, non-credible threats are not
limited to games with rich subgame structures. They can occur even in the
absence of (proper) subgames. The game in Figure 6.1(a) also illustrates this
point.

Example 6.5 The strategy combination (T, R) is a subgame perfect Nash
equilibrium where player 2 uses a weakly dominated strategy. In fact, conditional on player 2's information set being reached, the choice R is strictly
dominated. But how can player 1 ever expect her opponent to pick a strictly
dominated choice if it becomes her turn to move? The answer is that player 1
precludes player 2's information set from being reached (by choosing T) and,
hence, there is no constraint on what player 2 may do there. But the reason
why 1 precludes 2's information set from being reached is that 1 expects 2 to
punish by R if 2 is given the move. Yet, if 2 is indeed called upon to move she
will not want to execute the threat of playing R.

Player 2's behavior in Example 6.4 is not precisely irrational, because,
given that she will never have to decide, any decision is as good as any other.
But, *conditional on* being called upon to move, player 2 behaves suboptimally.
To exclude such "conditionally irrational" behavior requires a solution concept that takes explicit account of counterfactual events, like an "unreached
information set being reached". In the previous example it is obvious what
optimality at the unreached information set means. In general, however, the
optimal choice at an information set depends on the move (in the information set) at which the choice is taken—which the player does not know by
definition. Hence, *conditional optimality*, given that an information set has
been reached, requires the player to assign probabilities to the moves in the
information set. This is accomplished by the concept of *beliefs*.

But at this point "beliefs" do not refer to a reinterpretation of a mixed strategy combination among the opponents (as in Chapter 5). It is a new concept, absent from the theory up to now and peculiar to the refinements introduced below.

Given an extensive form game $G = (F, v)$, a *system of beliefs* is an n-tuple $\mu = (\mu_1, ..., \mu_n)$ of functions $\mu_i : X_i \to \mathbb{R}_+$ such that

$$\sum_{x \in h} \mu_i(x) = 1, \text{ for all } h \in H_i, \tag{6.4}$$

where H_i denotes the set of all information sets and X_i the set of all decision points, $X_i = \cup_{h \in H_i} h$, for player i, for $i = 1, ..., n$. The function μ_i is to be interpreted as the conditional probabilities that player i assigns to the moves in her information set $h \in H_i$, given that h has been reached.

Given a behavior strategy combination $b \in B$, a system of beliefs μ is *consistent with b* if

$$\mu(x) = \frac{\pi_b(x)}{\pi_b(h)} \text{ for all } x \in h \tag{6.5}$$

whenever $\pi_b(h) \equiv \sum_{x \in h} \pi_b(x) > 0$, for all information sets $h \in H_i$ and all players $i = 1, ..., n$.

Definition 6.2 *A Nash equilibrium (in behavior strategies) $b \in B$ of an extensive form game is a **perfect Bayesian equilibrium** if there is a system of beliefs $\mu = (\mu_1, ..., \mu_n)$ that is consistent with b and*

$$\sum_{x \in h} \mu_i(x) U_i(b|x) \geq \sum_{x \in h} \mu_i(x) U_i(b_{-i}, b_i'|x) \text{ for all } b_i' \in B_i \tag{6.6}$$

for all information sets $h \in H_i$ and all $i = 1, ..., n$, where $U_i(b|x)$ is defined in (6.1).

Condition (6.6) is already guaranteed to hold in a Nash equilibrium at information sets that can be reached, i.e., for which $\pi_b(h) > 0$, by consistency (6.5). Hence, perfect Bayesian equilibrium adds a conditional optimality condition (6.6) only at unreached ($\pi_b(h) = 0$) information sets. In Example 6.4 the equilibrium (T, R) fails to be a perfect Bayesian equilibrium, because the choice R at player 2's information set cannot be optimal for *any* system of beliefs. The equilibrium (M, L), on the other hand, *is* a perfect Bayesian equilibrium.

Since by (6.4) $\mu_i(x) = 1$ must hold at any information set that contains only one move, every perfect Bayesian equilibrium must be subgame perfect. It is stronger than subgame perfection, however, because it also restricts behavior at unreached information sets that contain several moves.

Perfect Bayesian equilibrium does not put any constraints on the beliefs at unreached information sets, beyond requiring them to be probability assignments. This can lead to inconsistencies between the beliefs of different players, giving rise to non-credible threats.

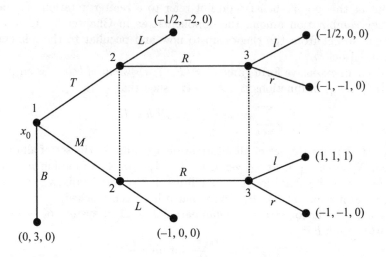

Figure 6.2: Inconsistent beliefs

Example 6.6 *Consider the three-player game in Figure 6.2. The combination* (B, L, r) *constitutes a perfect Bayesian equilibrium, supported by the following beliefs. Player 2, when called upon to choose, believes that this is the case because player 1 has chosen M. Player 3, when called upon to choose, believes that this has happened because player 1 has chosen T (and player 2 has chosen R). So, the out-of-equilibrium beliefs of players 2 and 3 are inconsistent. Again, this equilibrium involves a (sort of) non-credible threat, because player 3 uses a weakly dominated strategy. For, if player 3's beliefs assigned positive probability to player 1 having chosen M, then she would always choose l (rather than r). But if 3 chose l, then player 2 would always choose R (rather than L), in which case player 1 would choose M (rather than B).*

Sequential Equilibrium   To rule out such phenomena, one requires that the beliefs be robust against slight perturbations of the equilibrium strategy. If all players choose *completely mixed* behavior strategies, i.e. behavior strategies that assign (strictly) *positive* probabilities to *all* choices at *all* information sets, then all information sets will be reached with positive probability. But then, consistency (6.5) determines uniquely a system of beliefs.

Sequential equilibrium restricts beliefs at unreached information sets by requiring that beliefs be determined as the limit of beliefs consistent with some sequence of completely mixed behavior strategy combinations. Hence, it is a perfect Bayesian equilibrium where beliefs are justified by small perturbations of the equilibrium strategy combination.

Definition 6.3 *A Nash equilibrium (in behavior strategies) $b \in B$ for an extensive form game is a **sequential equilibrium** if it is a perfect Bayesian equilibrium such that, for the supporting system of beliefs μ, there is a sequence $\{b^t\}_{t=1}^{\infty}$ of completely mixed behavior strategy combinations which converges to b and satisfies $\mu = \lim_{t\to\infty} \mu^t$ where $\{\mu^t\}_{t=1}^{\infty}$ is the (unique) sequence of beliefs for which μ^t is consistent with b^t for all $t = 1, 2, ...$*

In Example 6.6 (B, L, r) fails to be a sequential equilibrium. The combination (M, R, l), on the other hand, *is* a sequential equilibrium. To see this, simply let $b_1^t(B) = b_1^t(T) = b_2^t(L) = b_3^t(r) = 1/(t+3)$ for all $t = 1, 2, ...$

While sequential equilibrium is stronger than perfect Bayesian equilibrium, it does not put any restriction on beliefs when there is only one (non-singleton) information set in the game. This would be different if the completely mixed strategy combinations b^t that justify the beliefs were themselves required to constitute equilibria of games where players are forced to put at least some minimum probabilities $\eta_c^t > 0$ (which converge to zero as t goes to infinity) on each of their (nontrivial) choices $c \in C_i$. Such a requirement would give rise to what is known as "extensive form (trembling hand) perfect equilibria", discussed in the next section.

In fact, the *existence* of sequential equilibrium (and, thereby, the existence of perfect Bayesian equilibrium) for all finite (extensive form) games follows from the existence of such (trembling hand) perfect equilibria for all finite games (see the next section). To make the latter apply to an extensive form game, it is sufficient to consider the agent normal form of the extensive form.

The way sequential equilibrium is constructed suggests that, upon reaching an off-equilibrium information set, players attempt to reconstruct what has happened and adjust their beliefs accordingly. Such a story would require that at every unreached information set players hold beliefs that are consistent with an alternative strategy combination that "explains" what has happened.

Accordingly, call a pair (b, μ), consisting of a behavior strategy combination and a system of beliefs, *structurally consistent* if μ is consistent with b and if, for all information sets $h \in H_i$ that are not reached by b (for which $\pi_b(h) = 0$), there is some alternative strategy combination $b' \in B$ which reaches h (for which $\pi_{b'}(h) > 0$) such that $\mu_i(x) = \pi_{b'}(x)/\pi_{b'}(h)$ for all $x \in h$, all $h \in H_i$, and all $i = 1, ..., n$. Surprisingly, sequential equilibrium does not guarantee structural consistency.

Example 6.7 *Consider the three-player game in Figure 6.3. The combination (T, L, m) constitutes a sequential equilibrium with supporting beliefs $\mu_3(x) = \mu_3(y) = 1/2$. To see that this is a perfect Bayesian equilibrium, observe first that, given μ, player 3 behaves optimally at her information set. (She is actually indifferent.) Given that 3 chooses m, it is weakly dominant for player 2 to choose L. Given this, player 1 is best off by choosing T. To see that this is a sequential equilibrium, let $b_1^t(T) = b_2^t(L) = b_3^t(m) = t/(t+1) \to_{t\to\infty} 1$ for*

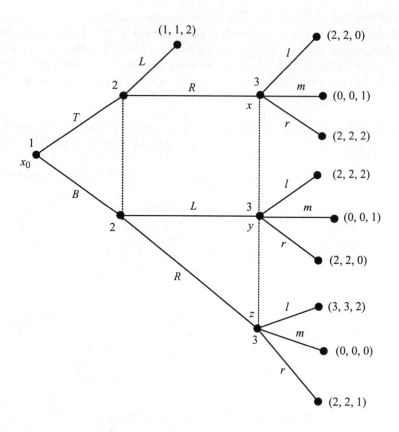

Figure 6.3: Structural (in)consistency

$t = 1, 2, ...$ *Then by Bayes' rule* $\mu_3^t(x) = \mu_3^t(y) = t/(2t+1) \to_{t\to\infty} 1/2$ *and* $\mu_3^t(z) = 1/(2t+1) \to_{t\to\infty} 0$.

    Yet, this sequential equilibrium is not structurally consistent. Suppose player 3, upon being reached, can find some $b' \in B$ *which reaches her information set* h. *Then* $\pi_{b'}(h) > 0$ *implies either* $b_1'(B) > 0$ *or* $b_2'(R) > 0$ *(or both), because if* $b_1'(B) = b_2'(R) = 0$ *then* $\pi_{b'}(h) = 0$. *Suppose further that* b' *is such that* $\pi_{b'}(z) = 0$. *Then either* $b_1'(B) = 0$ *or* $b_2'(R) = 0$, *because if both* $b_1'(B)$ *and* $b_2'(R)$ *are positive then* $\pi_{b'}(z) > 0$. *But* $b_1'(B) + b_2'(R) > 0$ *and* $b_1'(B)b_2'(R) = 0$ *imply that* $\pi_{b'}(x)\pi_{b'}(y) = 0$, *because either* $b_2'(R) = 0$ *and* $b_1'(B) > 0$ *implies* $\pi_{b'}(x) = 0$ *or* $b_2'(R) > 0$ *and* $b_1'(B) = 0$ *implies* $\pi_{b'}(y) = 0$. *But this is inconsistent with* $\mu_3(x)\mu_3(y) = 1/4 > 0$. *Hence, there exists no such* $b' \in B$.

    There is a sense in which the sequential equilibrium in this example also involves a non-credible threat. For player 3 to choose m at her information set is optimal *if and only if* $\mu_3(x) = \mu_3(y) = 1/2$. In particular, for any other belief, the choice m would be suboptimal. Yet, by "threatening" to choose m player 3 forces the other two players not to give her the move. And it is this threat that allows her to hold these beliefs. If, for instance, player 3 believed $\mu_3(z) = 1$, she would choose l. But then it is certainly better for player 2 to choose R, which would induce player 1 to choose B, and 3's information set would be reached. (This is, in fact, an alternative sequential equilibrium.)

Such examples have inspired a whole plethora of further "belief based" refinements of sequential equilibrium, some for general games and some for particular classes of games, most prominently "signalling games". A survey of those, however, is beyond the scope of the present text.

Summary

- Backwards induction suggests "building" solutions for large games from solutions of smaller subproblems.

- A subgame perfect Nash equilibrium for an extensive form game induces a Nash equilibrium in every subgame.

- A perfect Bayesian equilibrium requires the behavior at every information set to be optimal with respect to some beliefs that are consistent with the equilibrium.

- A sequential equilibrium is a perfect Bayesian equilibrium where the beliefs are justified by small perturbations of the equilibrium strategy combination.

Exercises

Exercise 6.1 Consider the following extensive form game with perfect information. Player 1 at the root can choose an outside option which gives her 100 dollars and nothing to player 2. If she chooses not to take the 100 dollars, player 2 learns this and gets to choose between two possibilities. She can either terminate the game, which yields her 8 dollars and gives 2 dollars to player 1, or she can give the move back to player 1. If she does the latter, player 1 (learns this and) gets to choose between the following options: either no player receives any payment, or player 1 receives 1 dollar and player 2 the remaining 9 dollars. Assume that the players' utilities are accurately measured by monetary payoffs and discuss the subgame perfect equilibrium of this game.

Exercise 6.2 (Centipede game) Consider the extensive form game where plays are $W = \{w_1, ..., w_K\}$, nodes are $N = \left\{ (\{w_k, ..., w_K\})_{k=1}^{K}, (\{w_k\})_{k=1}^{K} \right\}$ for some odd integer $K > 1$, nontrivial choices are

$$
\begin{aligned}
C_1 &= (\{w_{2j+1}\}, \{w_{2j+2}, ..., w_K\})_{j=0}^{(K-3)/2}, \\
C_2 &= (\{w_{2j}\}, \{w_{2j+1}, ..., w_K\})_{j=1}^{(K-1)/2},
\end{aligned}
$$

and payoffs are $v(w_k) = (v_1(w_k), v_2(w_k)) = ((k+1)/2, (k-1)/2)$ if k is even and $v(w_k) = ((k-2)/2, (k+2)/2)$ if k is odd and $k < K$, and at the very end $v(w_K) = ((K+1)/2, (K-1)/2)$. Determine the subgame perfect equilibrium for this game and discuss its properties.

Exercise 6.3 (Shubik's dollar auction) An amount of 1.1 dollars is sold to two potential buyers by a sequential open oral all-pay auction. That is to say, two participants bid for 1.1 dollars by publicly stating their bids during an unlimited number of rounds. Each of them has a budget of 2 dollars. Bids have to be made in multiples of 1 dollar, $b \in \{0, 1, 2\}$, and can only be raised or maintained from round to round. Bidder $i = 1$ makes the bid in the first round, $i = 2$ in the second, $i = 1$ in the third, and so on. The 1.1 dollars will go to the bidder who is not (strictly) overbid by the other in the next round, but every participant will have to pay whatever she bid last. For which price will the 1.1 dollars be sold in the subgame perfect equilibrium?

Exercise 6.4 (Bertrand oligopoly with different costs) Consider a market for a homogeneous good served by $n \geq 2$ price-setting firms which have constant marginal costs $0 < c_1 < c_2 \leq ... \leq c_n$. There is a continuum of consumers whose behavior is represented by a demand function $D(p)$ with $D(c_n) > 0$ and $D'(p) < 0$ whenever $D(p) > 0$. Each consumer wants to buy from the firm offering the lowest price. If several firms ask the lowest price, consumers choose among those. Show that in a subgame perfect equilibrium firm 1 will take the whole market by asking $p_1 = c_2$.

Exercise 6.5 (Divide and choose) Consider the following procedure to share a surplus among $n \geq 2$ parties. First, player 1 proposes a division of the surplus (a vector of non-negative real numbers which sum to the size of the surplus). All players learn this proposal. Then player 2 can pick one of the pieces, then player 3 picks one of the remaining, then player 4, and so on until player n. Player 1, finally, receives the share that no other player has picked. What is the distribution of the surplus in (any one) subgame perfect equilibrium of the associated game?

Exercise 6.6 (Chain store game) Consider a potential entrant to a monopolist market. If she does not enter, the monopolist earns 2 units of payoff, and the entrant nothing. If the potential entrant enters the market, the monopolist (learns this and) has two responses available. She can either accommodate the entrant, which will result in both players obtaining one unit of payoff, or she may fight, which will cause the entrant a loss of 1 unit of payoff and yield the monopolist a profit of κ units of payoff. Assume that κ is common knowledge among the players, and discuss how the subgame perfect equilibrium of this game depends on κ.

Exercise 6.7 Modify Exercise 6.6 by assuming that the monopolist knows κ, but the potential entrant does not. Formulate this as a Bayesian game and discuss the perfect Bayesian equilibria of it. Does subgame perfection yield a different prediction from that of perfect Bayesian equilibrium?

Exercise 6.8 Reinterpret the game in Example 6.1 as a bilateral bargaining game under incomplete information as follows. Instead of having a continuum of consumers with valuations $v \in [0,1]$, there is only one single buyer with a valuation $v \in [0,1]$, which is known to her, but unknown to the seller. The seller's prior beliefs about the type v of the buyer are given by the uniform distribution on the unit interval. Again, there are two periods, $t = 0, 1$, and the seller can make price offers in each of those (provided the buyer has not already bought in the first period). The seller's expected profit is now $\delta^t p_t$ times the probability that the buyer will accept in period $t = 0, 1$. Show that a sequential equilibrium for this game corresponds to the equilibrium described in Example 6.2.

Exercise 6.9 (Beer-quiche game) Consider the following two-player game with incomplete information. First, chance chooses whether player 1 is a "strong" type (with probability 9/10) or a "weak" type (with probability 1/10). Player 1 learns her type (but player 2 does not) and then decides whether to have a "beer" or a "quiche" for breakfast. Player 2 sees the breakfast and has to decide whether to "fight" or not. For the "strong" type of player 1, having "beer" adds 1 unit to her payoff; for the "weak" type, having "quiche" adds 1 unit of payoff. For both types of player 1 not being "fought" adds 2 units of payoff. For player 2 "fighting" the "weak" type yields 1 unit of payoff and "not fighting" the "weak" type yields 0. "Fighting" the "strong" type of player 1 yields player 2 zero payoff, and "not fighting" the "strong" type yields 1 unit of payoff. Argue that there is a sequential equilibrium where both types of player 1 have "quiche" and player 2 does "not fight" if she sees "quiche" (but "fights" upon observing "beer" with positive probability).

Notes on the Literature: Subgame perfect Nash equilibrium was introduced by Selten [162] and shown to exist for all finite games by Selten [163]. Existence of subgame perfect Nash equilibrium in infinite games with perfect information has been studied by Harris [78], Hellwig, Leininger, Reny, and Robson [85], and Hellwig and Leininger [84], among others. Examples 6.1, 6.2, and 6.3 are based on Güth and Ritzberger [75]. Ultimatum bargaining was first experimentally explored by Güth, Schmittberger, and Schwarze [76]. Problems with subgame perfection's maintained hypothesis of rational behavior in the light of evidence to the contrary are discussed by Basu [13] and Reny [146], [147]. Example 6.4 is adapted from Selten [163]. Exercise 6.6 is based on Selten [164]. Sequential equilibrium was introduced by Kreps and Wilson [102]. An early implicit use of perfect Bayesian equilibrium was made by Akerlof [2] and Spence [171]. A first formal application was given by Milgrom and Roberts [119]. Fudenberg and Tirole [64] study the relation between perfect Bayesian and sequential equilibrium. Example 6.7 is based on Kreps and Ramey [101]. On "belief based" refinements of sequential equilibrium there is an extensive literature which cannot be reviewed here (see, e.g., Sections 10.4-5 of van Damme [37]). An example of such a concept is the "intuitive criterion" by Cho and Kreps [32]. (Exercise 6.9 is from this paper.) Mailath, Okuno-Fujiwara, and Postlewaite [107] review such concepts in the context of signaling games.

6.2 Strategy Trembles

Perfection A somewhat different approach to ruling out implausible equilibria is to require robustness against strategy perturbations. Suppose that players—while perfectly rational in making decisions—may make mistakes when *implementing* the strategy of their choice. Not every Nash equilibrium is robust against such "strategy trembles".

Example 6.8 *The two-player game below has two Nash equilibria. Both players using their first pure strategies is clearly a strict equilibrium. But both using their second (weakly dominated) pure strategies is also an equilibrium. Yet, if any one player expects her opponent to make a mistake with small positive probability when implementing her second pure strategy, she would certainly not want to choose her own second pure strategy. This is despite the fact that both players are better off at the second equilibrium. The second equilibrium is not robust against small doubts about what the opponent will do.*

	s_2^1	s_2^2
s_1^1	1 1	5 0
s_1^2	0 5	5 5

Even if a player is perfectly rational in computing her best strategy, she may face a "bound" on rationality when executing the strategy of her choice. Once she has decided on a particular strategy, there is a small but positive probability that "her hand may tremble" and some other strategy may result. This represents a "procedural" bound on rationality. And we could ask for equilibria to be robust against such (sufficiently) small error probabilities.

More precisely, suppose player i would like to implement strategy $s_i \in S_i$ with certainty. Yet, because her actual behavior is error ridden, she will actually behave according to s_i only with a probability less than one. With the remaining probability a chance move takes over and picks a strategy at random, e.g. according to some fixed completely mixed strategy. Formally this is equivalent to a small, but *positive lower bound* on the probability that player i can assign to each of her pure strategies. This gives rise to a "perturbed game" in which players are forced to choose any one of their strategies with some positive minimal probability.

Definition 6.4 *Consider an n-player normal form game* $\Gamma = (S, u)$. *For each* $i = 1, ..., n$ *let* $\eta_i : S_i \to \mathbb{R}_{++}$ *be a (positive) function which satisfies* $\sum_{s_i \in S_i} \eta_i(s_i) < 1$, *let* $\eta = (\eta_1, ..., \eta_n)$, *and*

$$\Delta_i(\eta) = \{\sigma \in \Delta_i \, | \, \sigma_i(s_i) \geq \eta_i(s_i), \text{ for all } s_i \in S_i \}. \tag{6.7}$$

The **perturbed game** $\Gamma(\eta)$ *is the infinite normal form game*

$$\Gamma(\eta) = (\Delta_1(\eta) \times ... \times \Delta_n(\eta), U)$$

where $U = (U_1, ..., U_n)$ *is the extension of the payoff function* $u = (u_1, ..., u_n)$ *to mixed strategies (see (3.11) in Section 3.3).*

The function η gives the probabilities of "trembles" for the players $i = 1, ..., n$, i.e., the (minimal) error probabilities. The renormalized η's (dividing by their sum for each player) can be thought of as the conditional probabilities with which chance picks pure strategies, given that the player has failed to implement her desired strategy. Since those are part of the data of the perturbed game, they are commonly known among players.

The presence of trembles forces players in $\Gamma(\eta)$ to play *completely mixed* strategies, i.e. mixed strategies that assign (strictly) positive probability to *all* pure strategies. The set of all completely mixed strategy combinations is the (relative) interior of the strategy space,

$$\text{int}(\Theta) = \times_{i=1}^{n} \text{int}(\Delta_i) = \times_{i=1}^{n} \{\sigma_i \in \Delta_i \, | \, \sigma_i(s_i) > 0, \text{ for all } s_i \in S_i \}. \tag{6.8}$$

To approximate the original game, trembles must be small. If we wish to rule out equilibria that are vulnerable to all conceivable trembles, we need to require that the equilibrium survives (approximately) at least one sequence of trembles that are sufficiently small. Doing so will, for instance, eliminate the second equilibrium in the previous example. This is largely the idea of *perfect* equilibrium.

Definition 6.5 *A (possibly mixed) Nash equilibrium* $\sigma \in \Theta$ *of a game* $\Gamma = (S, u)$ *is ("trembling hand")* **perfect** *if there is a sequence* $\{\eta^t\}_{t=1}^{\infty}$ *of trembles and a sequence* $\{\sigma^t\}_{t=1}^{\infty}$ *of (completely mixed) strategy combinations such that*

(a) *for every* $\varepsilon > 0$ *there is some* t_ε *such that* $\eta_i^t(s_i) < \varepsilon$, *for all* $s_i \in S_i$ *and all* $i = 1, ..., n$, *for all* $t > t_\varepsilon$,

(b) *for every* $\varepsilon > 0$ *there is some* t_ε *such that* $\|\sigma - \sigma^t\| < \varepsilon$ *for all* $t > t_\varepsilon$, *and*

(c) *each* $\sigma^t \in \Theta$ *is a Nash equilibrium of* $\Gamma(\eta^t)$, *for all* $t = 1, 2, ...$

Hence, perfection requires the following. (a) There exists a sequence of trembles that converges to zero, i.e. such that the limit of η^t as t goes to infinity is zero. (b) There is a sequence of completely mixed strategy combinations that converges to the (perfect) equilibrium. And (c) each completely mixed strategy combination in this sequence is an equilibrium for the corresponding perturbed game $\Gamma(\eta^t)$. Since equilibria are defined by weak inequalities and payoff functions are continuous, a limit point for such a sequence of completely mixed equilibria (for the perturbed games) must itself be a Nash equilibrium.

If Γ is finite, a perfect equilibrium can be shown to exist for Γ (see Theorem 6.2 below). So, perfect equilibrium is a solution concept that is selective and exists (for finite games).

A perfect equilibrium for the agent normal form of an extensive form game is called an *extensive form* (trembling hand) *perfect* equilibrium. If the extensive form is finite, so is its agent normal form. Thus, every finite extensive form game has an extensive form perfect equilibrium. Since the approximating equilibria of perturbed games are in completely mixed strategies, consistent beliefs are guaranteed at all information sets. Hence, every extensive form perfect equilibrium is sequential and, henceforth, subgame perfect. By this inclusion the existence of perfect equilibria (for the agent normal form) ensures the existence of sequential equilibria for all finite extensive form games.

But note that a perfect equilibrium of the normal form is not necessarily an extensive form perfect equilibrium for an extensive form game with the given normal form (see Exercise 6.11 below). Moreover, in exceptional cases an extensive form perfect equilibrium may not be a perfect equilibrium for the associated normal form (see Exercise 6.12 below). In generic cases, however, an extensive form perfect equilibrium can be shown to be normal form perfect.

Finding perfect equilibria in concrete games is often facilitated by the following characterization of perfect equilibrium. (A proof is given in the appendix to this section.)

Proposition 6.1 *For any finite n-player game* $\Gamma = (S, u)$ *the following three statements are equivalent:*

(a) $\sigma \in \Theta$ *is a perfect Nash equilibrium;*

(b) $\sigma \in \Theta$ *is a limit point of a sequence* $\{\sigma^t\}_{t=1}^{\infty}$ *of completely mixed strategy combinations that satisfy*

$$\text{if } U_i(\sigma^t_{-i}, s_i) < U_i(\sigma^t_{-i}, s'_i), \text{ then } \sigma^t_i(s_i) \leq \varepsilon_t \qquad (6.9)$$

for all $s_i, s'_i \in S_i$, *all* $i = 1, ..., n$, *and all* $t = 1, 2, ...,$ *as* $\varepsilon_t \to_{t \to \infty} 0$;

(c) $\sigma \in \Theta$ *is a limit point of a sequence* $\{\sigma^t\}_{t=1}^{\infty}$ *of completely mixed strategy combinations such that* $\sigma \in \tilde{\beta}(\sigma^t)$ *for all* $t = 1, 2, ...$

The proposition states, in (b), that any perfect equilibrium may equivalently be thought of as a limit point of ε-perfect equilibria as $\varepsilon \to 0$. An *ε-perfect equilibrium* is a completely mixed strategy combination where any strategy that is not a best reply is used with probability at most $\varepsilon > 0$. Moreover, in (c), it states that any perfect equilibrium can equivalently be thought of as a best reply (not only against itself, but) against an approximating sequence of completely mixed strategy combinations (which converges to the perfect equilibrium).

Proposition 6.1(c) implies that all perfect equilibria are admissible, i.e. do not use weakly dominated strategies. For, if $s_i \in S_i$ is weakly dominated by $s_i' \in S_i$, then $U_i(\sigma_{-i}, s_i) < U_i(\sigma_{-i}, s_i')$ for all *completely mixed* strategy combinations $\sigma \in \Theta$. Hence, s_i cannot be a best reply against completely mixed strategy combinations. With more than two players, however, it is *not* true that any admissible Nash equilibrium is perfect (see Exercise 6.10 below). An equivalence of perfect and admissible Nash equilibria holds only for (finite) two-player games.

Strict Perfection   Given the justification of perfect equilibrium, it is possibly surprising that it requires robustness only against *one* sequence of trembles, and not against *all* possible trembles. But the latter is too much to ask for, if existence is desired.

Call a Nash equilibrium $\sigma \in \Theta$ for $\Gamma = (S, u)$ *strictly perfect* if for every $\varepsilon > 0$ there is some $\delta > 0$ such that, for all trembles η that satisfy $\eta_i(s_i) < \delta$, for all $s_i \in S_i$ and all $i = 1, ..., n$, there is a Nash equilibrium σ^η for $\Gamma(\eta)$ within ε from σ (i.e., $\|\sigma - \sigma^\eta\| < \varepsilon$). Hence, strictly perfect equilibrium replaces the existence-quantifier in the definition of perfect equilibrium by an all-quantifier, thus asking for robustness against *arbitrary* strategy trembles.

As a point-valued solution concept, strictly perfect equilibrium may fail to exist, as the next example illustrates.

Example 6.9 *In the two-player game below, any mixed strategy combination where player 2 uses only her first pure strategy is a Nash equilibrium.*

	s_2^1		s_2^2		s_2^3	
s_1^1	1		1		0	
		1		0		0
s_1^2	1		0		1	
		1		0		1

And there are no other Nash equilibria. If $\eta_2^t\left(s_2^2\right) > \eta_2^t\left(s_2^3\right)$ for all $t = 1, 2, ...,$ then all equilibria σ^t for $\Gamma(\eta^t)$ converge to $\left(s_1^1, s_2^1\right)$. If $\eta_2^t\left(s_2^2\right) < \eta_2^t\left(s_2^3\right)$ for all t, then the equilibria of $\Gamma(\eta^t)$ converge to $\left(s_1^2, s_2^1\right)$. Hence, there is no strictly perfect equilibrium.

If it exists, a strictly perfect equilibrium is clearly perfect. It will later be shown that the example illustrating nonexistence of strictly perfect equilibrium is somewhat unusual. Indeed, for generic (finite) games there exist equilibria that satisfy even stronger requirements than strict perfection. Hence, "most" (finite) games do have a strictly perfect equilibrium.

Proper Equilibrium Another aspect of perfect equilibrium is that the trembles do not have any structure. In particular, one might argue that if players make mistakes they will try to avoid more "costly" mistakes. Hence, worse strategies should occur with smaller probability than better strategies. This idea of more or less "rational" mistakes motivates *proper* Nash equilibrium.

Definition 6.6 *A Nash equilibrium* $\sigma \in \Theta$ *of a game* $\Gamma = (S, u)$ *is* **proper** *if there is a sequence* $\{\sigma^t\}_{t=1}^{\infty}$ *of completely mixed strategy combinations which converges to* σ *and satisfies*

$$\text{if } U_i\left(\sigma_{-i}^t, s_i\right) < U_i\left(\sigma_{-i}^t, s_i'\right), \text{ then } \sigma_i^t(s_i) \leq \varepsilon_t \sigma_i^t(s_i') \qquad (6.10)$$

for all $s_i, s_i' \in S_i$ *and all* $i = 1, ..., n$, *where again* $\varepsilon_t > 0$ *converges to zero as* $t \to \infty$.

Condition (6.10) defines an *ε-proper equilibrium*. Hence, a proper equilibrium is a limit point of a sequence of ε-proper equilibria as $\varepsilon \to 0$. Since every ε-proper equilibrium is ε-perfect, every proper equilibrium is perfect by Proposition 6.1(b). But, of course, "proper"ness is stronger than perfection in general.[2] Still, because every proper equilibrium is perfect, the following result guarantees the existence of proper *and* perfect equilibria for all finite games. (A proof is provided in the appendix to this section.)

Theorem 6.2 *Every finite game* $\Gamma = (S, u)$ *has a proper (and, hence, perfect) Nash equilibrium.*

The significance of proper equilibrium, however, derives not (only) from refining perfect equilibrium. Rather, it is due to the fact that a proper equilibrium of a normal form game induces a sequential equilibrium in any extensive form game with this normal form. This does not require the agent normal form—by contrast to perfect equilibrium. (A proof, following [97], is given in the appendix to this section.)

Proposition 6.2 *If* $\sigma \in \Theta$ *is a proper equilibrium of a finite normal form game* $\Gamma = (S, u)$, *then for any extensive form game with normal form* Γ *the mixed strategy combination* σ *induces a behavior strategy combination which constitutes a sequential equilibrium for the extensive form.*

---

[2] Whether a strictly perfect equilibrium is proper is not clear at all (see [37], p. 32).

That the (mixed) proper equilibrium σ "induces" a behavior strategy combination in any extensive form with normal form Γ deserves explanation. Recall that a proper equilibrium is a limit point of ε-proper equilibria. Since the latter are completely mixed strategy combinations, there is a unique behavior strategy combination that is consistent with an ε-proper equilibrium (see the appendix to Section 3.4). A limit point of these behavior strategy combinations as $\varepsilon \to 0$ is the behavior strategy combination "induced" by the proper equilibrium. And this can be shown to constitute a sequential equilibrium.

The converse of Proposition 6.2 is not true. Not every sequential equilibrium of an extensive form game is proper in the associated normal form. Nor can "sequential" be replaced by (extensive form) "perfect" (see [97], p. 1033).

Still, Proposition 6.2 indicates that backwards induction does *not* have to rely on the extensive form. Via proper equilibrium, the normal form is sufficient to identify backwards induction equilibria.

Yet, neither proper nor perfect equilibrium are without problems. A drawback that these concepts share, for instance, is that they are not invariant with respect to adding strictly dominated strategies. This seems undesirable, because strictly dominated strategies appear to be strategically irrelevant.

Example 6.10 *Consider the three-player game below. Player 3's second strategy is strictly dominated. If it were not present, the only perfect and hence (by Theorem 6.2) proper equilibrium would be for all players to choose their first pure strategies. Yet, if player 3's strategy s_3^2 is available, then also $\left(s_1^2, s_2^2, s_3^1\right)$ is a proper and, hence, perfect equilibrium.*

s_3^1	s_2^1	s_2^2
s_1^1	1 / 1 / 1	0 / 0 / 1
s_1^2	0 / 0 / 1	0 / 0 / 1

s_3^2	s_2^1	s_2^2
s_1^1	0 / 0 / 0	0 / 0 / 0
s_1^2	0 / 0 / 0	1 / 1 / 0

To see this, let $\sigma_i^t \left(s_i^1\right) = (t+1)^{-2}$ for $i = 1, 2$ and $\sigma_3^t \left(s_3^2\right) = 1/(t+1)$. Since for $i = 1, 2$ we have

$$U_i \left(\sigma_{-i}^t, s_i^1\right) - U_i \left(\sigma_{-i}^t, s_i^2\right) = \sigma_{3-i}^t \left(s_{3-i}^1\right) - \sigma_3^t \left(s_3^2\right) = \frac{-t}{(1+t)^2} < 0,$$

the sequence $\{\sigma^t\}_{t=1}^{\infty}$ constitutes a sequence of ε_t-proper equilibria with $\varepsilon_t = 1/t > 0$ which converges to $\left(s_1^2, s_2^2, s_3^1\right)$ as $t \to \infty$.

So, while perfect and proper equilibria can be seen as capturing robustness against small imperfections in implementing strategies, the effects of these trembles do depend on the presence of strategies that will not be used by rational players.

Summary

- A perfect equilibrium is a Nash equilibrium that is robust against at least one sequence of sufficiently small strategy trembles.

- A proper equilibrium is a perfect equilibrium where more "costly" mistakes are made with (an order) smaller probability.

- Every finite game has a proper and, hence, perfect Nash equilibrium.

- Every proper equilibrium induces a sequential equilibrium in any extensive form game with this normal form.

Appendix

Proof of Proposition 6.1 "(a) implies (b)": Let $\sigma \in \Theta$ be a perfect equilibrium and define $\varepsilon_t = \max_{i=1,\ldots,n} \max_{s_i \in S_i} \eta_i^t(s_i)$ for all $t = 1, 2, \ldots$ Since σ^t is an equilibrium for $\Gamma(\eta^t)$, any strategy $s_i \in S_i$ that is not a best reply against σ^t is used with (the minimal) probability $\sigma_i^t(s_i) = \eta_i^t(s_i) \leq \varepsilon_t$. Hence, condition (6.9) holds.

"(b) implies (c)": Let $\{\sigma^t\}_{t=1}^{\infty}$ be a sequence of ε_t-perfect equilibria and assume $\sigma^t \to_{t\to\infty} \sigma$ (without loss of generality). As $\varepsilon_t \to 0$ it follows from (6.9) that every $s_i \in \operatorname{supp}(\sigma_i)$ is a (pure) best reply against σ^t for t sufficiently large. Hence, $\sigma \in \tilde{\beta}(\sigma^t)$ for all t sufficiently large.

"(c) implies (a)": Let $\{\sigma^t\}_{t=1}^{\infty}$ be as in (c) and σ be a limit point of this sequence. Define η^t by $\eta_i^t(s_i) = \sigma_i^t(s_i)$ if $s_i \notin \operatorname{supp}(\sigma_i)$ and $\eta_i^t(s_i) = \varepsilon_t$ if $s_i \in \operatorname{supp}(\sigma_i)$ for all $s_i \in S_i$ and all $i = 1, \ldots, n$. Then $\eta^t \gg 0$ for all $t = 1, 2, \ldots$, $\eta^t \to_{t\to\infty} 0$ and $\Gamma(\eta^t)$ is well defined. By construction, for sufficiently large t we have $\sigma_i^t \in \Delta_i(\eta^t)$ for all i, and σ^t constitutes an equilibrium for $\Gamma(\eta^t)$, because only best replies against σ^t are used with more than the minimal probabilities. This completes the proof of Proposition 6.1.

Proof of Theorem 6.2 Here we show the existence of proper equilibria for all finite games $\Gamma = (S, u)$. The technique uses perturbed games, but different perturbed games from perfection.[3]

---

[3] The technique of the proof can also be used to show the existence of merely perfect equilibria, simply by replacing the perturbed games with strategy spaces $\Theta(\varepsilon)$ with those relevant for perfection, i.e. those with strategy spaces $\Theta(\eta) = \times_{i=1}^n \Delta_i(\eta)$.

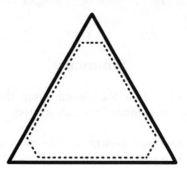

Figure 6.4: Perturbations for proper equilibria

For all $i = 1, ..., n$ denote by $K_i = |S_i|$ the number of pure strategies for player i and define $\Delta_i(\varepsilon)$ as the convex hull of the $K_i!$ permutations of the vector

$$\frac{1 - \varepsilon}{1 - \varepsilon^{K_i}} \left(1, \varepsilon, \varepsilon^2, ..., \varepsilon^{K_i - 1}\right),$$

for all $\varepsilon \in (0, 1)$. Every vector in $\Delta_i(\varepsilon)$ can be written as a convex combination of (no more than K_i of) these vectors that span $\Delta_i(\varepsilon)$. Figures 6.4 and 6.5 illustrate the difference between $\Delta_i(\varepsilon)$, depicted in Figure 6.4 for three pure strategies, and $\Delta_i(\eta)$ (as in Definition 6.4), depicted in Figure 6.5. The bold triangles are Δ_i and the dotted polyhedra are $\Delta_i(\varepsilon)$ and $\Delta_i(\eta)$, respectively.

Set $\Theta(\varepsilon) = \Delta_1(\varepsilon) \times ... \times \Delta_n(\varepsilon)$ for $\varepsilon \in (0, 1)$. By construction, each $\Delta_i(\varepsilon)$ is nonempty, compact, and convex. Define the game $\Gamma(\varepsilon)$ as the infinite normal form game $\Gamma(\varepsilon) = (\Theta(\varepsilon), U)$, where $U = (U_1, ..., U_n)$ is the extension of the payoff function $u = (u_1, ..., u_n)$ to mixed strategies (see (3.11) in Section 3.3).

Since the payoff function U_i is (linear and, therefore, weakly) quasi-concave in player i's (mixed) strategy, the game $\Gamma(\varepsilon)$ meets the conditions for existence of a Nash equilibrium (see Theorem 5.6 in Section 5.3). Hence, for any $\varepsilon \in (0, 1)$ the game $\Gamma(\varepsilon)$ has a Nash equilibrium. Since $\Theta(\varepsilon)$ is a subset of the compact set Θ, any sequence of equilibria for games $\Gamma(\varepsilon)$ contains a convergent subsequence as $\varepsilon \to 0$. Let $\sigma \in \Theta$ be a limit point for such a convergent subsequence.

We claim that σ is a proper equilibrium for Γ. To show this, it suffices to show that all equilibria for $\Gamma(\varepsilon)$ are ε-proper. Let σ^ε be an equilibrium for $\Gamma(\varepsilon)$ for some $\varepsilon \in (0, 1)$. If $U_i\left(\sigma^\varepsilon_{-i}, s_i\right) < U_i\left(\sigma^\varepsilon_{-i}, s'_i\right)$ and $\sigma^\varepsilon_i(s_i) > \varepsilon\sigma^\varepsilon_i(s'_i)$, then any of the vectors in the convex combination that makes up σ^ε_i and gives (relative) weight ε^k to s_i and (relative) weight ε^l to s'_i with $k < l$, for $k, l = 0, ..., K_i - 1$ (which must exist by hypothesis), can be replaced by a

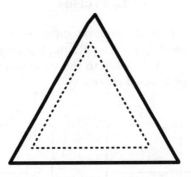

Figure 6.5: Perturbations for perfect equilibria

vector that gives weight ε^l to s_i and weight ε^k to s_i' (and keeps the relative weight of all other strategies constant). Since this will improve i's payoff, it contradicts the equilibrium hypothesis. Therefore, $U_i\left(\sigma_{-i}^\varepsilon, s_i\right) < U_i\left(\sigma_{-i}^\varepsilon, s_i'\right)$ implies $\sigma_i^\varepsilon(s_i) \leq \varepsilon\sigma_i^\varepsilon(s_i')$, demonstrating that σ^ε is an ε-proper equilibrium for Γ. This completes the proof of Theorem 6.2.

Proof of Proposition 6.2 Let $\{\sigma^\varepsilon\}_{\varepsilon>0}$ be a sequence of ε-proper equilibria, $\{b_\varepsilon\}_{\varepsilon>0}$ the induced sequence of behavior strategy combinations, and $\{\mu_\varepsilon\}_{\varepsilon>0}$ the associated sequence of beliefs consistent with b_ε. Extract a subsequence along which all these objects converge. We need to show that $b = \lim_{\varepsilon\to0} b_\varepsilon$ is such that each agent maximizes her payoff given $\mu = \lim_{\varepsilon\to0} \mu_\varepsilon$ and the other agents' strategies.

Suppose not. Then there is some player i and an information set $h \in H_i$ such that b_i assigns positive probability to a choice $c \in C_i$ available at h, whose expected payoff (given μ and b) is less than for some other choice $c' \in C_i \setminus \{c\}$ available at h. Since player i's agents at information sets after h are maximizing i's payoff, player i's expected payoff (starting at h and given μ and b_{-i}) of choosing c' and then continuing as in b_i is larger than that of choosing c, irrespective of the continuation. By the convergence assumption the same is true given μ_ε and b_ε, provided $\varepsilon > 0$ is sufficiently small.

It follows that every (pure normal form) strategy for i that does not avoid h and chooses c has smaller expected payoff, given σ_{-i}^ε, than a modification of that strategy that chooses c' and then continues as in b_i. Since σ^ε is ε-proper, σ_i^ε assigns the first strategy at most ε times the probability of the second strategy. It follows that σ_i^ε assigns to c a probability of at most εK_i. Letting $\varepsilon \to 0$ we see that $\sigma_i = \lim_{\varepsilon\to0} \sigma_i^\varepsilon$ assigns zero probability to c—a contradiction. This completes the proof of Proposition 6.2.

Exercises

Exercise 6.10 Consider the three-player game below. Verify that $\left(s_1^1, s_2^1, s_3^1\right)$ and $\left(s_1^2, s_2^1, s_3^1\right)$ are admissible Nash equilibria, but only $\left(s_1^1, s_2^1, s_3^1\right)$ is perfect. (Hint: What would player 1 do, close to the second equilibrium, if there were some chance that player 2 would choose s_2^2?)

s_3^1:

	s_2^1	s_2^2
s_1^1	1, 1, 1	1, 0, 1
s_1^2	1, 1, 1	0, 0, 1

s_3^2:

	s_2^1	s_2^2
s_1^1	1, 1, 0	0, 0, 0
s_1^2	0, 1, 0	1, 0, 0

Exercise 6.11 Consider the following two-player extensive form game. Player 1 starts by either choosing an outside option which gives both players payoff 2, or choosing either L or R. Player 2 gets to move only if 1 chooses L or R, but she does not learn which of these two has been chosen. At her information set player 2 has two choices, say, l and r.

If L and l materialize, the payoff vector $(4, 1)$ accrues to the players (where the first entry is player 1's payoff and the second 2's). If L and r materialize, players earn $(1, 0)$; if R and l are played, neither of them gets anything; and for R and r they earn $(0, 1)$.

Show that the unique sequential and, hence, extensive form perfect equilibrium is (L, l). But also show that in the normal form of this game there is a perfect equilibrium where player 1 chooses her outside option. (Hint: Consider the case where player 2 expects her opponent to make the mistake R with a larger probability than the mistake L.)

Exercise 6.12 Consider the following (degenerate) two-player extensive form game. Player 1 starts by either giving the move to player 2 or retaining the move and then deciding between the payoff vectors $(1, 1)$ or $(0, 0)$ (where the first entry is 1's payoff and the second 2's). If player 2 gets to move, she (learns that and) also decides between the payoff vectors $(1, 1)$ and $(0, 0)$.

Show that the unique perfect equilibrium for the normal form of this game is for player 1 to retain the move and then choose the payoff vector $(1, 1)$ (and for player 2 to choose the same payoff vector if she gets to choose). Moreover, show that in the extensive form it is also a perfect equilibrium for player 1 to give the move to player 2 who subsequently chooses $(1, 1)$ (and player 1 at her unreached information set also chooses $(1, 1)$). (Hint: First, for finite two-player games every admissible equilibrium is perfect. Second, extensive form perfect equilibria are perfect equilibria of the agent normal form.)

Exercise 6.13 Argue that for a finite n-player game any strict Nash equilibrium is strictly perfect (and proper and perfect).

Exercise 6.14 Show that any Nash equilibrium of a finite n-player game that is in completely mixed strategies is a proper and, hence, perfect equilibrium.

Exercise 6.15 Argue that in Example 6.9 there is a connected set of Nash equilibria such that every perturbed game has a Nash equilibrium close to this set; i.e., there is a set of Nash equilibria which meets the definition of strict perfection as a set.

> **Notes on the Literature:** Perfect Nash equilibrium was introduced by Selten [163]. Proposition 6.1 summarizes a result by Selten [163] (part (c)) and one by Myerson [126] (part (b)). Myerson's [126] paper is also the source of proper Nash equilibrium, where Theorem 6.2 was first demonstrated. The concept of strictly perfect equilibrium is due to Okada [135]. Perfect equilibria for games with infinite strategy spaces have been studied by Simon and Stinchcombe [168] . That for finite two-player games every admissible Nash equilibrium is perfect is shown by van Damme ([37], Theorem 3.2.2). Examples 6.9 and 6.10 are from this source (pp.16 and 30, respectively), as are Exercises 6.11 and 6.12 (p. 114). Proposition 6.2 is due to van Damme [36] and Kohlberg and Mertens ([97], Proposition 0).

6.3   Strategic Stability

All the concepts introduced so far have certain deficiencies. Strictly perfect equilibria may fail to exist. Perfect equilibrium in the normal form may fail to induce a solution consistent with backwards induction in an associated extensive form. Moreover, it shares with proper equilibrium a sensitivity to adding strictly dominated strategies. Subgame perfect and sequential equilibrium, as well as extensive form perfect equilibrium, are sensitive to inessential transformations of the extensive form. Moreover, sequential equilibria and extensive form perfect equilibria may be inadmissible (in the normal form).

Set-Valued Solutions   This raises the issue of whether all "desirable" properties for a "self-enforcing" solution can be combined in a single refinement concept. Of course, a prior question is what the "desirable" properties are. Some of them are obvious, like the three basic requirements of existence, selectiveness, and the Nash property. But these are already combined in Nash equilibrium itself. The other desirable properties follow from the findings in the previous sections.

Those suggest that a self-enforcing solution should also satisfy *backwards induction* and should be *independent of irrelevant alternatives*, like strategies that rational players will certainly not use, or inessential transformations of the extensive form.

Since backwards induction solutions may not be invariant to the latter, there is a potential conflict between the two. To avoid this, backwards induction should be qualified as "backwards induction in all (Thompson-) equivalent extensive forms". From Section 4.2 we already know what that means. It means that a self-enforcing solution should depend only on the *semi-reduced* normal form (see Section 4.1).

Moreover, when looking at the mixed extension of a (finite) game, any player has all probability distributions over her pure strategies available as alternatives. Hence, if she is given another pure strategy which is equivalent to an already available mixed strategy, then this additional strategy can be considered an irrelevant alternative. It follows that a self-enforcing solution should not depend on whether or not a pure strategy is available which is equivalent to a mixture over other pure strategies. This implies that such a solution should depend only on the (mixed strategy) *reduced* normal form.[4]

But backwards induction, and the requirement that only the reduced normal form matters may be impossible to satisfy simultaneously, as the following example shows.

---

[4]   Recall, however, that the transition to the reduced normal form may destroy certain types of extensive form information; see Section 4.3.

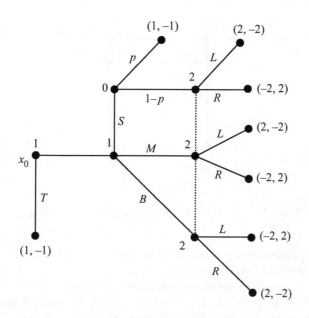

Figure 6.6: Backwards induction depends on p

Example 6.11 *Consider the two-player extensive form game in Figure 6.6 for $p \in [0,1]$. In its unique sequential equilibrium player 2 chooses L with probability $\sigma_2(L) = (4 - 3p)/(8 - 4p)$; i.e., the unique backwards induction solution depends on p (see Exercise 6.16 below). But for all p the reduced normal form is as given below. Since the reduced normal form is independent of p, no single equilibrium point can correspond to the backwards induction solution for all games in Figure 6.6 with some arbitrary $p \in [0,1]$.*

	L	R
T	1	1
	-1	-1
IM	2	-2
	-2	2
IB	-2	2
	2	-2

It follows that a solution that depends only on the reduced normal form and satisfies (a version of) backwards induction in all extensive form games with that reduced normal form cannot be a single point, if existence is to be preserved. Rather *all* equilibria (of the reduced normal form) with payoffs $(1, -1)$ have to be considered equivalent in the above example. In other words,

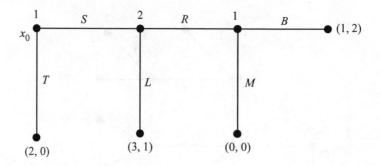

Figure 6.7: (T, R) is sensitive against eliminating SB

insisting on existence, a self-enforcing solution has to be a *set of equilibria*, rather than a single equilibrium point.

The independence requirement also has implications concerning dominated strategies. Since no rational player will use strictly dominated strategies (see Section 5.1), those are certainly irrelevant. Hence, a self-enforcing solution should not depend on whether or not a strictly dominated strategy has been deleted from the game.

It follows that such a solution should be *invariant* with respect to *adding strictly dominated strategies*. (Example 6.10 shows that perfect and proper equilibria fail this test.)

One can go further. If it is taken for granted that no player will use weakly dominated strategies, then those should be considered irrelevant. It follows that self-enforcing solutions should survive iterated elimination of weakly dominated strategies. In fact, many have argued that decision theory in itself should imply that no rational player will use a weakly dominated strategy. (See, however, the discussion in Section 5.1.) Under this presumption, one is led to require that a self-enforcing solution be *admissible*.

Moreover, independence of irrelevant alternatives is a "global" property, applicable independently of the specific game under consideration. Once a specific candidate solution has been found, it can be argued that no player will use a strategy that is never a best reply against the solution. Then, such strategies that are never best replies are to be considered irrelevant. Consequently, a self-enforcing solution should remain so when such a strategy is eliminated. In short, a solution should be *invariant* with respect to *eliminating strategies that are never best replies*.

Jointly adopting all these invariance requirements may, however, again conflict with existence when the solution is taken to be a single strategy combination. The following simple example illustrates this.

Example 6.12 *Consider the two-player extensive form game in Figure 6.7. In its unique subgame perfect equilibrium, player 1 at the beginning decides to terminate the game (T) and player 2, if called upon to move, would give the move back to player 1 (R), who would then choose B.*

Against the (backwards induction) equilibrium strategy combination (T, R), player 1's strategy SB is never a best reply (see the semi-reduced normal form below). If this strategy is eliminated, however, player 1 is forced to pick M at her second decision point. This amounts to an extensive form similar to Figure 6.7, but where player 1's second decision point has disappeared and after R (by player 2) the game ends with payoffs (0,0). The unique subgame perfect equilibrium of this new game is (S, L). It follows that there is no (single) strategy combination which both satisfies backwards induction and is invariant with respect to elimination of strategies that are never best replies.

	L	R
T	2 0	2 0
SM	3 1	0 0
SB	3 1	1 2

The example illustrates once again that combining backwards induction with invariance (with respect to eliminating strategies that are never best replies) conflicts with existence when the solution is meant to be single strategy combination. But it also reveals the meaning of the alternative, a *set-valued* solution.

Take a closer look at the example. In particular, to support (T, R) it is not necessary that player 2 choose R with certainty. That $\sigma_2(R) \geq 1/2$ is sufficient. But against $\sigma_2(R) = 1/2$, player 1's strategy SB *is* a best reply— and should *not*, therefore, count as never being a best reply.

Yet, one can argue that $\sigma_2(R) = 1/2$ should not be allowed for under backwards induction. After all, subgame perfection pins down $\sigma_2(R) = 1$ in Figure 6.7. On the other hand, if the backwards induction solution *is* played in this game, then player 2 will never get to choose. If she gets to choose, this very fact constitutes evidence of a violation of backwards induction on the part of player 1. Should, in the light of this evidence, player 2 expect her opponent to choose B if given another move?

There is probably no good answer to this question, because what is optimal for player 2 depends on how she interprets the event of being reached. Is it because player 1 made a "mistake" or because she is going for her payoff 3 (threatening to choose M at her last decision point)? Both are inconsistent with backwards induction "rationality". But no model of "irrationality" has been provided. A theory that assumes rationality of all players should be content to remain silent on this question.

Luckily, however, the theory does not have to be very precise. In order to know her best choice, player 1 does not need to know precisely what player 2 will choose. It is enough for her to consider R more likely than L, i.e., $\sigma_2(R) \geq 1/2$ (where σ_2 is interpreted as player 1's belief about what 2 will do). Given that 1 assesses the situation in that way, player 2 indeed has no basis for preferring one of her strategies over the other. Hence, the equilibrium outcome T gets supported in strategy space by a *set* of strategy combinations ($\sigma_1(T) = 1$ and $1/2 \leq \sigma_2(R) \leq 1$).

This suggests a qualification to the meaning of backwards induction. Recall that identifying a backwards induction solution relied on the assumption that a player is rational whatever she has previously done. But that player 2, upon being reached, will "assume" that player 1 is rational is only a *test* of the assumption that player 1 will not play S. Once this is confirmed, one cannot continue to maintain the assumption that, in the event of being reached, player 2 will continue to assume that 1 is rational—indeed she need not, and perhaps cannot.

Short of a theory of "irrationality", we have to allow for all (equilibrium) strategy combinations consistent with the backwards induction outcome. Once this is accepted, the conflict between invariance (with respect to eliminating strategies that are never best replies) and backwards induction disappears. Strategy SB now becomes a best reply against one of the elements in the solution.

Three conclusions emerge from this discussion. First, as noted before, a self-enforcing solution may (in strategy space) correspond to a *set* of strategy combinations (rather than a single point). Second, not all points in this set may be backwards induction equilibria; rather we have to be content with the solution *containing* a backwards induction equilibrium. Third, "independence of irrelevant alternatives" has to be applied in view of the whole set of strategy combinations; that is, a strategy may be deleted only if it is never a best reply against *any one* of the elements in the solution.

Similar conflicts arise between backwards induction and admissibility. Examples can be constructed (see [114] in which *all* extensive form perfect equilibria are inadmissible. Once inadmissible strategies get eliminated, this attributes more "rationality" to players, making them more predictable, and, hence, produces smaller solutions. Backwards induction equilibria may not survive that. But again, this conflict disappears once *sets* of strategy combinations are allowed as solutions.

The price for this reconciliation is that we have to be satisfied if we can advise players what to do in circumstances that are consistent with all players being rational. So, a self-enforcing solution does not necessarily pin down behavior in parts of the game tree that are inconsistent with the solution. Intuitively, a self-enforcing solution has to be set-valued because it must incorporate the possibility of "irrational" play.

The question remains what mathematical objects should be candidates for solutions, now that it is understood that single strategy combinations do not qualify. Using the set of *all* Nash equilibria may allow patently implausible equilibria to show up in the solution. So, something more narrow is required. In fact, the resolution of this is not entirely clear. As a preliminary qualification, *connected* sets of Nash equilibria may be required. This, at least, ensures that the outcome varies continuously with beliefs or presentation effects captured by the solution. (For almost all games the outcome will then not vary at all; see Theorem 6.4 in the next section.)

Desiderata Recall that a solution concept is a mapping. Now it maps into *sets* of (mixed) strategy combinations. And it assigns self-enforcing solutions which will be referred to as "strategically stable sets". Formally, let $E(\Gamma)$ assign the set of all Nash equilibria for a game Γ and denote by E^* the mapping from (finite) games to sets of strategy combinations that assigns *strategically stable sets* of Nash equilibria. Then, the desirable properties for self-enforcingness are as follows.

(EX) (Existence): $E^*(\Gamma) \neq \emptyset$ for all games Γ.

(NE) (Nash property): $E^*(\Gamma) \subseteq E(\Gamma)$ for all games Γ.

(CO) (Connectedness): If $Q \in E^*(\Gamma)$ then Q is connected,[5] for all games Γ.

The first two are the basic requirements. Clearly, (NE) takes care of selectiveness, because Nash equilibrium in itself is selective. (CO) defines the range of the solution concept by asking that solutions are connected sets (in the space of mixed strategies).

The formalization of backwards induction is somewhat cautious. Its idea is to require that a self-enforcing solution contains an equilibrium which will induce a sequential equilibrium in any extensive form game with this normal form. Yet, there is no characterization of such equilibria. So, we have to be content to appeal to Proposition 6.2 and ask for the solution to contain a proper equilibrium. As Example 6.12 shows, we cannot ask for more than inclusion.

(BI) (Backwards induction): If $Q \in E^*(\Gamma)$, then Q contains a proper equilibrium, for all games Γ.

The first "ordinality" requirement is that a self-enforcing solution depends only on the reduced normal form. That is, if two games have the same reduced normal form, then they should have the same solutions. This is stated, somewhat informally (because the strategy spaces of the unreduced games may differ), as follows.

---

[5] A set is *connected* if it cannot be covered by two disjoint open sets.

(IV) (Invariance): If Γ and Γ' have the same reduced normal form, then $Q \in E^*(\Gamma')$ if and only if $Q \in E^*(\Gamma)$, for all games Γ and Γ'.

Next, we turn to admissibility.[6] We want to require that at all points in a solution only admissible strategies are used. There are several ways to formalize this. A strong one is to allow only admissible best replies.

Call $\sigma_i' \in \Delta_i$ an *admissible best reply* for player i against $\sigma \in \Theta$ if there is a sequence $\{\sigma^k\}_{k=1}^{\infty}$ of completely mixed strategy combinations which converges to σ such that σ_i' is a best reply against any element in this sequence; i.e., $\sigma_i' \in \tilde{\beta}_i(\sigma^k)$ for all $k = 1, 2, \ldots$

Write $\tilde{\beta}_i^a : \Theta \to \Delta_i$ (resp. $\beta_i^a = \tilde{\beta}_i^a \cap S_i$) for the correspondence assigning i's admissible best replies (resp. admissible pure best replies), so that $\tilde{\beta}_i^a(\sigma) \subseteq \tilde{\beta}_i(\sigma)$ (resp. $\beta_i^a(\sigma) \subseteq \beta_i(\sigma)$) for all $\sigma \in \Theta$ and all i. For the product mapping write $\tilde{\beta}^a = \tilde{\beta}_1^a \times \ldots \times \tilde{\beta}_n^a$ (resp. $\beta^a = \beta_1^a \times \ldots \times \beta_n^a$), and for any subset $Q \subseteq \Theta$ let $\tilde{\beta}^a(Q) = \cup_{\sigma \in Q} \tilde{\beta}^a(\sigma)$ (resp. $\beta^a(Q) = \cup_{\sigma \in Q} \beta^a(\sigma)$) be the admissible best replies (resp. admissible pure best replies) against some strategy combination in Q. The admissibility requirement can now be written as follows.

(AD) (Admissibility): If $Q \in E^*(\Gamma)$ then $Q \subseteq \tilde{\beta}^a(\Theta)$ for all games Γ.

Accepting (AD) implies that a strategy that is not an admissible best reply against the solution will certainly not be used. Hence, such a strategy counts as irrelevant, and, accordingly, one is led to accept the following.

(IA) (Independence of irrelevant alternatives): If $Q \in E^*(\Gamma)$ and $s_i \notin \beta_i^a(Q)$ for some player i, then Q contains a solution of the game where $s_i \in S_i$ has been deleted.

The statement of (IA) allows only pure strategies (viewed as particular mixed strategies) to be eliminated. But doing so deletes all mixed strategies with the deleted pure strategy in their support. That (IA) only asks for Q to *contain* a solution of the game where s_i has been eliminated makes sure that not all of Q is destroyed by this test.

Property (IA) is strong, because it allows both inadmissible strategies *and* strategies that are not best replies against any element in the solution to be eliminated. (If $s_i \in S_i$ is not a best reply against the solution, then it cannot be an admissible pure best reply against the solution.) Hence, it represents a combination of admissibility and invariance with respect to elimination of strategies that are never best replies.

---

[6] There is an issue as to whether admissibility is indeed a fundamental requirement (see the discussion in Section 5.1). But at this point we follow the literature on strategic stability in adopting admissibility.

Definitions In an attempt to provide an explicit definition of a solution that satisfies (versions of) the above seven desiderata, Kohlberg and Mertens [97] have proposed a provisional definition of strategically stable sets. It boils down to a minimal set-valued version of strict perfection.

Definition 6.7 *For a (finite) game Γ a set of Nash equilibria is **KM-stable** if it is minimal[7] with respect to the following property:*

(S) $Q \subseteq E(\Gamma)$ *is closed and for every $\varepsilon > 0$ there is $\delta_0 > 0$ such that for any completely mixed strategy combination $\tilde{\sigma} \in \text{int}(\Theta)$ and any vector $(\delta_1, ..., \delta_n)$ with $0 < \delta_i < \delta_0$ for all $i = 1, ..., n$, the game where every strategy combination $\sigma \in \Theta$ is replaced by $(1 - \delta_i)\sigma_i + \delta_i\tilde{\sigma}_i$ for all i has an equilibrium σ^δ which is ε-close to Q (i.e., $\min_{\sigma \in Q} \|\sigma^\delta - \sigma\| < \varepsilon$).*

The perturbed games in property (S) are constructed by using a completely mixed strategy combination $\tilde{\sigma}$ in order to satisfy invariance (IV). Every pure strategy is perturbed by the same amount towards the same completely mixed strategy. An arbitrary convex combination of pure strategies will then also be perturbed in the same way. Thus, KM-stable sets are invariant under the addition or deletion of mixed strategies as extra pure strategies.

It turns out that KM-stable sets exist for all finite games[8] and consist entirely of normal form perfect equilibria. The latter follows from minimality. Every element of a KM-stable set is a limit of equilibria of perturbed games, in which weakly dominated strategies become strictly dominated. Thus, KM-stable sets satisfy admissibility. Moreover, every KM-stable set contains a KM-stable set of a game that is obtained (from the original game) by deleting a (weakly) dominated strategy. And every KM-stable set contains a KM-stable set of a game that is obtained (from the original game) by deleting a strategy that is not a best reply against any one of the equilibria in the original KM-stable set (see [97], Proposition 6).

So, KM-stable sets satisfy existence (EX), the Nash property (NE), invariance (IV), admissibility (AD), and independence (IA). But they may fail connectedness (CO) and backwards induction (BI). The following example illustrates both failures.

Example 6.13 *Consider the following extensive form game with three players. Player 1 starts by either taking an outside option (s_1^1) which yields payoffs $(2, 0, 0)$ or moving into the simultaneous move subgame depicted below, where each of the three players has two choices. The unique sequential equilibrium of this game has player 1 moving into the subgame playing $\sigma_1(s_1^2) = \sigma_1(s_1^3) = 1/2$ and $\sigma_i(s_i^1) = 1/2$ for $i = 2, 3$ with payoffs $(11/4, 5/4, 5/4)$. But the union*

---

[7]   "Minimality" is meant with respect to set inclusion; i.e., the set under scrutiny does not properly contain another set which also satisfies the defining property.

[8]   A proof of this relies on Proposition 6.5 in the next section.

of the two points $\left(s_1^1, s_2^1, s_3^1\right)$ and $\left(s_1^1, s_2^2, s_3^2\right)$ is a KM-stable set: if player 1 reaches the subgame, then, conditional on such a mistake, players 2 and 3 face a game which, depending on the conditional probabilities for s_1^2 and s_1^3, has $\left(s_2^1, s_3^1\right)$ or $\left(s_2^2, s_3^2\right)$ or both as strict equilibria. Against any of those, player 1 is strictly better off taking the outside option s_1^1. Hence, this does indeed constitute a KM-stable set. Yet, it does not contain the sequential equilibrium. It even yields different payoffs from the sequential equilibrium. Moreover, this KM-stable set is not connected.

Left matrix (columns s_3^1, s_3^2; bottom label s_1^2):

	s_3^1		s_3^2	
s_2^1	0 3 3		5 1 1	
s_2^2	5 1 1		1 0 0	

Right matrix (columns s_3^1, s_3^2; bottom label s_1^3):

	s_3^1		s_3^2	
s_2^1	1 0 0		5 1 1	
s_2^2	5 1 1		0 3 3	

These failures explain why Definition 6.7 is preliminary. Kohlberg and Mertens [97] also consider other preliminary concepts, besides KM-stable sets. A *hyperstable set* is minimal in the class of closed sets of Nash equilibria Q that satisfy the following: for any game with the same reduced normal form and any sufficiently small perturbation of the payoffs of that game, there is an equilibrium close to Q. A *fully stable set* is minimal in the class of closed sets of Nash equilibria Q that satisfy the following: for any restriction of each player's strategy set to some compact convex polyhedron, in the interior of the mixed strategy space and at sufficiently small (Hausdorff) distance from it, the resulting game has an equilibrium close to Q.

Both these concepts perform quite well. They satisfy existence (EX), the Nash property (NE), versions of connectedness (CO), backwards induction (BI), invariance (IV), and independence (IA). But both fail admissibility. On that ground, Kohlberg and Mertens reject these two concepts.

Insisting on admissibility, the ordinality requirement (IV) may actually be strengthened. A self-enforcing solution be "ordinal"; i.e., it should depend only on those aspects of the game that are relevant for the players' individual decision problems. That the solution depends only on the reduced normal form is one part of that. If admissibility is accepted, then one can also argue that admissible best replies capture all aspects that are relevant for "rational" decision making of players. Hence, one is led to require that two games have the same solutions if their admissible best reply correspondences are the same.

(BR) (Best reply invariance): If admissible best replies $\tilde{\beta}_\Gamma^a$ for the game Γ and admissible best replies $\tilde{\beta}_{\Gamma'}^a$ for the game Γ' are the same, then $E^*\left(\Gamma'\right) = E^*\left(\Gamma\right)$.

The application of (BR) is restricted to games with the same player and strategy sets. Hence, this requirement should be supplemented with requirements that player names and strategy labels do not matter.

Hillas [87] proposes a definition of strategic stability which satisfies existence (EX), the Nash property (NE), connectedness (CO), backwards induction (BI), admissibility (AD), and best reply invariance (BR). Roughly, a closed set of equilibria is H(illas)-*stable* if it is minimal among all closed sets of equilibria $Q \subseteq E(\Gamma)$ with the following property: each upper hemicontinuous compact convex valued correspondence that is pointwise close to the best reply correspondence of a game with the same reduced normal form as Γ has a fixed point close to Q.

But H-stable sets are sensitive to certain irrelevant changes in the player set. And this seems undesirable. Changes in the set of players that do not affect the individual decision problems should not affect a self-enforcing solution.

For instance, irrelevant players should have no influence on the solution. Say that a subset of players $J \subseteq I \equiv \{1, ..., n\}$ constitutes a *small world* if their payoffs do not depend on the strategies of "outside" players not in J. Let $\Gamma_J = \left(\times_{i \in J} S_i, (u_i)_{i \in J} \right)$ be the game played by "insiders". For a small world "outside" players are to be considered irrelevant and should have no impact on the solution inside of it.

(SW) (Small worlds): If J is a small world in Γ, then Q_J is a solution for Γ_J if and only if it is the projection of a solution Q for Γ.

Similarly, a game that consists of two small worlds should decompose. If two disjoint player sets play different games in different rooms, then it should not matter whether the games are analyzed jointly or separately.

(DC) (Decomposition): If for a game Γ both J and $I \setminus J$ constitute small worlds, then $Q \in E^* (\Gamma)$ if and only if $Q = Q_J \times Q_{I \setminus J}$ with $Q_J \in E^* (\Gamma_J)$ and $Q_{I \setminus J} \in E^* (\Gamma_{I \setminus J})$.

Next, recall that a player may be split into agents (see Section 4.1). Still, some of those agents may be dependent in the sense that there is a play which belongs to (a move in) the information sets of both. Call two agents (of the same player) *independent* if the plays passing through their information sets are disjoint. A player is *split into independent agents* if she is replaced by a collection of agents, each controlling possibly several information sets, such that no play passes through information sets of different agents (of the same player). Such a splitting of a player into independent agents should be irrelevant for a self-enforcing solution.

(PS) (Player splitting): If Γ' is obtained from Γ by splitting some player into independent agents, then $E^* (\Gamma') = E^* (\Gamma)$.[9]

---

[9] The equality sign here is a slight abuse of notation, because the underlying strategy

The player splitting property implies that the Harsanyi form and the Selten form (see Section 3.5) of an incomplete information game will have the same solutions. And this seems certainly desirable.

Adding these three requirements about the player set and best reply invariance brings the total number of desiderata for strategic stability up to eleven properties. This in itself suggests that an explicit definition of strategically stable sets which satisfy all these will be quite involved—and it is! Mertens ([113] and [114]) gives such a definition; but its statement is involved and beyond the scope of the present text.

In concrete applications, strategically stable sets can often be identified by their properties. For instance, if only one set of equilibria survives the elimination of dominated strategies, this remaining set must be strategically stable. Hence, the important piece of information is captured by the following result, which states that all ten properties listed can be combined with existence (EX). (For a proof see Mertens [113], [114].)

Theorem 6.3 *For every finite game Γ there exists a closed set of Nash equilibria which satisfies connectedness (CO), backwards induction (BI), invariance (IV), admissibility (AD), independence (IA), best reply invariance (BR), small worlds (SW), decomposition (DC), and player splitting (PS).*

Of course, requirements (EX) and (NE) are taken care of by the statement of the theorem. The strategically stable sets alluded to in Theorem 6.3 are called *M-stable sets*. Note, however, that Mertens in his definition does not insist on minimality. In fact, he identifies a conflict between minimality and the ordinality requirements.

Strategic stability occasionally has surprising implications. For instance, it implies that the possibility of inflicting damage on oneself confers power. Suppose that before playing a one-shot "Battle of the Sexes" game (like in Example 6.14 below) player 1 has the opportunity to "burn" one unit of utility such that player 2 can observe that. Then the only strategically stable outcome has player 1 not burning utility and players playing the most preferred outcome for player 1. The argument is that the game can be reduced to this outcome by independence (IA). With both players having the opportunity to "burn" utility the situation is even worse. This insight is illustrated by the concluding example.

Example 6.14 *(Burning money) Consider the following two-player extensive form game. Initially both players choose simultaneously whether or not to burn one unit of money. Then they observe each other's choices and play the 2×2 simultaneous move subgames below. Here, $x_i = 1$ if player i has burned money and $x_i = 0$ otherwise, for $i = 1, 2$.*

spaces of the two games are different. But there is a natural projection from the strategy space for Γ' to the one for Γ to which this equality sign refers.

By not burning money and then playing L_i with probability $3/4$ player i can guarantee herself a positive payoff. Therefore, no player i will burn money and then play L_i, because this yields at best zero. Thus, all those strategies are never best replies to any point in the solution and can be deleted.

But in the reduced game any strategy that involves not burning money and playing R_i if the opponent burns money is weakly dominated by not burning money, playing L_i if the opponent burns money, and otherwise doing the same as in the dominated strategy. This is so, because the two strategies differ only when the opponent burns money, in which case she will choose R_{3-i}. Applying admissibility to delete those strategies yields a game where each player has three strategies, because now not burning money must be followed by L_i at the information set where the opponent has burned money.

Now suppose there is a stable set in which player 2, say, does not burn money. Then player 1 can guarantee herself payoff 2 by burning money. Since playing L_1 if 2 does not burn money yields player 1 at most payoff 1, this strategy can never be a best reply against any element in the stable set. A symmetric argument applies to player 2. Hence, we are left with a game where each player i can only either burn or not burn money (and then play R_i in either case).

This game has only three equilibria. In the two pure equilibria one of the players burns money with certainty; in the mixed equilibrium both players burn money with positive probability. So, any equilibrium in the strategically stable set involves a waste of resources.

	L_2	R_2
L_1	$-x_1$ $\quad -x_2$	$1-x_1$ $\quad 3-x_2$
R_1	$3-x_1$ $\quad 1-x_2$	$-x_1$ $\quad -x_2$

Summary

- A strategically stable set is a closed, nonempty, and connected set of Nash equilibria which satisfies a number of desirable properties.

- Desirable are ordinality (the solution depends only on the reduced normal form and admissible best replies), backwards induction, admissibility, independence of irrelevant alternatives, and invariances with respect to irrelevant changes in the player set.

- Set-valuedness of strategic stability is forced by the combination of the desirable properties when existence is required.

- Intuitively, strategic stability has to yield set-valued solutions in order to incorporate the possibility of play inconsistent with its premises.

Exercises

Exercise 6.16 Determine the (unique) sequential equilibrium for the game in Figure 6.6. (Hint: First solve the proper subgame and then the full game.)

Exercise 6.17 Determine the set of all (mixed) Nash equilibria for the game from Example 6.12. (Hint: It consists of two linear pieces in Θ.)

Exercise 6.18 Find an extensive form game with perfect information which has the semi-reduced normal form given below. Then argue that a self-enforcing solution cannot consist only of the strategy combination (T_1, T_2). (Hint: For the first part see Section 4.3.)

	T_2	L_2
T_1	1 0	1 0
$L_1 t$	0 2	4 0
$L_1 l$	0 2	3 3

Exercise 6.19 Argue that for any (finite) game every strict equilibrium and every completely mixed equilibrium are both (singleton) KM-stable sets.

Exercise 6.20 Consider the following extensive form game with two players. At its start player 1 can take an outside option which yields her payoff 2 and nothing to player 2. Or she can move into the simultaneous move subgame below. Argue that a solution that depends only on the reduced normal form and induces a backwards induction solution in any extensive form with this reduced normal form must yield player 1 payoff 3. (Hint: Add a mixture between the outside option and T for player 1 where T is played with some probability below $1/2$, reinterpret this as a new extensive form game where player 1 has three strategies in the only proper subgame, and find the backwards induction solution for the latter.)

	L	R
T	3 1	0 0
B	0 0	1 3

Notes on the Literature: The seminal paper on strategic stability is by Kohlberg and Mertens [97]. Definition 6.7 for KM-stable sets, Example 6.11 and Example 6.13 (the latter due to

Gül), are also from this source. The presentation in this section is largely inspired by van Damme's [39] discussion of strategic stability. Mertens ([113], [114], and [115]) has introduced the full list of desirable properties for strategic stability, has given the definition of M-stable sets, and has proved Theorem 6.3. Hillas [87] has provided a modification of strategic stability, H-stable sets. Example 6.14 is from van Damme [38]. For applications see Ben-Porath and Dekel [16] and Glazer and Weiss [68].

6.4   Payoff Perturbations

Strategy trembles are but one way to perturb the data of a game. Equally important information is obtained from a more global viewpoint, by variation of the payoff function. In particular, there are a few basic results that concern the structure of Nash equilibria for *all* (finite) games. These are obtained by taking seriously the idea that Nash equilibrium defines a solution mapping. The *Nash equilibrium correspondence* maps the space of games to (those) strategy combinations (that constitute Nash equilibria).

The Equilibrium Correspondence   More precisely, fix (finite) strategy spaces S_i for all players $i = 1, ..., n$ and vary the payoff function $u = (u_1, ..., u_n)$. Since any function on a finite domain can be represented by a list of vectors, recording its value at each point of the function's domain, the space of all games can be identified with some Euclidean space $\mathbb{R}^{nK}$, where $K = |S|$ is the number of pure strategy combinations (see Section 4.4). That is, each point $u \in \mathbb{R}^{nK}$ represents a particular game $\Gamma(u) = (S, u)$. Likewise, each point $u \in \mathbb{R}^{nK}$ is associated with a particular (mixed) best reply correspondence

$$\tilde{\beta}_u(\sigma) = \times_{i=1}^{n} \arg \max_{\sigma_i' \in \Delta_i} U_i(\sigma_{-i}, \sigma_i'),$$

where U_i is the extension of u_i to mixed strategies (see (3.11) in Section 3.3). The Nash equilibrium correspondence $E : \mathbb{R}^{nK} \to \Theta$ is then formally defined by

$$E(u) = \left\{ \sigma \in \Theta \,\middle|\, \sigma \in \tilde{\beta}_u(\sigma) \right\}. \tag{6.11}$$

A first, immediate observation about the Nash equilibrium correspondence is that it is *upper hemi-continuous* (u.h.c.; for a definition see the appendix to Section 5.3). This follows from the fact that the extended payoff function U is (linear and, therefore) continuous in $u \in \mathbb{R}^{nK}$ and the best reply correspondence $\tilde{\beta}_u$ is u.h.c. (by the maximum theorem). Hence, if $\{u^t\}_{t=1}^{\infty}$ is a sequence of games which converges to $u \in \mathbb{R}^{nK}$ (i.e., $u^t \to_{t \to \infty} u$) and $\{\sigma^t\}_{t=1}^{\infty}$ a sequence of associated Nash equilibria, $\sigma^t \in E(u^t)$ for all $t = 1, 2, ...$, which converges to $\sigma \in \Theta$ (i.e., $\sigma^t \to_{t \to \infty} \sigma$), then $\sigma \in E(u)$. Intuitively, this means that the set of equilibria "does not implode" as payoffs vary. (Yet, it may "explode".)

An equivalent way to state this property is that the *graph*

$$\mathcal{G}(E) = \left\{ (u, \sigma) \in \mathbb{R}^{nK} \times \Theta \,\middle|\, \sigma \in E(u) \right\} \tag{6.12}$$

of the Nash equilibrium correspondence is closed. This is so, because u.h.c. is equivalent to the correspondence having a closed graph (because of its compact range).

The Structure of Equilibria A second observation is that each Nash equilibrium $\sigma \in E(u)$ constitutes a solution to a system of finitely many inequalities of the form

$$U_i(\sigma_{-i}, s_i) \leq U_i(\sigma) \text{ for all } s_i \in S_i \text{ and all } i = 1, ..., n. \qquad (6.13)$$

Every U_i is linear in each player's mixed strategy $\sigma_j \in \Delta_j$ for all $i, j = 1, ..., n$. Therefore, every U_i is multilinear in the strategy combination $\sigma \in \Theta$. In particular, all inequalities in (6.13) are *polynomial* inequalities. Hence, the equilibrium correspondence can be thought of as assigning solutions of polynomial inequalities to the (linear) coefficients $u \in \mathbb{R}^{nK}$.

Polynomials have very convenient properties. Recall that if a system of polynomial equations vanishes on an open set, then it vanishes everywhere. Similarly, recall that a (non-vanishing) polynomial of order r has at most r distinct roots (zeros). These simple observations give rise to an important insight about the structure of Nash equilibria for any given game. (For a proof of the following result see [97], Appendix B.)

Proposition 6.3 *For any finite game* $\Gamma = (S, u)$, *the set of all Nash equilibria consists of finitely many connected components.*

So, for all games the set of Nash equilibria "breaks" into finitely many (connected) pieces, each of which consists entirely of solutions to the system (6.13) of polynomial inequalities. Moreover, each of these components is a finite union of smooth manifolds. The latter follows from the fact that they are "semi-algebraic" sets.[10]

This, in a sense, says that Nash equilibrium as a solution concept is very selective. For every (finite) game, it singles out finitely many closed (because the best reply correspondence is u.h.c.) and connected subsets which constitute (jointly) the set of *all* equilibria.

Example 6.15 *The two-player game below has two components of Nash equilibria: the singleton set, which consists of the strict equilibrium* (s_1^2, s_2^1), *and a one-dimensional component* $\{\sigma \in \Theta \,|\, \sigma_1(s_1^1) = 1 \text{ and } 0 \leq \sigma_2(s_2^1) \leq 1/2\}$. *Figure 6.8 illustrates. The square is* Θ, *the thick broken lines indicate player 1's best replies, and the solid bold lines indicate player 2's best replies. Where the best replies intersect, the Nash equilibria are located. But note that, in general, higher dimensional components need not be linear, as happens in this example.*

	s_2^1		s_2^2	
s_1^1	0		1	
		1		1
s_1^2	1		0	
		1		0

---

[10] Semi-algebraic sets are sets defined by finitely many polynomial equalities and (strict) inequalities.

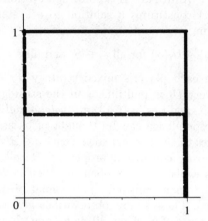

Figure 6.8: One-dimensional component

Proposition 6.3 does *not* necessarily say that for every game there are only finitely many equilibrium *payoffs*. But it comes close to that. In fact, for "almost all" games there are only finitely many equilibrium payoffs. The "almost all" here is very strong, because it refers to *extensive form* games.

In analogy to the definition of the space of games for a normal form, the space of all games consistent with a fixed extensive form can be defined. Fix the extensive form (tree, choices, and probabilities for chance moves) and vary the payoff function on plays. If there are only finitely many plays, the resulting space can again be identified with some Euclidean vector space. In fact, it will be a linear subspace of $\mathbb{R}^{nK}$ (see Section 4.4). The equilibrium correspondence E can now be restricted to this space of extensive form games.

Say that a property holds for *almost all* extensive form games if the closure of the set of games, for which the property is true, covers the whole space (of extensive form games) and its complement is of (Lebesgue) measure zero. (This is stronger than "almost all" on $\mathbb{R}^{nK}$, because the extensive form space is of lower dimension.[11]) Intuitively, if a property holds for almost all games and a game is drawn at random (from the space), then the game (that is drawn) will satisfy the property with probability one.

Now recall (from Section 3.3) that every (mixed) strategy combination $\sigma \in \Theta$ for an extensive form game gives rise to a probability distribution π_σ on plays. Call such a distribution π_σ an *outcome* of the game. The next result says that, from the continuum of possible outcomes (induced by mixed

---

[11]   In fact, this is a very strong notion of "almost all" even for a fixed space. For instance, the closure of the rational numbers covers the reals, but the rationals have measure zero in the reals (so, their complement has positive measure).

strategy combinations) Nash equilibrium singles out finitely many, for almost all extensive form games. (The proof is based on Sard's theorem (Theorem 5.8 in the appendix to Section 5.3); see [102], Appendix A3.)

Theorem 6.4 *For almost all finite extensive form games, the set of Nash equilibrium outcomes is finite.*

That is, for any finite extensive form, the set $\{\pi_\sigma \,|\, \sigma \in E(u)\}$ is finite, for almost all payoff functions u from the space of extensive form games. Intuitively, for almost all extensive form games, Nash equilibrium is extremely selective.

Combining this with Proposition 6.3 yields that for almost all extensive form games the (equilibrium) payoffs are *constant* across any connected component of Nash equilibria. This follows from continuity of the payoff function U and Theorem 6.4. (If payoffs varied over a connected component, they would give rise to a continuum of equilibrium payoffs.) Hence, from the possible continuum of payoffs (induced by mixed strategy combinations), Nash equilibrium almost always selects finitely many.

Corollary 6.1 *For almost all (extensive form) games, there are only finitely many Nash equilibrium payoffs.*

Hence, by Proposition 6.3 and Theorem 6.4, the graph $\mathcal{G}(E)$ (defined in (6.12)) of the Nash equilibrium correspondence "above" each game $u \in \mathbb{R}^{nK}$ breaks into finitely many connected pieces. And almost always payoffs are constant across each such piece. But the graph $\mathcal{G}(E)$ has even more structure.

In particular, it turns out that the graph $\mathcal{G}(E)$ is itself a connected set which pretty much "looks like" the space of games, i.e. like $\mathbb{R}^{nK}$. From this (which will formally be stated below) it follows that for every game there is at least one of the (finitely many) connected components (of Nash equilibria) which is such that every nearby game has a Nash equilibrium close to the component.

Essentiality The latter adds a *lower hemi-continuity* aspect to the properties of (parts of) the Nash equilibrium correspondence. For at least one component and every sequence of approximating games, there is an associated sequence of equilibria (for the approximating games) that converges to the component. (But beware: this does *not* say that the equilibrium correspondence is continuous.)

Definition 6.8 *For a (finite) game $\Gamma(u)$, a closed set of equilibria $C \subseteq E(u)$ is* **essential** *if for every $\varepsilon > 0$ there is some $\delta > 0$ such that, whenever $u' \in \mathbb{R}^{nK}$ is δ-close to u (i.e., $\|u - u'\| < \delta$), then there is an equilibrium $\sigma' \in E(u')$ for $\Gamma(u')$ that is ε-close to C (i.e., $\min_{\sigma \in C} \|\sigma - \sigma'\| < \varepsilon$).*

If the set C in this definition is a connected component of equilibria, it is called an *essential component*. If a single Nash equilibrium $\sigma \in E(u)$ satisfies the robustness property in Definition 6.8, i.e. if $C = \{\sigma\}$, it is called an *essential Nash equilibrium*. Every essential equilibrium is strictly perfect and, hence, perfect. This is so, because strategy trembles are specific payoff perturbations.

To see this, let $\eta = (\eta_i)_{i=1}^n = \left((\eta_i(s_i))_{s_i \in S_i}\right)_{i=1}^n \gg 0$ be a vector of strategy trembles with $\lambda_i(\eta_i) \equiv \sum_{s_i \in S_i} \eta_i(s_i) < 1$ for all i (see Definition 6.4). For all i and $\sigma_i \in \Delta_i$ let $r_i(\sigma_i, \eta_i) = (1 - \lambda_i(\eta_i))\sigma_i + \eta_i \in \text{int}(\Delta_i)$ (where the latter denotes the interior of Δ_i, i.e. the completely mixed strategies) such that the entries of the vector r_i are given by

$$r_i(s_i | \sigma_i, \eta_i) = (1 - \lambda_i(\eta_i))\sigma_i(s_i) + \eta_i(s_i) \text{ for all } s_i \in S_i. \qquad (6.14)$$

The vector $r_i(\sigma_i, \eta_i)$ is the (mixed) strategy actually implemented in the perturbed game (with trembles η) when player i decides for $\sigma_i \in \Delta_i$. Viewing each pure strategy as a vector (vertex) in Δ_i, one obtains, for every i, all $s_i \in S_i$ and all $\sigma_i \in \Delta_i$,

$$\sum_{s_i' \in S_i} \sigma_i(s_i') r_i(s_i | s_i', \eta_i) = [1 - \lambda_i(\eta_i) + \eta_i(s_i)]\sigma_i(s_i) \qquad (6.15)$$

$$+ \sum_{s_i' \in S_i \setminus \{s_i\}} \sigma_i(s_i')\eta_i(s_i) = r_i(s_i | \sigma_i, \eta_i).$$

Now define, for all i and all $s \in S$, with $r(\sigma, \eta) = (r_i(\sigma_i, \eta_i))_{i=1}^n \in \text{int}(\Theta)$ for all $\sigma \in \Theta$,

$$u_i(s | \eta) = U_i(r(s, \eta)) = \sum_{s' \in S} \prod_{j=1}^n r_j(s_j' | s_j, \eta_j) u_i(s') \qquad (6.16)$$

and observe that $u(\eta) = \left((u_i(s | \eta))_{s \in S}\right)_{i=1}^n$ converges to u as η goes to zero. Using (6.15) yields that

$$
\begin{aligned}
U_i(\sigma | \eta) &\equiv \sum_{s \in S} \prod_{j=1}^n \sigma_j(s_j) u_i(s | \eta) \\
&= \sum_{s' \in S} \left[\sum_{s \in S} \prod_{j=1}^n \sigma_j(s_j) r_j(s_j' | s_j, \eta_j) \right] u_i(s') \\
&= \sum_{s' \in S} \prod_{j=1}^n r_j(s_j' | \sigma_j, \eta_j) u_i(s') = U_i(r(\sigma, \eta))
\end{aligned}
$$

for all $\sigma \in \Theta$ and all i. Hence, playing the perturbed game $\Gamma(\eta)$ (see Definition 6.4) is equivalent to playing a game with (perturbed) payoffs $u(\eta)$. We have proved the following proposition.

Proposition 6.4 *For any finite game, every essential equilibrium is strictly perfect (and, hence, perfect).*

Since, moreover, essential equilibria are robust to *all* payoff perturbations, they seem a strong candidate for a refinement concept. But Example 6.9 from Section 6.2 can be used to show that there are games that do *not* have an essential equilibrium (see Exercise 6.21 below). So, essential equilibrium fails (at least for nongeneric games) the existence requirement on solutions.

Yet, it can be shown that every game does have an essential component of Nash equilibria. (A proof of this claim will follow from a result in the next section, which states that every component with non-zero index is essential. Since the index sum across all equilibrium components is +1, it follows that there always is a component with a non-zero index.)

Proposition 6.5 *Every finite game has an essential component of Nash equilibria.*

That is, applying the essentiality requirement (Definition 6.8) to the connected components of Nash equilibria from Proposition 6.3 restores existence. By Theorem 6.4, for almost all extensive form games an essential component will uniquely determine an outcome of the game, even if the supporting strategy combinations are not unique. So, for almost all relevant cases there is an essential equilibrium *outcome*.

Geometry of Nash Equilibria The original proof of Proposition 6.5 (see Kohlberg and Mertens [97], Appendix B) is based on a topological description of the graph $\mathcal{G}(E)$ of the Nash equilibrium correspondence. It states that $\mathcal{G}(E)$, compactified by adding a point "infinity", looks like a mildly deformed sphere around the sphere of games $\mathbb{R}^{nK}$ (similarly compactified). In short, $\mathcal{G}(E)$ is a (continuous) deformation of $\mathbb{R}^{nK}$.

Now if for some game there were not an essential component, then all pieces of $\mathcal{G}(E)$ above that game could be pulled away by a small deformation of $\mathcal{G}(E)$. But this would leave a hole in the (deformed) graph, contradicting that $\mathcal{G}(E)$ itself is a twisted copy of the space of games $\mathbb{R}^{nK}$. If you like, "drilling a hole" into a sphere is not a continuous deformation.

In the appendix to this section a result will be proved which ensures that $\mathcal{G}(E)$ is a deformation of $\mathbb{R}^{nK}$. For the moment it will be illustrated how such a "topological" description of $\mathcal{G}(E)$ works and what it means for the simplest possible game.

Example 6.16 *Consider the one-player game with two pure strategies. By expected utility, for each game $(u_1(s_1^1), u_1(s_1^2)) \in \mathbb{R}^2$ a constant can be subtracted from both payoffs without affecting Nash equilibria. By subtracting the*

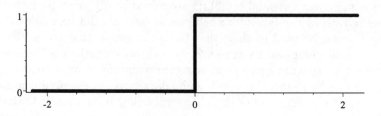

Figure 6.9: Graph of the equilibrium correspondence

second coordinate from the first, we obtain the following representation:

s_1^1	u
s_1^2	0

This is sufficient to identify the space of games with the real line, i.e. $u \in \mathbb{R}$. Clearly, if $u > 0$ (resp. $u < 0$) the player will optimally choose $\sigma_1\left(s_1^1\right) = 1$ (resp. $\sigma_1\left(s_1^1\right) = 0$). If $u = 0$, then any mixed strategy is optimal. Therefore, the graph of the Nash equilibrium correspondence is as shown in Figure 6.9. The space of games is the horizontal axis and Θ is represented by the vertical unit interval which depicts the value of $\sigma = \sigma_1\left(s_1^1\right) \in [0,1]$. Note that for all $u \in \mathbb{R}$ the unique Nash equilibrium component is essential.

The graph of the equilibrium correspondence in the example (Figure 6.9) is closed, connected, and is the union of three (linear) one-dimensional manifolds (with boundary).[12] Hence, it has the same dimension as the space of games. And by "stretching and twisting" it can be continuously deformed into a line. In fact, if appropriately stretched, $\mathcal{G}(E)$ will look like the graph of a continuous function. (This is what Theorem 6.6(b) in the appendix shows.) Surprisingly, these properties generalize to arbitrary finite games.

The precise statement of the fundamental result on the geometry of the Nash equilibrium correspondence is as follows. Let $q : \mathcal{G}(E) \to \mathbb{R}^{nK}$ be the projection map from the graph to the space of games, defined by $q(u, \sigma) = u$. (This is clearly a continuous mapping.)

---

[12]   On the notion of a *manifold*, see the appendix to Section 5.3.

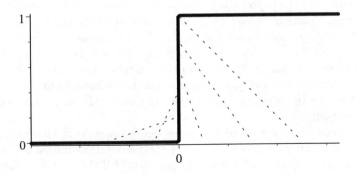

Figure 6.10: How the homeomorphism maps to the line

Theorem 6.5 *There exists a homeomorphism* $\phi : \mathbb{R}^{nK} \to \mathcal{G}(E)$ *such that the projection of* ϕ *to the space of games, i.e.,* $q \circ \phi$, *is homotopic to the identity on* $\mathbb{R}^{nK}$ *under a homotopy that extends to the one-point compactification of* $\mathbb{R}^{nK}$ *by continuity.*

This theorem will now be explained in detail. First, a *homeomorphism* is a continuous function which is onto and one-to-one and has a continuous inverse.[13] If there is a homeomorphism between two sets, they are *homeomorphic*. This means that one set can be continuously deformed into the other. In this sense homeomorphic sets are topologically equivalent. So, the graph of the equilibrium correspondence $\mathcal{G}(E)$ and the space of games $\mathbb{R}^{nK}$ are homeomorphic (via the homeomorphism ϕ) and, thus, topologically "look alike".

This implies that the graph $\mathcal{G}(E)$ has no "fat" parts; i.e., there is no open set of games (in $\mathbb{R}^{nK}$) such that all games in this set have an equilibrium component which contains the same set (in Θ) with positive dimension. This indicates that Nash equilibrium components of *positive* dimension (as allowed for in Proposition 6.3) are somewhat exceptional. They result from "folds" in the graph $\mathcal{G}(E)$ which otherwise looks like the space of games.

For Example 6.16, the homeomorphism ϕ boils down to

$$\phi(u) = \left(u + 1 - 2 \max \left\{ 0, \min \left\{ 1, \frac{u+1}{2} \right\} \right\}, \max \left\{ 0, \min \left\{ 1, \frac{u+1}{2} \right\} \right\} \right),$$

---

[13] Recall: a mapping is *onto* (surjective) if the image of the domain is the whole range. It is *one-to-one* (injective) if all points in the domain have distinct images.

for all $u \in \mathbb{R}$, with continuous inverse $\phi^{-1}(u,\sigma) = u+\sigma$, for all $(u,\sigma) \in \mathcal{G}(E)$. Indeed, $u \leq -1$ implies $\phi(u) = (u+1,0) \in \mathcal{G}(E)$, $-1 < u < 1$ implies $\phi(u) = (0,(u+1)/2) \in \mathcal{G}(E)$, and $u \geq 1$ implies $\phi(u) = (u-1,1) \in \mathcal{G}(E)$.

In Figure 6.10 the payoff u is on the horizontal and the strategy σ is on the vertical axis. Points u smaller than -1 are moved to the right by one payoff unit, with strategy coordinate $\sigma = 0$. Points larger than 1 are moved to the left by one payoff unit, with strategy coordinate $\sigma = 1$. Points $u \in [-1,1]$ are mapped along the thin broken lines to the strategy coordinate at zero.

Furthermore, the payoff coordinate of the homeomorphism ϕ (i.e., $q \circ \phi$) is *homotopic* to the identity on the space of games (the map that takes any game into itself).

"Homotopy" is a notion of two maps being *equivalent*. That one map is "homotopic" to another says that one is a continuous deformation of the other. A great many properties of a map are not altered if the map is deformed in a continuous way.[14] Abstractly, two functions g_0 and g_1 from X to Y are *homotopic* if there is a continuous function (the *homotopy*) $G: X \times [0,1] \to Y$ such that $G(x,0) = g_0(x)$ and $G(x,1) = g_1(x)$ for all $x \in X$.

Hence, while the "payoff coordinate" of the homeomorphism ϕ is not directly "above" the argument u, it can be continuously deformed in such a way that it is. For Example 6.16 the homotopy is given by

$$\Phi(u,t) = u + 1 - t - 2(1-t)\max\left\{0, \min\left\{1, \frac{u+1}{2}\right\}\right\},$$

for all $(u,t) \in \mathbb{R} \times [0,1]$, so that $\Phi(u,0) = q(\phi(u)) = \phi_1(u)$ and $\Phi(u,1) = u$, for all $u \in \mathbb{R}$, as required.

Finally, the homotopy Φ extends to the one-point compactification of the space of games by continuity. This ensures that the homotopy maps points "near infinity" in its domain to points "near infinity" in its range.[15] So, in a sense, it is continuous even "at infinity". Of course, this is an obscure statement, because infinity does not belong to the real numbers. To make sense, it requires a way of adding "infinity" to the domain. The one-point compactification does that. It consists of adding a single point "infinity". This turns the space of games into an nK-dimensional sphere (which is a compact manifold). To extend the projection q and the homotopy Φ to this added point, simply set $q(\infty) = (\infty)$ and $\Phi(\infty,t) = (\infty)$ for all $t \in [0,1]$.

In Example 6.16 the one-point compactification is easiest to understand. There the space of games is the real line. Now bend the line so as to bring its two (open) ends ($\pm\infty \notin \mathbb{R}$) together and, where they are about to touch,

---

[14] In the real world no functional relationship is ever perfectly determined. The only meaningful properties of a mapping, consequently, are those that remain valid when the map is slightly deformed. Such properties are called *homotopy stable* properties.

[15] Technically, such a map is called "proper". (Do not confuse with proper Nash equilibrium.) A function is *proper* if the preimage of every compact set (in the range) is compact (in the domain).

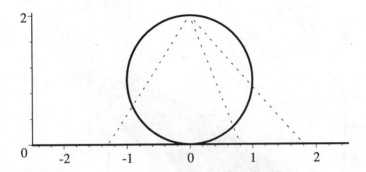

Figure 6.11: One-point compactification

"glue" a point called "infinity" (∞). The resulting "glued" object looks like a circle (the 1-sphere) which is indeed a compact manifold. Figure 6.11 illustrates the procedure. Each point on the real line is mapped along a (dashed) ray, through the point and the "north pole" of the circle, to where the ray first intersects the circle. The "north pole" is the added point "infinity".

Once the real line is compactified, the homotopy Φ can be conceived as a mapping from the (unit) circle to itself. Then, continuity "at infinity" boils down to continuity at a particular point of the circle. For instance, parametrize the circle by the angle x (in radians) between the lines from the circle's center to the image of $u = 2$ and any point on the circle. Then, each point on the circle (and, thereby, each $u \in \mathbb{R} \cup \{\infty\}$) can be identified with some angle $x \in [-\pi, \pi]$; e.g., $u = -2$ is represented by $x = \pm\pi = \pm 3.1416$, $u = 0$ is represented by $x = -\pi/2$, $u = 2$ is represented by $x = 0$, and $u = \infty$ is represented by $x = \pi/2$. (Beware: the points $x = \pi$ and $x = -\pi$ are the *same* point on the circle.) With this parametrization, Figure 6.12 visualizes continuity of the homotopy Φ on the compactified space of games for Example 6.16.

In Figure 6.12 the angle $x \in [-\pi, \pi]$ (which parametrizes the one-point compactification of the real line) is on the horizontal and the vertical axes. On the third axis is the homotopy parameter $t \in [0, 1]$. The sloping surface is the (graph of the) homotopy Φ. The deformation for small values of t is

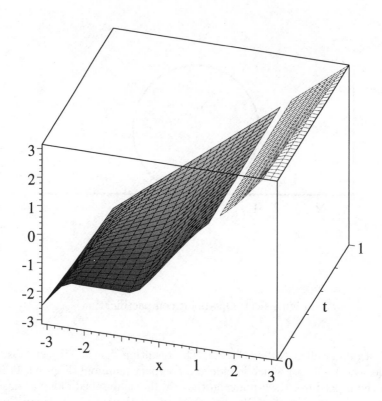

Figure 6.12: Continuity "at infinity"

because here Φ is close to $q \circ \phi$. As t grows, the deformation disappears and Φ approximates the diagonal, i.e. the identity. Potentially, a continuity problem could arise at the image of "infinity", $x = \pi/2 = 1.5708$. But visual inspection of Figure 6.12 shows that Φ is continuous even "at infinity" (see the little "gap" in the surface).

Thus, we conclude that the projection from $\mathcal{G}(E)$ to the space of games behaves like the identity not only *locally*, but also *globally*. The space of games can now be identified with a compact manifold (a sphere), so "globally" has an unambiguous meaning. To appreciate this point, it is instructive to consider the following (counter)example.

Example 6.17 *Compare the Nash equilibrium correspondence E for Example 6.16 with a correspondence* $A : \mathbb{R} \to [0,1]$ *which is* empty-valued *for* $u < 0$, *assigns to* $u = 0$ *the full unit interval* $[0,1]$, *and to each* $u > 0$ *the two*

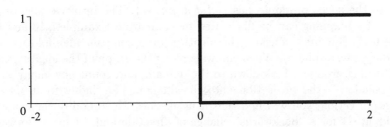

Figure 6.13: Correspondence with global degree zero

values, $\sigma = 0$ and $\sigma = 1$. Figure 6.13 illustrates (with u on the horizontal and σ on the vertical axis). This correspondence is also u.h.c. and its graph $\mathcal{G}(A) = \{(u, \sigma) \in \mathbb{R} \times [0, 1] \,|\, \sigma \in A(u)\}$ is also homeomorphic to the line $\mathbb{R}$. The homeomorphism for A is given by

$$\varphi(u) = (|u| - \max\{0, \min\{u, 1\}\}, \max\{0, \min\{u, 1\}\})$$

with (continuous) inverse $\varphi^{-1}(u, \sigma) = \sigma + u$ if $\sigma > 0$ and $\varphi^{-1}(u, \sigma) = -u$ if $\sigma = 0$.

But now there is no proper homotopy. Suppose there is a homotopy $\Psi : \mathbb{R} \times [0, 1] \to \mathbb{R}$ such that $\Psi(u, 0) = q(\varphi(u)) = \varphi_1(u)$ and $\Psi(u, 1) = u$ for all $u \in \mathbb{R}$. (Indeed there exists such a homotopy, but what will be shown is that it cannot extend continuously to the one-point compactification.) Let $\{u_k\}_{k=1}^{\infty}$ be a sequence from $\mathbb{R}_-$ which diverges to $-\infty$. Since $u_k < 0$ for all $k = 1, 2, ...,$ $\Psi(u_k, 0) = q(\varphi(u_k)) = |u_k| > 0$ and $\Psi(u_k, 1) = u_k < 0$ for all k. Since Ψ is continuous in $t \in [0, 1]$, for each k there exists some $t_k \in [0, 1]$ such that $\Psi(u_k, t_k) = 0$. Because t_k is from the compact interval $[0, 1]$, it can be assumed (by extracting a subsequence, if necessary) that $t_k \to_{k\to\infty} t_0 \in [0, 1]$. Hence, $(u_k, t_k) \to_{k\to\infty} ((\infty), t_0)$, but $\Psi(u_k, t_k) = 0$ does not converge to $\Psi((\infty), t_0) = (\infty)$. Intuitively, the argument $u_k < 0$ of Ψ and its value $|u_k| > 0$ run off in different directions as u_k diverges. Therefore, Ψ cannot extend to the one-point compactification by continuity.

That the projection from $\mathcal{G}(E)$ to the space of games is *globally* homotopic to the identity is like an implicit existence theorem of Nash equilibrium for all games. It says that the projection from $\mathcal{G}(E)$ to $\mathbb{R}^{nK}$ is globally *onto* (surjective). So, "above" each point $u \in \mathbb{R}^{nK}$ in the space of games there is a point in the graph $\mathcal{G}(E)$ of the Nash equilibrium correspondence.

Technically, this second part of Theorem 6.5 says that the projection from $\mathcal{G}(E)$ to the space of games has *global degree* $+1$. The intuitive idea of the degree of a mapping can be illustrated by comparing Example 6.16 and Example 6.17. But first, consider the identity (on a compact manifold without boundary, say, a sphere). When you walk along the graph of the identity without "turning around", look down to the domain, and count how many times your "shadow" (the projection onto the domain, i.e. "lights from above") changes direction, you will find that it never does. Now modify the counting by adding -1 for a "backwards" change of direction and $+1$ for a "forward" change, starting with $+1$ for the initial "forward" movement.[16] Perform the same exercise on the graph of some other mapping and call the (modified) sum of direction change the "degree" of the mapping.

If you do that on the graph of the Nash equilibrium correspondence for Example 6.16 (Figure 6.8), your (modified) count of direction changes will give $+1$ (a single forward movement), i.e. precisely the same as for the identity. If you do it on the graph of A from Example 6.17 (Figure 6.12), you will find that your "shadow" reverses direction once (while you do not) when you cross the image of $u = 0$. This adds -1, so the "degree" of A is 0.

In this low-dimensional environment geometric intuition should convince you that with "degree" 0 your "shadow" will *not* visit the whole domain, even if your own journey takes you along the whole graph of A. By contrast, with "degree" $+1$ your "shadow" *will* meet the whole domain, as long as you do the whole graph. Hence, with "degree" $+1$ the projection (from the graph to the domain) is onto (surjective)—which says that above every point of the domain *there is* an image point.

This intuitive idea can be generalized and the (global) *degree* can be defined for arbitrary (continuously differentiable) functions from a compact manifold without boundary to a connected manifold with the same dimension. Moreover, a result known as the *homotopy lemma* (see e.g. [121], p. 28) states that any two such maps that are homotopic have the same degree. These results are what the second part of Theorem 6.5 implicitly refers to.

Summary

- The Nash equilibrium correspondence assigns to every game the set of all its Nash equilibria. It is an upper hemi-continuous correspondence.

---

[16]   This, of course, assumes that "direction" is defined. Technically this is done by introducing an *orientation* on a manifold. If a manifold is at all *orientable*, then there are precisely two ways to do this.

- For every finite game, the set of Nash equilibria consists of finitely many connected components.

- For almost all finite extensive form games, the set of Nash equilibrium outcomes is finite.

- A Nash equilibrium component is essential if every nearby game has an equilibrium close to the component. Every finite game has an essential component of Nash equilibria.

- The graph of the Nash equilibrium correspondence is a twisted, but unknotted, copy of the space of games and projects surjectively onto the latter.

Appendix

This Appendix is devoted to an extension of Theorem 6.5, due to Demichelis and Germano [45], and, thereby, provides a proof of it (statement (c) below). The extension concerns the following. The graph of a correspondence may be a twisted copy of the space of games, yet it may contain a "knot". In such a case the homeomorphism that stretches the graph to the underlying space will not extend to the ambient space (games times strategies), because if it did, it would "tear apart" the ambient space when "unknotting" the graph. The following theorem shows that this is *not* so for the graph of the Nash equilibrium correspondence; i.e., $\mathcal{G}(E)$ is *unknotted*.

This implies (statement (b) below) that the graph of the Nash equilibrium correspondence is a continuous deformation of the graph of a continuous function (from games to strategies). A solution concept that induces a continuous function (from games to strategies) would, of course, be the bliss point from the viewpoint of theory. But the Nash equilibrium correspondence is not a continuous function. Yet, its graph looks (topologically) *as if* it would be a continuous function, at least after some mild plastic surgery.

Technically, the following theorem shows that the homotopy from Theorem 6.5 is really an "isotopy" for the ambient space. An *isotopy* is a homotopy that maps, for each fixed value of the parameter $t \in [0,1]$, the domain homeomorphically onto its range. Finally, a function is *proper* if the preimage of every compact set (in the range) is compact (in the domain). Properness is equivalent to the existence of an extension to the one-point compactification by continuity.

Theorem 6.6 *There exists a continuous function $f : \mathbb{R}^{nK} \to \Theta$ and an isotopy $H : \mathbb{R}^{nK} \times \Theta \times [0,1] \to \mathbb{R}^{nK} \times \Theta$ such that*

(a) H_1 is the identity on $\mathbb{R}^{nK} \times \Theta$,

(b) $\mathcal{G}(E) = H_0^{-1}(\mathcal{G}(f))$, and

(c) the projection of $H_t^{-1} \circ F$ onto $\mathbb{R}^{nK}$ is homotopic to the identity on $\mathbb{R}^{nK}$ under a proper homotopy, for all $t \in [0,1]$,

where $H_t(u,\sigma) \equiv H(u,\sigma,t)$, for all $(u,\sigma) \in \mathbb{R}^{nK} \times \Theta$, is the evaluation map for the isotopy H, for each $t \in [0,1]$, $\mathcal{G}(f)$ is the graph of f, and $F : \mathbb{R}^{nK} \to \mathbb{R}^{nK} \times \Theta$ assigns the point in $\mathcal{G}(f)$ above u, i.e. $F(u) = (u, f(u))$, for all $u \in \mathbb{R}^{nK}$.

Proof Let $K_i = |S_i|$ be the number of (pure) strategies for player $i = 1,...,n$ and $K = \prod_{i=1}^{n} K_i = |S|$ be the number of (pure) strategy combinations. For each $i = 1,...,n$ and each $u \in \mathbb{R}^{nK}$, let $w_i(u) = (w_i(s_i|u))_{s_i \in S_i} \in \mathbb{R}^{K_i}$ with

$$w_i(s_i|u) = \frac{K_i}{K} \sum_{s_{-i} \in S_{-i}} u_i(s_{-i}, s_i), \text{ for all } s_i \in S_i, \tag{6.17}$$

the average payoff from strategy s_i, and

$$v_i(s|u) = u_i(s) - w_i(s_i|u), \text{ for all } s \in S, \text{ and} \tag{6.18}$$
$$V_i(\sigma_{-i}, s_i|u) = \sum_{s'_{-i} \in S_{-i}} v_i(s'_{-i}, s_i|u) \prod_{j \neq i} \sigma_j(s'_j)$$

for all $s_i \in S_i$ and all $\sigma \in \Theta$. Set $w(u) = (w_i(u))_{i=1}^{n} \in \mathbb{R}^{\sum_i K_i}$. These definitions imply that, for all i and all $\sigma \in \Theta$,

$$U_i(\sigma_{-i}, s_i) = w_i(s_i|u) + V_i(\sigma_{-i}, s_i|u) \text{ for all } s_i \in S_i, \tag{6.19}$$

where U_i denotes the extension of u_i to mixed strategies. Next, define the function $x : \mathbb{R}^{\sum_i K_i} \to \Theta$ by $x(y) = (x_i(y_i))_{i=1}^{n}$, where $x_i(y_i) = (x_i(s_i|y_i))_{s_i \in S_i}$ for all $y_i = (y_i(s_i))_{s_i \in S_i} \in \mathbb{R}^{K_i}$ is defined by

$$x_i(s_i|y_i) = \max\{0, y_i(s_i) - z_i(y_i)\}, \text{ for all } s_i \in S_i, \text{ with} \tag{6.20}$$

$$z_i(y_i) = \min\left\{ z \in \mathbb{R} \ \middle| \ \sum_{s_i \in S_i} \max\{0, y_i(s_i) - z\} \leq 1 \right\}$$

for all $i = 1,...,n$. Since $\sum_{s_i \in S_i} \max\{0, y_i(s_i) - z\}$ is a continuous piecewise linear non-increasing function of z (which is strictly decreasing as long as it is positive),

$$\sum_{s_i \in S_i} \max\{0, y_i(s_i) - z_i(y_i)\} = 1. \tag{6.21}$$

Together with $x_i(s_i|y_i) \geq 0$, for all $s_i \in S_i$, this implies $x_i(y_i) \in \Delta_i$ for all $y_i \in \mathbb{R}^{K_i}$. If $y_i \in \Delta_i$, then $z_i(y_i) \geq 0$, because otherwise the sum exceeds 1, so that $\sum_{s_i \in S_i} \max\{0, y_i(s_i)\} = 1$ implies $z_i(y_i) = 0$ and $x_i(y_i) = y_i$ for all $y_i \in \Delta_i$; i.e., on Δ_i the function x_i is the identity. Other vectors $y_i \in \mathbb{R}^{K_i} \setminus \Delta_i$ are projected in a continuous piecewise linear fashion onto Δ_i. In fact, using the Kuhn-Tucker conditions, it can be shown that $x_i(y_i) = \arg\min_{x \in \Delta_i} \|x - y_i\|$.

To define the continuous function $f : \mathbb{R}^{nK} \to \Theta$, let $f(u) = (f_i(u))_{i=1}^n$ be given by $f_i(u) = x_i(w_i(u))$ for all $u \in \mathbb{R}^{nK}$ and all $i = 1, ..., n$. This is continuous as the composition of two continuous functions, $f_i = x_i \circ w_i$. Note that all these functions are defined player-wise. (The argument u is for convenience.)

The mapping $H : \mathbb{R}^{nK} \times \Theta \times [0, 1] \to \mathbb{R}^{nK} \times \Theta$ is defined via its evaluation maps $H_t : \mathbb{R}^{nK} \times \Theta \to \mathbb{R}^{nK} \times \Theta$ for each $t \in [0, 1]$. For each $(u, \sigma) \in \mathbb{R}^{nK} \times \Theta$, let $H_t(u, \sigma) = (h_t(u, \sigma), \sigma)$ where

$$h_t(u, \sigma) = \left[[u_i(s) + (1 - t)(\sigma_i(s_i) + V_i(\sigma_{-i}, s_i | u))]_{s \in S} \right]_{i=1}^n . \qquad (6.22)$$

Similarly, define $H_t^{-1} : \mathbb{R}^{nK} \times \Theta \to \mathbb{R}^{nK} \times \Theta$ by $H_t^{-1}(u, \sigma) = (h_t^{-1}(u, \sigma), \sigma)$ for all $(u, \sigma) \in \mathbb{R}^{nK} \times \Theta$ where

$$h_t^{-1}(u, \sigma) = \left[[u_i(s) - (1 - t)(\sigma_i(s_i) + V_i(\sigma_{-i}, s_i | u))]_{s \in S} \right]_{i=1}^n . \qquad (6.23)$$

To show that H_t and H_t^{-1} are inverses of each other, it suffices to show that for each fixed $\sigma \in \Theta$ the functions h_t and h_t^{-1} are inverses of each other. To see this, observe that for all i and all $s_i \in S_i$ from (6.17), (6.22), and (6.23),

$$w_i(s_i | h_t(u, \sigma)) = w_i(s_i | u) + (1 - t)(\sigma_i(s_i) + V_i(\sigma_{-i}, s_i | u)),$$
$$w_i(s_i | h_t^{-1}(u, \sigma)) = w_i(s_i | u) - (1 - t)(\sigma_i(s_i) + V_i(\sigma_{-i}, s_i | u)),$$

so from (6.18) $v_i(s | h_t(u, \sigma)) = v_i(s | u) = v_i(s | h_t^{-1}(u, \sigma))$ for all $s \in S$. Hence, $V_i(\sigma_{-i}, s_i | h_t(u, \sigma)) = V_i(\sigma_{-i}, s_i | u) = V_i(\sigma_{-i}, s_i | h_t^{-1}(u, \sigma))$ for all $\sigma \in \Theta$, all $s_i \in S_i$, and all $i = 1, ..., n$ implies that $h_t^{-1}(h_t(u, \sigma), \sigma) = u$ and $h_t(h_t^{-1}(u, \sigma), \sigma) = u$ for all $u \in \mathbb{R}^{nK}$.

Since continuity is evident (in fact, H_t is smooth for all $t \in [0, 1]$), it follows that for each fixed $t \in [0, 1]$ the evaluation map H_t is a homeomorphism, so H is the isotopy in the statement. By (6.22), at $t = 1$ the function H_1 is the identity on $\mathbb{R}^{nK} \times \Theta$, i.e. $H_1(u, \sigma) = (u, \sigma)$ for all $(u, \sigma) \in \mathbb{R}^{nK} \times \Theta$, which demonstrates statement (a).

To see (b), let $u \in \mathbb{R}^{nK}$ and denote $u' = h_0^{-1}(u, f(u))$. For all i and all $s_i \in S_i$, we have from (6.17)

$$w_i(s_i | u') = w_i(s_i | u) - x_i(s_i | w_i(u)) - V_i(x_{-i}(w_{-i}(u)), s_i | u),$$
$$v_i(s | u') = u_i(s) - w_i(s_i | u) = v_i(s | u) \text{ for all } s \in S.$$

Hence, $w_i(s_i | u') + V_i(x_{-i}(w_{-i}(u)), s_i | u') = w_i(s_i | u) - x_i(s_i | w_i(u))$ is i's expected payoff from using s_i against $x_{-i}(w_{-i}(u)) \in \Theta_{-i}$, for all $s_i \in S_i$, and

$$\sum_{s_i \in S_i} x_i(s_i | w_i(u)) [w_i(s_i | u) - x_i(s_i | w_i(u))] = z_i(w_i(u))$$

is i's expected payoff from $f(u) = x\,(w(u))$, because $x_i\,(w_i(u)) \in \Delta_i$. Since from (6.20) $x_i\,(s_i\,|w_i(u)) > 0$ if and only if $w_i\,(s_i\,|u) > z_i\,(w_i(u))$, we have $w_i\,(s_i\,|u) - x_i\,(s_i\,|w_i(u)) \leq z_i\,(w_i(u))$ for all $s_i \in S_i$ and all i, from (6.20); i.e., $f(u) = x\,(w(u)) \in \Theta$ is a Nash equilibrium for the game with payoff vector $u' = h_0^{-1}\,(u, f(u))$. Hence, $H_0^{-1}\,(\mathcal{G}(f)) \subseteq \mathcal{G}(E)$.

Conversely, let $(u, \sigma) \in \mathcal{G}(E)$ and denote now $u' = h_0(u, \sigma)$. We show that $\sigma = f(u')$. For all i and $s_i \in S_i$ we have $w_i\,(s_i\,|u') = \sigma_i(s_i) + U_i\,(\sigma_{-i}, s_i)$ from (6.22), (6.17), and (6.19). Since $(u, \sigma) \in \mathcal{G}(E)$, equation (6.21) implies that $z_i\,(w_i(u')) = U_i(\sigma)$ and, therefore,

$$ x_i\,(s_i\,|w_i(u')) = \max\{0, \sigma_i(s_i) + U_i(\sigma_{-i}, s_i) - U_i(\sigma)\} = \sigma_i(s_i) $$

for all $s_i \in S_i$ and all i. Hence, $\sigma = f(u')$ and, consequently, for each $(u, \sigma) \in \mathcal{G}(E)$ there is $u' \in \mathbb{R}^{nK}$ such that $(u, \sigma) = H_0^{-1}\,(u', f(u'))$, so that $\mathcal{G}(E) \subseteq H_0^{-1}\,(\mathcal{G}(f))$. This completes the proof of statement (b).

Note that $H_0^{-1} \circ F = \phi$ is precisely the homeomorphism introduced by Kohlberg and Mertens [97] for Theorem 6.5, i.e. $\phi(u) = H_0^{-1}\,(u, f(u)) = H_0^{-1}\,(F(u))$ for all $u \in \mathbb{R}^{nK}$. They show that $h_t^{-1} \circ F$ extends to the one-point compactification of $\mathbb{R}^{nK}$ by continuity (see [97], p. 1022). Consequently, $h_t^{-1} \circ F$ is a proper homotopy with $h_1^{-1}\,(F(u)) = h_1^{-1}\,(u, f(u)) = u$, for all $u \in \mathbb{R}^{nK}$, by (6.23). Introducing a function from $\tau \in [0, 1]$ to the unit interval which takes the value 0 for all $\tau \leq t < 1$ and the value 1 for $\tau = 1$ (see Exercise 6.23 below), this implies for all $t \in [0, 1]$ that $h_t^{-1} \circ F$ is homotopic to the identity under a proper homotopy, viz. statement (c).

This completes the proof of Theorem 6.6.

The isotopy H actually operates only on the mean payoffs w_i and leaves the deviations v_i from the mean unchanged. Combine this with the observation that strategy trembles can be perceived as particular payoff perturbations (see (6.16) in this section). Then it is clear that a similar result holds on the space of *strategy perturbations* for every fixed game. That is, the graph of the equilibrium correspondence on strategy perturbations for a fixed game also has nice properties.

Exercises

Exercise 6.21 Show that the game from Example 6.9 does not have an essential Nash equilibrium.

Exercise 6.22 Show that any strict Nash equilibrium is essential.

Exercise 6.23 Show that homotopy is an equivalence relation. (Hint: To prove transitivity in the smooth case the following smooth functions on the reals will be useful. Let $g(x) = e^{-1/x}$ for all $x \geq 0$ and $g(x) = 0$ for all $x < 0$ and define

$$\rho(x) = g\,(x - c_1)\,/\,[g\,(x - c_1) + g(c_2 - x)]$$

for any two real constants c_1, c_2 with $c_1 < c_2$.)

Exercise 6.24 Consider the two-player game below. Show that the six strategy combinations (s_1^1, s_2^1), (s_1^3, s_2^1), (s_1^3, s_2^2), (s_1^2, s_2^2), (s_1^2, s_2^3), and (s_1^1, s_2^3) all belong to a single connected component of Nash equilibria.

	s_2^1	s_2^2	s_2^3
s_1^1	3 3	2 1	1 3
s_1^2	1 2	2 2	1 2
s_1^3	3 1	2 1	0 0

Exercise 6.25 Determine the full set of all Nash equilibria for the following extensive form game with two players. Two players have to share 5 dollars (which come in 1 dollar bills). Player 1 can initially make a "fair" offer, namely 3 dollars for herself and 2 dollars for player 2. Such a fair offer will be accepted by player 2 with certainty. Alternatively, player 1 may take the risk of claiming 4 dollars for herself, leaving only 1 dollar for player 2. If she does so, it is then the turn of player 2, who can either accept the proposed division or reject it. If $(4, 1)$ is accepted, it is implemented and the game ends. If $(4, 1)$ is rejected, then it becomes again player 1's turn, and she can now either "accommodate" or "take revenge". In both cases player 1 earns 2 dollars. But if player 1 "accommodates", player 2 also gains 2 dollars. If player 1 "takes revenge", player 2 obtains nothing. (Assume that monetary payoffs accurately reflect preferences.) Is this the kind of game alluded to in Theorem 6.4?

Exercise 6.26 Draw the graph of the Nash equilibrium correspondence for the simple two-player games below on the subspace defined by $-1 < x < 1$. This will require a three-dimensional picture:

	s_2^1	s_2^2
s_1^1	x 1	0 0
s_1^2	0 0	1 1

Exercise 6.27 Draw the Nash equilibrium correspondence of the games below which are parametrized by $x \in R$. Note that all Nash equilibria of these

games are symmetric, for all $x \in R$. Therefore, replacing Θ by the unit interval, so as to plot the graph as a two-dimensional picture, will be sufficient.

	s_2^1	s_2^2
s_1^1	x $\quad$ x	0 $\quad$ 0
s_1^2	0 $\quad$ 0	$1-x$ $\quad$ $1-x$

Notes on the Literature: The approach viewing Nash equilibrium as a correspondence on the space of payoff vectors was pioneered by Harsanyi [81]. Blume and Zame [25] exploit the polynomial structure of the Nash equilibrium correspondence and show that it is a semi-algebraic mapping. Proposition 6.3 is due to Kohlberg and Mertens [97], who also proved Theorem 6.5 and introduced the homeomorphism extended in the appendix. (Exercise 6.24 is also from this source.) Theorem 6.4 is due to Kreps and Wilson [102]. The concept of an essential equilibrium point in game theory was introduced by Wu Wen-Tsün and Jiang Jia-He [183]. Its extension to compact sets of Nash equilibria was first mentioned by Jiang Jia-He [92] and is also discussed by van Damme ([37], p. 266) and (implicitly) by Kohlberg and Mertens [97]. In mathematics the notion of an essential fixed point was introduced by Fort [61] and extended to compact sets of fixed points by Kinoshita [95] and O'Neill [137] (see McLennan [111]). Theorem 6.6 is due to Demichelis and Germano [45]. A good introduction to homotopy theory is given e.g. by Guillemin and Pollack ([77], Section 1.6). Degree theory is discussed e.g. by Milnor ([121], Chapter 5).

6.5 Index Theory

Most refinements are *robustness* criteria, either with respect to strategy per-turbations or with respect to behavior in unreached parts of the game tree. When strategy trembles are considered, the philosophy seems to be that we trust our specification of the game, but not necessarily the players' ability to implement the intended strategy. But, why this particular robustness test? Maybe there are more general robustness notions available.

And indeed, there are. (But beware: requiring more robustness will, in general, produce larger solution sets.) In mathematics such notions of robust-ness go under such names as "regularity", "essentiality", or "transversality". These names basically refer to properties that cannot be destroyed by small perturbations of the underlying equation system.

These concepts allow for a *comparison* of, say, the various solutions to a given system of equations. Such an approach, in principle, should be applicable also to Nash equilibria, since the last section showed that the Nash equilibria of a given game are little less than the solutions to a system of equations (or inequalities). An abstract but simple example will illustrate these ideas.

Example 6.18 *Consider a smooth function $g : [0,1] \rightarrow \mathbb{R}$ which satisfies $g(0) > 0$ and $g(1) < 0$. By continuity there must be at least one "zero", i.e., a point $\bar{p} \in (0,1)$ such that $g(\bar{p}) = 0$. Figure 6.14 illustrates a case with three zeros (the bold curve). To produce a zero, the graph of g must intersect the (horizontal) p-axis. By the boundary assumptions it does so globally "from above to below". Yet, at the middle zero it locally cuts the p-axis "from below to above"; i.e., in the reverse direction. Moreover, it is easy to imagine that a (sufficiently strong) deformation of g will erase the middle zero, where g is "perversely" sloped ($\partial g/\partial p > 0$), so as to leave only a single zero, where g is "correctly" ($\partial g/\partial p < 0$) sloped (see the broken curve in Figure 6.14). Furthermore, if g defined a dynamics on the unit interval by $dp/dt = g(p(t))$ for "time" $t \geq 0$, then it is easy to see that the rightmost and leftmost zero would be stable, while the middle zero would be unstable. Hence, it makes intuitive sense to assign to any zero where g is negatively sloped an "index" $+1$ and to any zero where g is positively sloped an "index" -1. From the boundary assumptions, it is intuitively clear that the "index" sum across all zeros must be a constant, namely $+1$. This is so because to go to the other side of a river you must cross it an odd number of times. Hence, this assignment of an "index" to zeros of g provides a natural comparison among them.*

The difference of this example to Nash equilibria in games is that the latter do not have a differentiable structure. Yet, this is a minor technical

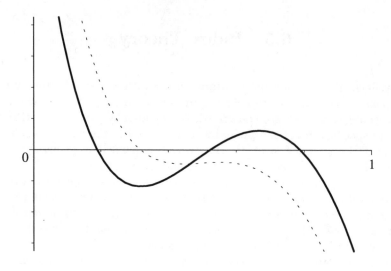

Figure 6.14: Function with three zeros

obstacle, not one of substance. Since, according to the findings in the last section, Nash equilibria are governed by polynomial inequalities (i.e., Nash equilibrium components are semi-algebraic sets), they have enough structure to apply reasoning from a differentiable viewpoint.

The point of this section, therefore, is to introduce an index assignment for the (connected) components of Nash equilibria in a given game (see Proposition 6.3). Such an index assignment provides a *classification* of equilibrium components that allows a comparison of all equilibrium outcomes of the game. (Recall from Theorem 6.4 that generically outcomes are constant across each component.)

This is not necessarily an exercise in refining Nash equilibrium. Refinements do not, at least not explicitly, *compare* equilibria across the whole strategy space. Rather, they look at a given equilibrium and ask whether it satisfies certain criteria. Classifying equilibrium components is more modest. Yet, it does not preclude the application of stronger criteria *after* the comparison has been made. It only makes it a more informed venture by supplying a prior comparison of the available equilibrium components.

Index of a Point   To introduce index theory, a brief review of concepts from differential topology is helpful (see also the appendix to Section 5.3). If $g : X \to Y$ is a smooth (infinitely often continuously differentiable) mapping between compact manifolds $X \subseteq \mathbb{R}^l$ and $Y \subseteq \mathbb{R}^m$ with the same dimension, the Jacobian $D_x g(\bar{x})$ at a point $\bar{x} \in X$ is the matrix of first-order partial derivatives, and its determinant is denoted by $|D_x g(\bar{x})|$. Call a point $y \in Y$ a *critical value* for g if $|D_x g(\bar{x})| = 0$, i.e. if the Jacobian is singular, for some $\bar{x} \in g^{-1}(y) = \{x \in X \,|\, g(x) = y\}$ in the preimage. All points $y \in Y$ that are not critical values are called *regular values*. Hence, for a regular value the Jacobian must be nonsingular at all points in its preimage.

For such a mapping, the (Brouwer) *degree* of g is defined as

$$\deg(g) = \sum_{\bar{x} \in g^{-1}(y)} \text{sign} \, |D_x g(\bar{x})| \tag{6.24}$$

for any regular value $y \in Y$. It can be shown that the degree of g does not depend on the particular regular value chosen to compute it.

If $\bar{x} \in g^{-1}(y)$ is an isolated[17] point in the preimage of a particular value, say, $y = 0$, the *index* of $\bar{x}$, $\text{ind}_g(\bar{x})$, is defined as the degree of the mapping $x \mapsto g(x)/\|g(x)\|$ that maps the boundary of a small ball centered at $\bar{x}$ into the unit sphere. If $y = 0 \in Y$ happens to be a regular value, this amounts to

$$\text{ind}_g(\bar{x}) = \text{sign} \, |D_x g(\bar{x})| \, . \tag{6.25}$$

Intuitively, the index counts how often the images (of points at the boundary of a small sphere around $\bar{x}$) wander in the "same" or the "opposite" direction as their argument (where the counting contributes $+1$ if it is the "same" and -1 if it is the "opposite" direction). Hence, it gives qualitative information on the directional changes of g around a point $\bar{x}$ where $g(\bar{x}) = 0$. In the above example the index is simply the (-1 times) slope of the function.

In the plane it is also easy to understand. Here, the index simply counts the number of times that $g(x)/\|g(x)\|$ rotates completely while we walk counterclockwise around a small circle about $\bar{x}$. (Again, counterclockwise rotation adds $+1$ and clockwise rotation contributes -1.)

Figure 6.15 illustrates. The upper two points, a "sink" and a "source", have index $+1$. The lower left point, a "saddle point", has index -1. The lower right point has index $+2$. You may think of the latter as a source and a sink that have "melted" into each other.

Next, a *vector field* on a manifold X is a smooth mapping that maps X to its tangent space $T(X)$. The *tangent space* of X, intuitively, consists of the linear spaces that best approximate X at each point of it. A vector field g on X *points outwards* at the boundary ∂X of X if $x + \lambda g(x) \notin X$ for all $x \in \partial X$ and all $\lambda > 0$.[18]

---

[17] The point $\bar{x}$ is *isolated* if there is a small neighborhood around it such that $\bar{x}$ is the only solution to $g(x) = y$ in this neighborhood.

[18] Recall (from the appendix to Section 5.3) that the *boundary of a manifold* is the image of the boundary of the half space to which it is diffeomorphic, and, therefore, not the same as its topological boundary.

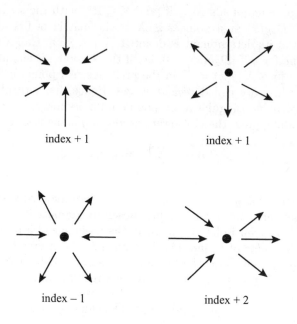

index + 1   index + 1

index − 1   index + 2

Figure 6.15: Index of a point

For vector fields on (compact) manifolds X (with boundary), differential topology supplies a most striking result, known as the *Poincaré-Hopf theorem* (see e.g. [121], p. 35, or [77], p. 134).

This theorem states that, if 0 is a regular value for a vector field that points outwards at the boundary, then the index sum across all zeros of the vector field is a constant which depends only on the topology of X (and not on the vector field). The constant is known as the *Euler characteristic* of X. Hence, the degree of such a vector field depends only on the topology of the underlying space.[19]

There is no point in going into details on what the Euler characteristic of a given space is. But its basic idea can be explained quite intuitively.

One way to understand the anatomy of an object is to dissect it into familiar pieces and examine the way the pieces fit together. Many objects (or spaces) can be decomposed such that each piece topologically "looks like" (is homeomorphic to) a simplex. This procedure is called "triangulation". Denote by r_k the resulting number of such simplicial pieces with dimension $k = 0, 1, ...$ (A simplex of dimension zero is a point or a "vertex", and it is an "edge" if

---

[19] This is often illustrated by pointing out that if your head were a doughnut you could comb your hair without leaving a bald spot. (The Euler characteristic of the torus is zero.) You cannot do that if your head is a sphere.

it is one-dimensional.) Then the Euler characteristic of the object is given by the alternating sum

$$\sum_{k \geq 0} (-1)^k r_k \qquad (6.26)$$

which can be shown to be independent of the particular triangulation that was used to compute it. For instance, the Euler characteristic of the circle or the torus ("doughnut") is zero, for a "figure eight" (two circles glued at one point) it is -1, for the surface of a cube it is $+2$, and so on. Of course, all objects that are homeomorphic have the same Euler characteristic (which, therefore, is a "topological invariant").

For the present purpose it is enough to know that all *contractible* manifolds have Euler characteristic $+1$. (And a space is contractible if its identity map is homotopic to a constant.) For instance, all convex spaces are contractible. The spaces that will be considered here are all contractible and, therefore, have Euler characteristic $+1$.

Regular Equilibria To apply all this to (finite) games, the space of mixed strategies needs to be reparametrized. For a game $\Gamma = (S, u)$ where player $i \ (= 1, ..., n)$ has $K_i = |S_i|$ pure strategies, express the probability of one of those, say, the K_i-th, by the residual probability over the other strategies, i.e. identify i's mixed strategy simplex Δ_i with the set of vectors

$$\Sigma_i = \left\{ \sigma_i \in \mathbb{R}_+^{K_i-1} \,\middle|\, \sum_{k=1}^{K_i-1} \sigma_i \left(s_i^k\right) \leq 1 \right\} \qquad (6.27)$$

so that expected payoffs can be written as

$$U_i(\sigma) = \sum_{k=1}^{K_i-1} \sigma_i \left(s_i^k\right) U_i \left(\sigma_{-i}, s_i^k\right) + \left(1 - \sum_{k=1}^{K_i-1} \sigma_i \left(s_i^k\right)\right) U_i \left(\sigma_{-i}, s_i^{K_i}\right)$$

for all $\sigma \in \Sigma \equiv \Sigma_1 \times ... \times \Sigma_n$ and all i.

Since a Nash equilibrium $\sigma \in \Sigma$ is a solution to the system of inequalities $U_i(\sigma_{-i}, s_i) \leq U_i(\sigma)$ for all $s_i \in S_i$ and all i, it must satisfy the system of equations

$$f_i^k(\sigma) \equiv \sigma_i \left(s_i^k\right) \left[U_i \left(\sigma_{-i}, s_i^k\right) - U_i \left(\sigma\right)\right] = 0 \text{ for all } k = 1, ..., K_i - 1 \quad (6.28)$$

and for all $i = 1, ..., n$. This system of equations emerges from applying the Kuhn-Tucker conditions to player i's expected payoff maximization problem. Its left hand side (denoted f) is well-known in evolutionary game theory, as the (multi-population) *replicator dynamics*, given by $d\sigma/dt = f(\sigma(t))$ where t denotes time (see e.g. [179], pp. 69-74).

The left hand side of (6.28) defines a smooth vector field $f : \Sigma \to \mathbb{R}^M$ (where $M = \sum_{i=1}^n K_i - n$) on Σ, because $\mathbb{R}^M$ is the tangent space to Σ.

Obviously, if $\sigma \in \Sigma$ is a Nash equilibrium, then $f(\sigma) = 0$; i.e., every Nash equilibrium is a zero for the vector field f, but *not* vice versa (e.g., all pure strategy combinations are zeros of f, but not necessarily Nash equilibria).

Yet, in general, zero $(0 \in \mathbb{R}^M)$ will not be a regular value for f. In fact, by Proposition 6.3, Nash equilibria may not even be isolated. But there are important exceptions to this rule.

In particular, call a Nash equilibrium $\bar{\sigma} \in \Sigma$ *regular* if the Jacobian $D_\sigma f(\bar{\sigma})$ is nonsingular at the Nash equilibrium. Regular Nash equilibria have all the properties that one can hope for—except for general existence. In particular, every regular equilibrium is essential, strictly perfect, perfect, proper, and (by strict perfection) a singleton KM-stable set. (For a proof see [149], Corollary 5, or [37], Theorems 2.4.7 and 2.4.8.)

Proposition 6.6 *If $\sigma \in \Sigma$ is a regular Nash equilibrium for a (finite) game $\Gamma = (S, u)$, then it is essential, strictly perfect, perfect, and proper.*

Behind this is a result which states that every regular Nash equilibrium is isolated and varies continuously in the payoffs on a neighborhood of Γ (see [149], Theorem 1). (Such Nash equilibria are called *strongly stable*; see [98]). Moreover, it is not difficult to see that every strict equilibrium is regular and a pure strategy Nash equilibrium is regular if and only if it is strict.

But examples can be constructed of games that do not have a regular Nash equilibrium. Therefore, regular Nash equilibrium fails the existence requirement. On the other hand, there is a sense in which the examples of games without regular equilibria are exceptional. In fact, almost all normal form games will have *only* regular Nash equilibria. Formally, the relevant result is as follows. (For a proof see [149], Theorem 2; [37], Section 2.6; [81]; or [181].)

Theorem 6.7 *For almost all[20] games $\Gamma = \Gamma(u)$ with $u \in \mathbb{R}^{nK}$, all Nash equilibria are regular.*

It seems as if this result makes all refinement exercises vacuous, because for almost all games all Nash equilibria satisfy all refinement criteria (by Proposition 6.6). And it is very easy to check that, because all that needs to be done is to calculate Jacobians. Yet, the "almost all" here refers to the space of normal form games. But from Section 4.4 it is known that every nontrivial extensive form corresponds to a lower-dimensional linear subspace of the space of normal form games. Hence, if the relevant games are given in extensive form, Theorem 6.7 has little to say. Therefore, it has to be interpreted with due care.

For a game with all Nash equilibria regular, an index assignment is straightforward. For a regular Nash equilibrium $\bar{\sigma} \in \Sigma$ of a game Γ the index of $\bar{\sigma}$ is defined by

$$\text{ind}_f(\bar{\sigma}) = \text{sign} \left| -D_\sigma f(\bar{\sigma}) \right|, \tag{6.29}$$

---

[20]   The phrase "almost all" here refers to an open dense set of games.

where the minus sign appears in order to let strict equilibria have index +1. Therefore, regular equilibria can only have index +1 or −1. It is not difficult to verify that any pure regular (hence, strict) Nash equilibrium must have index +1. Applying the Poincaré-Hopf theorem then yields the following result which describes the structure of equilibria for games that have exclusively regular equilibria. (For a proof see [149], Theorem 3; or, for the first part, [81]; or, for the second part, [73].)

Proposition 6.7 *For almost all (normal form) games,*
(a) there is a finite and odd number of Nash equilibria.
(b) if the game has $m \geq 1$ pure equilibria, then it has at least $m - 1$ mixed Nash equilibria.

The logic for the proof of this theorem is easy. If all equilibria are regular, then there can be only finitely many, because, roughly, each is contained in an isolating neighborhood, Σ is compact, and f is polynomial. Since the index sum across all equilibria must be +1, and regular equilibria can only have index +1 or −1, there must be an odd number. (In particular, there must be at least one, because zero counts as an even number.) Since pure equilibria have index +1, for $m \geq 1$ pure equilibria there must be at least $m-1$ equilibria with index −1, because otherwise the index sum would exceed +1. And equilibria with index −1 must be mixed.

The Index of a Component The drawback of the results on regular equilibria is that games with a nontrivial (but fixed) extensive form may not have regular equilibria. For the general case, we have to live with connected components of Nash equilibria that are higher-dimensional. For those, zero ($0 \in \mathbb{R}^M$) will *not* be a regular value for f and we have to find other ways to apply the logic of index theory.

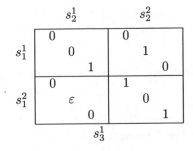

Example 6.19 *Consider the three-player game above for some small $\varepsilon > 0$. It has three components of Nash equilibria. Two consist of regular (and completely mixed) equilibria, but the third is higher-dimensional. Denoting*

$\sigma_1\left(s_1^1\right) = x$, $\sigma_2\left(s_2^1\right) = y$, and $\sigma_3\left(s_3^1\right) = z$, the equilibria are given by

$$\sigma^1 = \left(x^1, y^1, z^1\right) = \left(\frac{1}{2}, \frac{1}{2-\varepsilon}, \frac{1}{2-\varepsilon}\right), \ \sigma^2 = \left(x^2, y^2, z^2\right) = \left(\frac{1+\varepsilon}{2+\varepsilon}, \frac{1}{2}, \frac{1}{2}\right)$$

and the one-dimensional component $C = \{1\} \times [0, 1/2] \times \{0\}$. The vector field $f = (f_1, f_2, f_3)$ can be written as

$$
\begin{aligned}
f_1(\sigma) &= x\,(1-x)\,(y-z)\,, \\
f_2(\sigma) &= y\,(1-y)\,[1 - x - (1-\varepsilon)\,z - \varepsilon x z]\,, \\
f_3(\sigma) &= z\,(1-z)\,(1-2x)\,(1-2y)\,.
\end{aligned}
$$

For the two regular equilibria the index can be calculated directly by computing the determinant $|-D_\sigma f|$. For the equilibrium σ^1 this gives $\left|-D_\sigma f\left(\sigma^1\right)\right| = (2-\varepsilon)^{-4}\,(1-\varepsilon)^2\,\varepsilon/4 > 0$, and for σ^2 the computation gives

$$\left|-D_\sigma f\left(\sigma^2\right)\right| = -\frac{\varepsilon\,(1+\varepsilon)}{16\,(2+\varepsilon)^2} < 0.$$

Hence, $\text{ind}_f\left(\sigma^1\right) = +1$ and $\text{ind}_f\left(\sigma^2\right) = -1$. If an index can also be assigned to C such that the index sum across all equilibrium components equals $+1$, then the component C must obtain index $+1$. Note that as $\varepsilon \to 0$ the two regular equilibria σ^1 and σ^2 become a single equilibrium (with index zero) which is not regular anymore.

So, what is required is a way to extend the definition of an index from regular equilibria to (higher-dimensional) components of equilibria. The key idea is to slightly perturb the vector field.

Say that a vector field g on Σ *points inwards* at the boundary $\partial\Sigma$ of Σ if $\sigma_i\left(s_i^k\right) = 0$ implies $g_i^k(\sigma) > 0$ and $\sigma_i\left(s_i^k\right) = 1$ implies $g_i^k(\sigma) < 0$ for all $k = 1, ..., K_i - 1$ and $\sum_{k=1}^{K_i-1} \sigma_i\left(s_i^k\right) = 1$ implies $\sum_{k=1}^{K_i-1} g_i^k(\sigma) < 0$, for all $i = 1, ..., n$. (Clearly, if g points inwards, then $-g$ points outwards at the boundary.) The vector field f (from (6.28)) does not strictly point inwards at $\partial\Sigma$, but it does so in a weak sense. Hence, by slightly perturbing it, f can be turned inwards pointing.

Definition 6.9 *An **interior approximation** is a smooth mapping $F : \Sigma \times \mathbb{R}_+ \to \mathbb{R}^M$ such that*
(a) $F(\sigma, 0) = f(\sigma)$ for all $\sigma \in \Sigma$ and
(b) $f_\lambda \equiv F(\,.\,, \lambda)$ points inwards at the boundary $\partial\Sigma$ for all $\lambda > 0$.
*An interior approximation F is **regular** if there is some $\lambda_0 > 0$ such that $0 \in \mathbb{R}^M$ is a regular value for f_λ for all $\lambda \in (0, \lambda_0)$.*

Interior approximations slightly perturb f in such a way as to make it inwards pointing. *Regular* interior approximations make $0 \in \mathbb{R}^M$ a regular value for all sufficiently small perturbations of f. It can be shown that "almost all" interior approximations are regular, so there is no existence problem. (This, once again, is a consequence of Sard's theorem; see Theorem 5.8 in the appendix to Section 5.3.)

Now recall from Proposition 6.3 that for every game there are finitely many connected components of Nash equilibria. It follows that for each of those components $C \subseteq E(u)$ there is an isolating neighborhood $\mathcal{O}_C$ (an open set such that C contains all equilibria in this open set) and these isolating neighborhoods can be chosen disjoint for the different components. The idea underlying an index assignment to components of Nash equilibria is to look at the degree of a regular interior approximation on the isolating neighborhood for the component under scrutiny.

Definition 6.10 *For a (finite) normal form game* $\Gamma = (S, u)$ *and a connected component* $C \subseteq E(u)$ *of Nash equilibria, the **index** of C, denoted* $\mathrm{Ind}\,(C)$*, is defined by*

$$\mathrm{Ind}\,(C) = \lim_{\lambda \to 0} \sum_{\sigma \in f_\lambda^{-1}(0) \cap \mathcal{O}_C} \mathrm{ind}\, f_\lambda\,(\sigma)\,, \qquad (6.30)$$

where F is an interior approximation such that the associated evaluation maps f_λ have finitely many isolated zeros in the interior of Σ for all $\lambda > 0$ sufficiently small.

There are several aspects of this definition that deserve comment. First, the limit for $\lambda \to 0$ is, obviously, taken to approximate f. Since the index is a "homotopy invariant" there is no problem with the existence of a limit, since the sequence of numbers will effectively be constant.

Choosing a regular interior approximation F, the index can be computed from

$$\mathrm{Ind}\,(C) = \sum_{\bar{\sigma} \in f_\lambda^{-1}(0) \cap \mathcal{O}_C} \mathrm{sign}\,|-D_\sigma f_\lambda(\bar{\sigma})|\,,$$

for some $\lambda > 0$ sufficiently small, i.e. by summing signs of determinants. Hence, if the interior approximation is regular (as it always can be chosen), then it is easily seen that $\mathrm{Ind}\,(C)$ is the *degree* of the evaluation maps f_λ for the interior approximation F when these are restricted to the isolating neighborhood $\mathcal{O}_C$. Since the evaluation maps f_λ are all homotopic, they all have the same degree, for all $\lambda \geq 0$. (This follows from the "homotopy lemma".) Therefore, computing the index of a component of Nash equilibria boils down to calculating determinants.

Second, as notation suggests, it is of no relevance which particular interior approximation is chosen to compute the index. All interior approximations will give the same answer. More precisely, it can be shown that the index $\mathrm{Ind}\,(C)$ does *not* depend on the particular interior approximation F chosen to calculate it (see ([149], Lemma 4). The index is a property of the component of equilibria C and not of the particular vector field.

In fact, the index does not even depend on the vector field f (as required by Definition 6.9(a)). Any vector field that is Lipschitz continuous[21] in $\sigma \in \Sigma$, continuous in the payoff vector $u \in \mathbb{R}^{nK}$, is weakly inwards pointing at the boundary, and for which all Nash equilibria are zeros will do (see [45], Corollary 1). So, Definition 6.10 is but one of a multitude of possible definitions of the index—that will all give the same answers.

Moreover, the index of a component can be shown to be the same as the *local degree* of the projection map around a component from the graph of the Nash equilibrium correspondence, $\mathcal{G}(E)$, to the space of games, $\mathbb{R}^{nK}$. The "local degree" of the projection is the degree of the projection map from a neighborhood of the component in the graph $\mathcal{G}(E)$ (viewed as a space of its own) into a neighborhood of the game in $\mathbb{R}^{nK}$. So, the difference from the "global degree" is that the projection is considered only on a small part of the graph $\mathcal{G}(E)$ that contains the component. The following theorem, therefore, states that the index of a component is governed by the local geometry of the Nash equilibrium correspondence, rather than by the vector fields used to define it. (For the original proof of this result see [44], Theorem 1; an elementary version of the proof, based on Theorem 6.6, is provided in [45], Corollary 2.)

Theorem 6.8 *For any connected component C of Nash equilibria for a (finite normal form) game Γ, the local degree of the projection map from $\mathcal{G}(E)$ to a neighborhood of Γ agrees with* Ind (C).

In a sense this is a rather surprising result. After all, the index is a purely local concept which does not take into account payoff perturbations. The local degree, on the other hand, is defined on the graph of the equilibrium correspondence and, thus, presumes knowledge of the latter. Yet, as it turns out, the two give the same answers.

An immediate consequence of Theorem 6.8, together with Theorem 6.7, is yet another simple way to compute the index of a component. Since, like the index, the local degree is invariant with respect to payoff perturbations, the payoffs for Γ can be slightly perturbed is such a way as to obtain a game that has all equilibria regular. Then the index sum of all regular equilibria sufficiently close to the component is precisely the index of the component. Hence, perturbing payoffs in the right way can substitute for perturbing the vector field f.

Finally, a crucial property of Definition 6.10 is as follows. As f_λ is inwards pointing, $-f_\lambda$ points outwards at the boundary $\partial\Sigma$, so the Poincaré-Hopf

---

[21] A vector field f is (locally) *Lipschitz continuous* if for every compact subset C of its domain there is a real constant λ such that $\|f(\sigma) - f(\sigma')\| \leq \lambda \|\sigma - \sigma'\|$ for all $\sigma, \sigma' \in C$. Every Lipschitz continuous function is continuous, and every continuously differentiable function is Lipschitz continuous.

theorem is applicable to the index assignment. In other words, the index sum across all connected components of Nash equilibria is $+1$, as Σ is contractible and, therefore, has Euler characteristic $+1$. Given this additivity property, the index assignment does indeed provide a classification of the various components of equilibria, since it implicitly *compares* equilibrium components.

This, at times, allows even a comparison of regular equilibria. For instance, in a 2×2 "coordination game" both strict equilibria have index $+1$, while the completely mixed equilibrium has index -1. On the other hand, the completely mixed (and regular) equilibrium of a 2×2 "matching pennies" game has index $+1$. Both observations follow immediately from additivity.

Components with Nonzero Index A direct implication of the additivity property is that every game has a component of equilibria with positive index. (In particular, therefore, it must have at least one Nash equilibrium.) As will be seen, such components with nonzero index are good candidates to be selected as a solution. A first result in this direction states that every component with nonzero index is an essential component (see Definition 6.8). Hence, a component with nonzero index is robust to payoff perturbations—as is to be expected, in view of Theorem 6.8. (For the proof of the following result see [149], Theorem 4.)

Theorem 6.9 *If $C \subseteq E(u)$ is a connected component of Nash equilibria for a (finite normal form) game $\Gamma = (S, u)$ with $\mathrm{Ind}\,(C) \neq 0$, then C is an essential component.*

The proof of this theorem is technically somewhat complicated. Its intuition is, however, simple. First, if the degree of the evaluation maps f_λ restricted to the isolating neighborhood $\mathcal{O}_C$ is nonzero, they will have zeros in $\mathcal{O}_C$. Second, limits of such zeros as $\lambda \to 0$ are Nash equilibria. Third, as the degree is unchanged by payoff perturbations, the same holds for nearby games. Hence, all nearby games will have Nash equilibria close to the component with nonzero index.

Alternatively, Theorem 6.8 can be used to prove Theorem 6.9 directly. Since a nonzero index implies that the local degree is nonzero, the projection of a neighborhood of C in the graph $\mathcal{G}(E)$ onto the space of games is surjective. Hence, "above" every nearby game there is an equilibrium within the neighborhood of C in $\mathcal{G}(E)$.

The converse of Theorem 6.9 is *not* true; i.e., there may be essential components with index zero. This is illustrated by the next example.

Example 6.20 *Consider the following two-player extensive form game. Player 1 starts by either taking an outside option ("Out") which yields both players zero; or she moves into the one-shot subgame which corresponds to first three rows of the reduced normal form representation below, where, for the moment, $\varepsilon_1 = \varepsilon_2 = \varepsilon_3 = 0$. Denote $\sigma_1\left(s_1^1\right) = x$, $\sigma_1\left(s_1^2\right) = y$, $\sigma_1\left(s_1^3\right) = z$,*

$\sigma_2\left(s_2^1\right) = p$, and $\sigma_2\left(s_2^2\right) = q$. If in any one equilibrium of this game both x and y are (strictly) positive, then $p = 1/3$, $q \geq 1/3$, and $4p \geq q + 11/4$ must hold—which, however, is impossible. Hence in all equilibria $xy = 0$. If in any one equilibrium both x and z are (strictly) positive, then $p + q = 2/3$ and $1 \geq 2p + 22q$ must hold—which, however, is again impossible. Hence, in all equilibria $xz = 0$. Therefore, if in any one equilibrium either y or z is (strictly) positive, then $x = 0$.

Now consider an equilibrium where player 1 does not use her outside option; i.e., $x + y + z = 1$. If $y + z > 0$, then $x = 0$, so $y + z = 1$. But then also $p = 0$, because $(0, 1/10, 9/10)$ always does better for player 2 than s_2^1 if s_1^1 and the outside option are not used. Yet, if $p = 0$, then $z = 0$, because s_1^2 always does better for player 1 than s_1^3 if s_2^1 is not used. Hence, $y = 1$ then implies $q = 1$, against which player 1 can improve by choosing the outside option. It follows that there is only one equilibrium in which the outside option is not used, namely $x = p = 1$, i.e., $\left(s_1^1, s_2^1\right)$. This equilibrium is strict and, therefore, must have index $+1$.

	s_2^1		s_2^2		s_2^3	
s_1^1	5		-15		-11	
		4		0		2
s_1^2	-23		-1		3	
		0		8		1
s_1^3	-1		-21		1	
		2		-2		3
Out	0		0		0	
		ε_1		ε_2		ε_3

In any one equilibrium where player 1 uses the outside option she must obtain payoff zero. If player 1 used s_1^1 on top of the outside option, i.e. if $x > 0$ and $x + y + z < 1$, then $y = z = 0$, so that player 2's unique best reply is s_2^2. But that yields again the strict equilibrium $x = p = 1$. Hence, when the outside option is used, $x = 0$. If $0 < y + z < 1$, then $p = 0$ (by the same argument as above) implies $z = 0$. But then in all equilibria consistent with $x = z = p = 0$, $y = 0$ must also hold. Hence, all equilibria that use the outside option belong to a single connected component of Nash equilibria in which $x = y = z = 0$, i.e. where player 1 uses only the outside option.

Player 2's strategies in this "outside option" component are given by the conditions that make player 1 weakly prefer the outside option to any other strategy. This can be seen to be the convex hull of the six strategies $(p, q) = (3/4, 1/4)$, $(11/16, 0)$, $(1/2, 0)$, $(31/282, 10/282)$, $(0, 3/4)$, and $(0, 1)$. By the additivity property of the index, the index of the "outside option" component must be zero.

But it turns out that the "outside option" component is an essential component. Clearly, for essentiality only player 2's payoffs when 1 uses the outside option matter; i.e., all relevant payoff perturbations are captured by (sufficiently small) ε_1, ε_2, and ε_3 in the normal form above. Moreover, explicit computation of 2's payoffs reveals that only the differences $\delta_1 = \varepsilon_1 - \varepsilon_2$ and $\delta_3 = \varepsilon_3 - \varepsilon_2$ matter.

Now, if $\delta_1 < 0$ and $\delta_3 < 0$, then $x = y = z = p = 0$ and $q = 1$ clearly is an equilibrium of the perturbed game close to 2's strategy $(0,1)$ in the "outside option" component. If $\delta_3 > 0$ and $\delta_3 > \delta_1$, then $x = z = p = 0$, $y = \delta_3/(7+\delta_3)$, and $q = 3/4$ is an equilibrium of the perturbed game close to 2's strategy $(0, 3/4)$ in the "outside option" component. Finally, if $\delta_1 > 0$ and $\delta_3 < \delta_1$, then $x = y = q = 0$, $z = (\delta_1 - \delta_3)/(1 + \delta_1 - \delta_3)$, and $p = 1/2$ is an equilibrium of the perturbed game close to 2's strategy $(1/2, 0)$ in the "outside option" component. Since these choices of δ_1 and δ_3 cover all generic possibilities, invoking the u.h.c. property of the Nash equilibrium correspondence yields the fact that the "outside option" component is indeed essential.

So, an essential component of Nash equilibria *can* have index zero. Moreover, essentiality of a component is not an "ordinal" property, in the sense that it depends not only on the reduced normal form. In fact, the same example can be used to illustrate this point.

Example 6.21 *Consider the previous example, but for player 2 add the mixture $s_2^4 = \frac{3}{4}s_2^1 + \frac{1}{4}s_2^2$ as an extra pure strategy.*

In this enlarged game there is no equilibrium close to the "outside option" component when payoffs are slightly perturbed by some small $\varepsilon > 0$ as in the normal form below. In fact, the unique equilibrium of the enlarged and perturbed game is the strict equilibrium (s_1^1, s_2^1).

	s_2^1		s_2^2		s_2^3		s_2^4	
s_1^1	5		-15		-11		0	
		4		0		2		3
s_1^2	-23		-1		3		-17.5	
		0		8		1		2
s_1^3	-1		-21		1		-6	
		2		-2		3		1
Out	0		0		0		$-\varepsilon$	
		0		0		0		ε

To see why, consider first an equilibrium where s_2^3 is not used. Then player 1 will not use s_1^2 or s_1^3 either, because the outside option does better than these two if s_2^3 is not used. But then player 2 will not use s_2^2 either, because s_2^4 does better than s_2^2 if s_1^2 and s_1^3 are not used. Hence, player 2 in such an equilibrium will use only s_2^1 and s_2^4. But this implies that player 1's unique best reply is s_1^1, which yields the strict equilibrium (s_1^1, s_2^1). Hence, any equilibrium other than (s_1^1, s_2^1) must use s_2^3.

Again it can be shown, for $\varepsilon > 0$ sufficiently small, that neither s_1^2 nor s_1^3 can be used simultaneously with s_1^1 in any equilibrium. (If s_1^1 and s_1^2 were used simultaneously in equilibrium, this would imply a "probability" for s_2^3 that exceeds 1. If s_1^1 and s_1^3 were used simultaneously, this would imply a negative "probability" for s_2^2.) Hence, if in equilibrium $x > 0$, then $y = z = 0$. But in the reduced game where s_1^2 and s_1^3 have been deleted, player 2's strategies s_2^2 and s_2^3 are both strictly dominated by s_2^4. Yet, if the game is reduced further by deleting s_2^2 and s_2^3, then s_1^1 becomes strictly dominant for player 1, so that the strict equilibrium $\left(s_1^1, s_2^1\right)$ emerges again. Hence, in any equilibrium other than $\left(s_1^1, s_2^1\right)$ we must have $x = 0$. But then s_2^1 cannot be used, because $(0, 0, 2/3, 1/3)$ does better than s_2^1 for player 2 when s_1^1 is not used. If s_2^1 is not used either, then s_1^3 cannot be used, because $(0, 0, \rho, 1 - \rho)$ for $1/3 < \rho < (6 - \varepsilon) / (17.5 - \varepsilon)$ (which exists for $\varepsilon < 1/4$) does better than s_1^3 for player 1 when s_2^1 is not used. In the game where s_1^1, s_2^1, and s_1^3 have been deleted, s_2^3 is strictly dominated and, therefore, cannot be used. But this contradicts the above finding that any equilibrium other than $\left(s_1^1, s_2^1\right)$ must use s_2^3. Hence, $\left(s_1^1, s_2^1\right)$ is the unique equilibrium of the perturbed game and the "outside option" component fails to be essential in the game enlarged by the mixture s_2^4.

Hence, a component of Nash equilibria that is essential for a given game may *not* be essential for another game with the same reduced normal form. This failure of "ordinality" by merely essential components is not, however, shared by components with nonzero index. It will be seen below (Proposition 6.8(a)) that the index is invariant with respect to adding an extra strategy, which is precisely a mixture over other strategies. So, a component with nonzero index is a true refinement of essential components.

The next result should be compared with the properties of, say, KM-stable sets. It states that, on top of Theorem 6.9, components with nonzero index satisfy more desirable properties. They depend only on the reduced normal form, and they contain a backwards induction equilibrium, and a component for any game obtained by deleting a (weakly) dominated strategy or one that is never a best reply against the component. Hence, components with nonzero index satisfy existence (EX), the Nash property (NE), connectedness (CO), essentiality, invariance (IV), backwards induction (BI), and a version of independence of irrelevant alternatives (IA). (A proof of the following result is given in the appendix to this section.)

Proposition 6.8 *For any (finite normal form) game $\Gamma = (S, u)$, a component C of Nash equilibria with nonzero index contains a component of Nash equilibria with nonzero index for any other game obtained from Γ by deleting a pure strategy that is either*

    (a) equivalent to a mixture over other strategies, or

    (b) weakly dominated, or

(c) never a best reply against (any element in) C, and where
(d) C contains a proper equilibrium for Γ.

In the light of Proposition 6.8, components with nonzero index appear as good candidates for equilibrium refinements. For generic extensive form games (by Theorem 6.4) they identify an essential backwards induction outcome which satisfies a number of invariances. Moreover, Proposition 6.8(a) together with Theorem 6.9 imply that a component with nonzero index contains a KM-stable set.

In fact, it can even be shown that every Nash equilibrium component with nonzero index contains an M-stable set. Thus, components with nonzero index satisfy all desiderata for strategic stability—with one exception.

Proposition 6.9 *For every (finite normal form) game, every component of Nash equilibria with nonzero index contains an M-stable set.*

The proof of this Proposition (see [46], Theorem 2) uses the fact that the space of strategy trembles is a subspace of the space of payoff perturbations (see (6.16) in Section 6.4). The converse of Proposition 6.9 is not true. In Example 6.19 at $\varepsilon = 0$ the interior equilibrium has index zero, but is a (singleton) strategically stable set, because it is completely mixed.

What components with nonzero index fail is admissibility. That is, a component with nonzero index may well contain inadmissible equilibria. And they do not satisfy a minimality criterion, like (some versions of) strategically stable sets.

On the other hand, there is nothing that forbids an ex post application of criteria like admissibility *after* components have been classified by the index assignment. For instance, after identifying a component with nonzero index, one may want to perturb strategies (as in the definition of stable sets, see Definition 6.7) and repeat the index assignment for the perturbed games. This, then, selects (connected) subsets of the component under scrutiny that do satisfy admissibility.

By Theorem 6.4 and Corollary 6.1 (from Section 6.4), for almost all extensive form games a component with nonzero index will identify a strategically stable outcome. Moreover, determining the index is computationally a lot easier than verifying strategic stability. The latter takes a complicated test against all possible strategy perturbations. The former consists merely of calculating determinants.

We conclude with an example which illustrates that some games may not have a component of Nash equilibria with an index equal to the Euler characteristic +1 of the underlying space.

Example 6.22 *Consider again a two-player outside option game where player 1 can initially either choose an outside option (which yields both players 0) or move into the subgame given by the first three rows of the normal form below.*

The subgame has three equilibria. The first, $\sigma^1 = ((1/2, 1/2, 0), (1/2, 1/2, 0))$ yields payoff 1 to player 1; the second, $\sigma^2 = ((1/3, 1/3, 1/3), (1/3, 1/3, 1/3))$, yields $-1/3$ to 1; and the third, $\sigma^3 = (s_1^3, s_2^3)$ yields her -2. Hence, only σ^1 gives more to player 1 than the outside option.

	s_2^1		s_2^2		s_2^3	
s_1^1	3		-1		-3	
		6		2		7
s_1^2	-1		3		-3	
		2		6		0
s_1^3	-3		4		-2	
		0		0		1
Out	0		0		0	
		0		0		0

Therefore, the overall game has two components of equilibria. The first is to move into the subgame and play σ^1 there, and the second is the higher-dimensional component where player 1 chooses the outside option with certainty. Since moving into the subgame and playing σ^1 is a regular equilibrium, its index is readily calculated by computing the determinant for the Jacobian of $-f$ at this equilibrium. This gives $\left|-D_\sigma f\left(\sigma^1\right)\right| = -1$ which is also the index of this regular equilibrium. By additivity, the "outside option" component must have index $+2$. Since those two are the only components of equilibria, there is no component with index $+1$.

Summary

- A regular Nash equilibrium is one where the Jacobian of a particular vector field, the replicator dynamics, is nonsingular. Its index is the sign of the determinant of (-1 times) the Jacobian.

- The index of a connected component of equilibria is the index sum across all zeros of an interior approximation (to the vector field) sufficiently close to the component. With the index of a component defined in this way, the index sum across all components of Nash equilibria equals $+1$.

- The index of a component equals the local degree of the projection from the graph of the equilibrium correspondence to a neighborhood of the game.

- Every component with nonzero index is essential and its index is invariant with respect to adding mixed strategies as extra pure strategies.

- Every component with nonzero index contains a backwards induction equilibrium, satisfies (a version of) independence of irrelevant alternatives, and contains an M-stable set.

Appendix

Proof of Proposition 6.8 Let $s_i^m \in S_i$ for some player i be either (a) equivalent to a mixture $\hat{\sigma}_i \in \Sigma_i$, or (b) weakly dominated by $\hat{\sigma}_i$, or (c) such that $U_i(\sigma_{-i}, s_i^m) < U_i(\sigma)$ for all $\sigma \in C$. For $\varepsilon > 0$ let $\Gamma^\varepsilon = (S, u^\varepsilon)$ be a game with only regular equilibria such that $\|u - u^\varepsilon\| < \varepsilon$ and $u_i(s_{-i}, s_i^m) > u_i^\varepsilon(s_{-i}, s_i^m)$ for all $s_{-i} \in S_{-i}$. By Theorem 6.7 such a game Γ^ε exists for all $\varepsilon > 0$. Denote $\tilde{\Gamma}^\varepsilon$ the game obtained from Γ^ε by deleting strategy $s_i^m \in S_i$.

Since Γ^ε is a payoff perturbation of $\Gamma = \Gamma^o$, it has a Nash equilibrium close to C by Theorem 6.9 for all $\varepsilon > 0$ sufficiently small. By construction, any equilibrium of Γ^ε close to C must give zero weight to s_i^m, for all $\varepsilon > 0$. If (a) s_i^m is equivalent to $\hat{\sigma}_i$ in Γ, then $\hat{\sigma}_i$ does better than s_i^m in Γ^ε; if (b) s_i^m is weakly dominated by $\hat{\sigma}_i$ in Γ, then $\hat{\sigma}_i$ strictly dominates s_i^m in Γ^ε; finally, if (c) s_i^m is never a best reply against C, then $U_i^\varepsilon(\sigma_{-i}, s_i^m) < \max_{s_i \in S_i} U_i^\varepsilon(\sigma_{-i}, s_i)$ for all $\sigma \in \Sigma$ sufficiently close to C by continuity. Hence, as $\varepsilon \to 0$ we obtain equilibria of $\Gamma = \Gamma^o$ which do not use s_i^m, which, therefore, are also equilibria of $\tilde{\Gamma}^o$.

Next, it is shown that (the intersection of) the isolating neighborhood $\mathcal{O}_C$ (with the reduced strategy space of $\tilde{\Gamma}^o$) for C isolates equilibria of $\tilde{\Gamma}^o$. Suppose $\sigma \in \partial \mathcal{O}_C$ is on the boundary of $\mathcal{O}_C$ and an equilibrium for $\tilde{\Gamma}^o$. Then, because σ cannot be an equilibrium for $\Gamma = \Gamma^o$ and all strategies and payoffs in $\tilde{\Gamma}^o$ for all players other than i are as in Γ, $U_i(\sigma_{-i}, s_i^m) > U_i(\sigma)$ must hold. But if (a) s_i^m is equivalent to or (b) weakly dominated by $\hat{\sigma}_i$, then $U_i(\sigma_{-i}, \hat{\sigma}_i) > U_i(\sigma)$ contradicts the equilibrium hypothesis; if (c) s_i^m is never a best reply against C, then $U_i(\sigma_{-i}, s_i^m) > U_i(\sigma)$ cannot hold, because, if it did, s_i^m would be a best reply against some element of C by choosing $\mathcal{O}_C$ sufficiently small, by continuity. Hence, $\mathcal{O}_C$ must isolate equilibria for $\tilde{\Gamma}^o$. It follows that C contains at least one component of equilibria for $\tilde{\Gamma}^o$.

Let f_ε (resp. $\tilde{f}_\varepsilon$) be the vector field (6.28) for Γ^ε (resp. $\tilde{\Gamma}^\varepsilon$) for $\varepsilon > 0$. By Theorem 6.8,

$$\text{Ind}(C) = \sum_{\bar{\sigma} \in f_\varepsilon^{-1}(0) \cap E(u^\varepsilon) \cap \mathcal{O}_C} \text{sign} \left| -D_\sigma f_\varepsilon(\bar{\sigma}) \right|$$

for all $\varepsilon > 0$ sufficiently small. Elementary linear algebra (Laplace's rule for determinants) yields

$$\left| -D_\sigma f_\varepsilon(\bar{\sigma}) \right| = \left(U_i^\varepsilon(\bar{\sigma}) - U_i^\varepsilon(\sigma_{-i}, s_i^m) \right) \left| -D_\sigma \tilde{f}_\varepsilon(\bar{\sigma}) \right|$$

whenever $\bar{\sigma}_i(s_i^m) = 0$. Hence, the degree of f_ε on $\mathcal{O}_C$ remains unchanged by deleting the unused strategy s_i^m, because at any equilibrium $\bar{\sigma}$ for Γ^ε we have $U_i^\varepsilon(\bar{\sigma}) > U_i^\varepsilon(\sigma_{-i}, s_i^m)$. Therefore, C contains at least one component of equilibria for $\tilde{\Gamma}^o$ with nonzero index.

Statement (d) follows from Proposition 5 of [97], if it is taken into account that every component with nonzero index contains a hyperstable set. By statement (a) and Theorem 6.9 a component with nonzero index satisfies the defining property of hyperstability. Consider the collection of subsets of the component that satisfy this property. Since this collection is nonempty and every decreasing chain of its elements has a lower bound (by compactness the intersection belongs to it), it has a minimal element, by Zorn's lemma. Since every hyperstable set contains a fully stable set, Proposition 5 of [97] (which states that a fully stable set contains a proper equilibrium) is applicable. This completes the proof of Proposition 6.8.

Exercises

Exercise 6.28 Argue, based on index theory, that in the two-player game below the unique pure strategy equilibrium (s_1^3, s_2^3) is not a good solution for the game. Find another solution which appears more reasonable.

	s_2^1		s_2^2		s_2^3	
s_1^1	4		0		3	
		0		4		0
s_1^2	0		4		3	
		4		0		0
s_1^3	0		0		3	
		3		3		3

Exercise 6.29 Consider the following two-player extensive form game. Player 1 initially can either take an outside option which yields her payoff 2 and payoff 5 to player 2; or she may give the move to player 2. If player 2 is given the move, she can also either take an outside option, which now yields her payoff 2 and payoff 3 to player 1; or she can move into the one-shot 2×2 subgame below. Determine the reduced normal form and the index for each of the (two) components of Nash equilibria of this game. (Hint: Invoke Theorem 6.8 and perturb payoffs in the normal form so as to obtain regular equilibria.)

	s_2^1		s_2^2	
s_1^1	0		1	
		0		3
s_1^2	1		0	
		3		0

Exercise 6.30 The two-player game below is derived from a "coordination" game where the off-diagonal payoffs $(0,0)$ are generated from a subgame with a "matching pennies" structure and a unique mixed strategy equilibrium. Argue that in this game the two equilibria that are supported on two pure strategies (the first two or the last two for each player) are preferable to the completely mixed equilibrium.

	s_2^1	s_2^2	s_2^3	s_2^4
s_1^1	3 2	3 2	4 −1	−4 1
s_1^2	3 2	3 2	−4 1	4 −1
s_1^3	4 −1	−4 1	2 3	2 3
s_1^4	−4 1	4 −1	2 3	2 3

Exercise 6.31 Show that every pure strategy regular equilibrium of a finite normal form game is (locally) asymptotically stable in the replicator dynamics $d\sigma/dt = f(\sigma(t))$, where f is defined in (6.28) and t denotes time. (Hint: You will have to invoke the following result. If all eigenvalues of the Jacobian of a system of differential equations, at one of its rest points, have negative real parts, then the rest point is asymptotically stable. Asymptotic stability means that all trajectories starting near the rest point will remain close to it forever, and all trajectories starting sufficiently close will converge to the rest point.)

Exercise 6.32 The following three-player game has two components of Nash equilibria. Find the two components and assign the index to each of them.

	s_2^1	s_2^2
s_1^1	0 0 0	0 1 0
s_1^2	1 0 0	0 0 1

s_3^1

	s_2^1	s_2^2
s_1^1	0 0 1	1 0 0
s_1^2	0 1 0	0 0 0

s_3^2

Exercise 6.33 Return to Exercise 6.9 and argue that only the "intuitive" component of equilibria (where both types of player 1 have "beer" for breakfast) appears plausible in the light of index theory.

Exercise 6.34 A Nash equilibrium (of a finite normal form game) is called quasi-strict if all best replies against it are used with positive probability (see Harsanyi [81]). Show that every regular equilibrium is quasi-strict.

Exercise 6.35 Assign the index to every equilibrium component for each of the games below which are generated by varying the parameter $x \in R$. Relate the result to the local degree of the projection from the graph of the equilibrium correspondence to the space of games.

	s_2^1		s_2^2
s_1^1	x		0
		x		0
s_1^2	0		$1-x$
		0		$1-x$

Exercise 6.36 Consider the normal form of a generic perfect information extensive form game. Such a game has a unique component of equilibria that induce the backwards induction outcome. Show that this backwards induction component has index $+1$ and all other components have index 0. (Hint: combine Propositions 6.9 and 6.2 with Theorem 6.3.)

Notes on the Literature: Index theory for connected components of Nash equilibria has been introduced by Ritzberger [149]. (This is where Definition 6.10 originates, Theorem 6.9 has first been shown there, and this source also contains proofs for Proposition 6.6, Theorem 6.7, and Proposition 6.7). There are a few precursors in the literature. Shapley [167] defined an index for isolated equilibria of finite two-player games, based on the Lemke-Howson algorithm. And Hofbauer [88] considered an index for isolated rest points of evolutionary single-population dynamics, like the single-population replicator dynamics. In the theory of general competitive equilibrium for economies, regularity and index theory have been used by Debreu [42] and Dierker [48], among others. Proposition 6.6 has also been proved by van Damme [37]. Theorem 6.7 was originally shown by Wilson [181] and Harsanyi [81]. The latter also introduced the concept of a regular equilibrium, and provided a proof of Proposition 6.7(a). Proposition 6.7(b) was also shown by Gül, Pearce, and Stacchetti [73]. Theorem 6.8 is due to Demichelis and Germano [44] (see also Demichelis and Germano [45]). Govindan and Wilson [70] also prove Proposition 6.8(a). Examples 6.20, 6.21, and 6.22 are taken from Hauk and Hurkens [83], where "outside option games" as introduced by van Damme [38] are studied. Proposition 6.9 is proved by Demichelis and Ritzberger ([46], Theorem 2). The proof of Proposition 6.8 is based largely on Propositions 3-6 by Kohlberg and Mertens [97]. Exercise 6.32 is borrowed from Hofbauer and Swinkels [90].

Bibliography

[1] Abreu, D., and D. Pearce (1984): "On the Inconsistency of Certain Axioms of Solution Concepts for Non-Cooperative Games", *Journal of Economic Theory* **34**, 169-174.

[2] Akerlof, G. (1970): "The Market for 'Lemons'", *Quarterly Journal of Economics* **90**, 629-650.

[3] Allais, M. (1953): "Le comportement de l'homme rationnel devant le risque, critique des postulats et axiomes de l'école Américaine", *Econometrica* **21**, 503-546.

[4] Anscombe, F., and R. Aumann (1963): "A Definition of Subjective Probability", *Annals of Mathematical Statistics* **34**, 199-205.

[5] Armbruster, W., and W. Böge (1979): "Bayesian Game Theory", in O. Möschlin and D. Pallaschke (eds.), *Game Theory and Related Topics*, Amsterdam, North-Holland.

[6] Arrow, K. J. (1971): *Essays in the Theory of Risk Bearing*, Chicago, Markham.

[7] Aumann, R. (1974): "Subjectivity and Correlation in Randomized Strategies", *Journal of Mathematical Economics* **1**, 67-96.

[8] Aumann, R. (1976): "Agreeing to Disagree", *Annals of Statistics* **4**, 1236-1239.

[9] Aumann, R. (1987): "Correlated Equilibrium as an Expression of Bayesian Rationality", *Econometrica* **55**, 1-18.

[10] Bagwell, K. (1995): "Commitment and Observability in Games", *Games and Economic Behavior* **8**, 271-280.

[11] Banerjee, A. (1992): "A Simple Model of Herd Behavior", *Quarterly Journal of Economics* **107**, 797-818.

[12] Barany, I. (1987): "Fair Distribution Protocols or How the Players Replace Fortune", CORE discussion paper 8718, Louvain.

[13] Basu, K. (1988): "Strategic Irrationality in Extensive Games", *Mathematical Social Sciences* **15**, 247-260.

[14] Basu, K., and J. W. Weibull (1991): "Strategy Subsets Closed under Rational Behavior", *Economics Letters* **36**, 141-146.

[15] Bazerman, M. H., and W. Samuelson (1983): "The Winner's Curse: An Empirical Investigation", in R. Tietz (ed.), *Aspiration Levels in Bargaining and Economic Decision Making*, Berlin-Heidelberg-New York, Springer Verlag.

[16] Ben-Porath, E., and E. Dekel (1992): "Signalling Future Actions and the Potential for Sacrifice", *Journal of Economic Theory* **57**, 36-51.

[17] Berge, C. (1963): *Topological Spaces*, New York, Macmillan.

[18] Bernheim, B. D. (1984): "Rationalizable Strategic Behavior", *Econometrica* **52**, 1007-1028.

[19] Bernoulli, D. (1738): "Specimen theoriae novae de mensura sortis", *Commentarii Academiae Scientiarum Imperialis Petropolitanae* **5**, 175-192. Translated by L. Sommer in *Econometrica* **22** (1954), 23-36.

[20] Bester, H., and K. Ritzberger (forthcoming): "Strategic Pricing, Signalling, and Costly Information Acquisition", *International Journal of Industrial Organization*.

[21] Bikhchandani, S., D. Hirshleifer, and I. Welch (1992): "A Theory of Fads, Fashion, Custom and Cultural Change as Informational Cascades", *Journal of Political Economy* **100**, 992-1027.

[22] Binmore, K. (1990): *Essays on the Foundations of Game Theory*, Cambridge, Mass., Basil Blackwell.

[23] Binmore, K. G., and P. Dasgupta (eds.) (1987): *The Economics of Bargaining*, Oxford, Basil Blackwell.

[24] Binmore, K. G., A. Rubinstein, and A. Wolinsky (1986): "The Nash Bargaining Solution in Economic Modeling", *Rand Journal of Economics* **17**, 176-188.

[25] Blume, L. E., and W. R. Zame (1994): "The Algebraic Geometry of Perfect and Sequential Equilibrium", *Econometrica* **62**, 783-794.

[26] Böge, W., and T. Eisele (1979): "On Solutions of Bayesian Games", *International Journal of Game Theory* **8**, 193-215.

[27] Börgers, T. (1994): "Weak Dominance and Approximate Common Knowledge", *Journal of Economic Theory* **64**, 265-276.

[28] Border, K. C. (1985): *Fixed Point Theorems with Applications to Economics and Game Theory*, Cambridge, Cambridge University Press.

[29] Borel, E. (1921): "La théorie du jeu et les equations intégrales à noyau symétrique gauche", *Comptes Rendus de l'Adadémie des Sciences* **173**, 1304-1308.

[30] Brandenburger, A., and E. Dekel (1987): "Rationalizability and Correlated Equilibrium", *Econometrica* **55**, 1391-1402.

[31] Chew, S. H., E. Karni, and Z. Safra (1987): "Risk Aversion in the Theory of Expected Utility with Rank Dependent Probabilities", *Journal of Economic Theory* **42**, 370-381.

[32] Cho, I.-K., and D. Kreps (1987): "Signaling Games and Stable Equilibria", *Quarterly Journal of Economics* **102**, 179-222.

[33] Cournot, A. A. (1838): *Recherches sur les Principes Mathématiques de la Théorie des Richesses*, Paris, Hachette.

[34] Cramton, P. (1998): "The Efficiency of the FCC Spectrum Auctions", *Journal of Law and Economics* **41**, 727-136.

[35] Dalkey, N. (1953): "Equivalence of Information Patterns and Essentially Determinate Games", *Annals of Mathematics Studies* **28**, 217-243.

[36] van Damme, E. (1984): "A Relation between Perfect Equilibria in Extensive Form Games and Proper Equilibria in Normal Form Games", *International Journal of Game Theory* **13**, 1-13.

[37] van Damme, E. (1987): *Stability and Perfection of Nash Equilibria*, Berlin-Heidelberg-New York, Springer-Verlag.

[38] van Damme, E. (1989): "Stable Equilibria and Forward Induction", *Journal of Economic Theory* **48**, 476-496.

[39] van Damme, E. (1994): "Strategic Equilibrium", mimeo (prepared for the *Handbook of Game Theory with Economic Applications*), January 1994.

[40] Dasgupta, P., and E. Maskin (1986): "The Existence of Equilibrium in Discontinuous Economic Games", *Review of Economic Studies* **53**, 1-41.

[41] Debreu, G. (1952): "A Social Equilibrium Existence Theorem", *Proceedings of the National Academy of Sciences* **38**, 886-893.

[42] Debreu, G. (1970): "Economies with a Finite Set of Equilibria", *Econometrica* **38**, 387-392.

[43] Dekel, E., and D. Fudenberg (1990): "Rational Behavior with Payoff Uncertainty", *Journal of Economic Theory* **52**, 243-267.

[44] Demichelis, S., and F. Germano (2000): "On the Indices of Zeros of Nash Fields", *Journal of Economic Theory* **94**, 192-217. (First version: *University of California at San Diego Discussion Paper* 96-33, 1996.)

[45] Demichelis, S., and F. Germano (2000): "On Knots and Dynamics in Games", CORE Discussion Paper 2000/10, Louvain.

[46] Demichelis, S., and K. Ritzberger (2000): "From Evolutionary to Strategic Stability", *CORE Discussion Paper* 2000/59.

[47] Diamond, D. W., and P. H. Dybvig (1983): "Bank Runs, Deposit Insurance, and Liquidity", *Journal of Political Economy* **91**, 401-419.

[48] Dierker, E. (1972): "Two Remarks on the Number of Equilibria of an Economy", *Econometrica* **40**, 951-953.

[49] Dieudonné, J. (1960): *Foundations of Modern Algebra*, New York, Academic Press.

[50] Dow, J., and S. R. d. C. Werlang (1992): "Risk Aversion, Uncertainty Aversion and the Optimal Choice of Portfolio", *Econometrica* **60**, 197-204.

[51] Dow, J., and S. R. d. C. Werlang (1994): "Nash Equilibrium under Uncertainty: Breaking Down Backward Induction", *Journal of Economic Theory* **64**, 305-324.

[52] Dresher, M. (1970): "Probability of a Pure Equilibrium Point in n-Person Games", *Journal of Combinatorial Theory* **8**, 134-145.

[53] Dufwenberg, M., and M. Stegeman (1999): "When Order Matters for Iterated Strict Dominance", *Research Papers in Economics* 1999:2, University of Stockholm.

[54] Ellsberg, D. (1961): "Risk, Ambiguity, and the Savage Axioms", *Quarterly Journal of Economics* **75**, 643-669.

[55] Elmes, S., and P. Reny (1994): "On the Strategic Equivalence of Extensive Form Games", *Journal of Economic Theory* **62**, 1-23.

[56] Fan, K. (1952): "Fixed Point and Minimax Theorems in Locally Convex Topological Linear Spaces", *Proceedings of the National Academy of Sciences* **38**, 121-126.

[57] Fan, K. (1972): "A Minimax Inequality and Applications", in O. Shisha (ed.), *Inequalities,* Vol. III, New York, Academic Press.

[58] Feddersen, T. J., and W. Pesendorfer (1996): "The Swing Voter's Curse", *American Economic Review* **86**, 408-424.

[59] Fishburn, P. C. (1970): *Utility Theory for Decision Making,* New York, John Wiley.

[60] Forges, F. (1990): "Universal Mechanisms", *Econometrica* **58**, 1341-1364.

[61] Fort, M. K. (1950): "Essential and Nonessential Fixed Points", *American Journal of Mathematics* **72**, 315-322.

[62] Friedman, J.W. (1977): *Oligopoly and the Theory of Games,* Amsterdam, North-Holland.

[63] Friedman, J.W. (1983): *Oligopoly Theory,* Cambridge, Cambridge University Press.

[64] Fudenberg, D., and J. Tirole (1991): "Perfect Bayesian Equilibrium and Sequential Equilibrium", *Journal of Economic Theory* **53**, 236-260.

[65] Gilboa, I. (1987): "Expected Utility with Purely Subjective Non-Additive Probabilities", *Journal of Mathematical Economics* **16**, 65-88.

[66] Gilboa, I., E. Kalai, and E. Zemel (1990): "On The Order of Eliminating Dominated Strategies", *Operations Research Letters* **9**, 85-89.

[67] Gilboa, I., and D. Schmeidler (1989): "Maxmin Expected Utility with a Non-Unique Prior", *Journal of Mathematical Economics* **18**, 141-153.

[68] Glazer, J., and A. Weiss (1990): "Pricing and Coordination: Strategically Stable Equilibrium", *Games and Economic Behavior* **2**, 118-128.

[69] Glicksberg, I. L. (1952): "A Further Generalization of the Kakutani Fixed Point Theorem with Application to Nash Equilibrium Points", *Proceedings of the National Academy of Sciences* **38**, 170-174.

[70] Govindan, S., and R. Wilson (1997): "Equivalence and Invariance of the Index and Degree of Nash Equilibria", *Games and Economic Behavior* **21**, 56-61.

[71] Green, J. (1987): "'Making Book against Oneself', the Independence Axiom, and Nonlinear Utility Theory", *Quarterly Journal of Economics* **98**, 785-796.

[72] Grossman, S. J., and J. E. Stiglitz (1980): "On the Impossibility of Informationally Efficient Markets", *American Economic Review* **70**, 393-408.

[73] Gül, F., D. Pearce, and E. Stacchetti (1993): "A Bound on the Proportion of Pure Strategy Equilibria in Generic Games", *Mathematics of Operations Research* **18**, 548-552.

[74] Güth, W., G. Kirchsteiger, and K. Ritzberger (1998): "Imperfectly Observable Commitments in n-Player Games", *Games and Economic Behavior* **23**, 54-74.

[75] Güth, W., and K. Ritzberger (1998): "On Durable Goods Monopolies and the Coase-Conjecture", *Review of Economic Design* **3**, 215-236.

[76] Güth, W., R. Schmittberger, and B. Schwarze (1982): "An Experimental Analysis of Ultimatum Bargaining", *Journal of Economic Behavior and Organization* **3**, 367-388.

[77] Guillemin, V., and A. Pollack (1974): *Differential Topology*, Englewood Cliffs NJ, Prentice-Hall.

[78] Harris, C. J. (1985): "Existence and Characterization of Perfect Equilibrium in Games of Perfect Information", *Econometrica* **53**, 613-628.

[79] Harsanyi, J. C. (1967-8): "Games of Incomplete Information Played by Bayesian Players, I, II, and III", *Management Science* **14**, 159-182, 320-334, 486-502.

[80] Harsanyi, J. C. (1973): "Game with Randomly Disturbed Payoffs: A New Rationale for Mixed Strategy Equilibrium Points", *International Journal of Game Theory* **2**, 1-23.

[81] Harsanyi, J. C. (1973): "Oddness of the Number of Equilibrium Points: A New Proof", *International Journal of Game Theory* **2**, 235-250.

[82] Harsanyi, J. C., and R. Selten (1988): *A General Theory of Equilibrium Selection in Games*, Cambridge Mass., MIT Press.

[83] Hauk, E., and S. Hurkens (1999): "On Forward Induction and Evolutionary and Strategic Stability", mimeo, Universitat Pompeu Fabra (forthcoming in *Journal of Economic Theory*).

[84] Hellwig, M., and W. Leininger (1987): "On the Existence of Subgame-Perfect Equilibrium in Infinite-Action Games of Perfect Information", *Journal of Economic Theory* **43**, 55-75.

[85] Hellwig, M., W. Leininger, P. Reny, and A. Robson (1990): "Subgame Perfect Equilibrium in Continuous Games of Perfect Information: An Elementary Approach to Existence and Approximation by Discrete Games", *Journal of Economic Theory* **52**, 406-422.

[86] Hey J. D., and C. Orme (1994): "Investigating Generalizations of Expected Utility Theory Using Experimental Data", *Econometrica* **62**, 1291-1326.

[87] Hillas, J. (1990): "On the Definition of the Strategic Stability of Equilibria", *Econometrica* **58**, 1365-1390.

[88] Hofbauer, J. (1990): "An Index Theorem for Dissipative Semiflows", *Rocky Mountain Journal of Mathematics* **20**, 1017-1031.

[89] Hofbauer, J., and K. Sigmund (1988): *The Theory of Evolution and Dynamical Systems*, Cambridge, Cambridge University Press.

[90] Hofbauer, J., and J. Swinkels (1995): "A Universal Shapley Example", mimeo, University of Vienna and Northwestern University.

[91] Hurkens, S., and E. van Damme (1997): "Games with Imperfectly Observable Commitments", *Games and Economic Behavior* **21**, 282-308.

[92] Jiang Jia-He (1963): "Essential Component of the Set of Fixed Points of the Multivalued Mappings and its Application to the Theory of Games", *Scientia Sinica* **12**, 951-964.

[93] Kahneman, D., and A. Tversky (1984): "Choices, Values, and Frames", *American Psychologist* **39**, 341-350.

[94] Kelsey, D. (1994): "Maxmin Expected Utility and Weight of Evidence", *Oxford Economic Papers* **46**, 425-444.

[95] Kinoshita, S. (1952): "On Essential Components of the Set of Fixed Points", *Osaka Mathematical Journal* **4**, 19-22.

[96] Knight, F. (1921): *Risk, Uncertainty and Profit*, Boston, Mass., Houghton Mifflin.

[97] Kohlberg, E., and J.-F. Mertens (1986): "On the Strategic Stability of Equilibria", *Econometrica* **54**, 1003-1037.

[98] Kojima, M., A. Okada, and S. Shindoh (1985): "Strongly Stable Equilibrium Points of *n*-Person Non-Cooperative Games", *Mathematics of Operations Research* **10**, 650-663.

[99] Krentel, W. D., J. C. C. McKinsey, and W. V. Quine (1951): "A Simplification of Games in Extensive Form", *Duke Mathematical Journal* **18**, 885-900.

[100] Kreps, D. M. (1979): "A Representation Theorem for 'Preference for Flexibility'", *Econometrica* **47**, 565-577.

[101] Kreps, D. M., and G. Ramey (1987): "Structural Consistency, Consistency, and Sequential Rationality", *Econometrica* **55**, 1331-1348.

[102] Kreps, D. M., and R. Wilson (1982): "Sequential Equilibrium", *Econometrica* **50**, 863-894.

[103] Kreps, D. M.(1988): *Notes on the Theory of Choice*, Boulder, Colo., Westview Press.

[104] Kuhn, H. W. (1953): "Extensive Games and the Problem of Information", in H. W. Kuhn and A. W. Tucker (eds.), *Contributions to the Theory of Games*, Vol. II, Princeton, Princeton University Press, 193-216.

[105] Luce, D. R., and H. Raiffa (1957): *Games and Decisions*, New York, John Wiley.

[106] Machina, M. J. (1987): "Choice under Uncertainty: Problems Solved and Unsolved", *Journal of Economic Perspectives* **1**, 121-154.

[107] Mailath, G. J., M. Okuno-Fujiwara, and A. Postlewaite (1993): "Belief-Based Refinements in Signalling Games", *Journal of Economic Theory* **60**, 241-276.

[108] Mailath, G. J., L. Samuelson, and J. M. Swinkels (1993): "Extensive Form Reasoning in Normal Form Games", *Econometrica* **61**, 273-302.

[109] Mailath, G. J., L. Samuelson, and J. M. Swinkels (1994): "Normal Form Structures in Extensive Form Games", *Journal of Economic Theory* **64**, 325-371.

[110] McAfee, P. R., and J. McMillan (1987): "Auctions and Bidding", *Journal of Economic Literature* **25**, 699-738.

[111] McLennan, A. (1988): "Selected Topics in the Theory of Fixed Points", mimeo, University of Minnesota.

[112] McMillan, J. (1994): "Selling Spectrum Rights", *Journal of Economic Perspectives* **8**, 145-162.

[113] Mertens, J.-F. (1989): "Stable Equilibria—A Reformulation, Part I: Definition and Basic Properties", *Mathematics of Operations Research* **14**, 575-624.

[114] Mertens, J.-F. (1991): "Stable Equilibria—A Reformulation, Part II: Discussion of the Definition and Further Results", *Mathematics of Operations Research* **16**, 694-753.

[115] Mertens, J.-F. (1992): "The Small Worlds Axiom for Stable Equilibria", *Games and Economic Behavior* **4**, 553-564.

[116] Mertens, J.-F., and S. Zamir (1985): "Formulation of a Bayesian Analysis for Games with Incomplete Information", *International Journal of Game Theory* **14**, 1-29.

[117] Milgrom, P. (1989): "Auctions and Bidding: A Primer", *Journal of Economic Perspectives* **3**, 3-22.

[118] Milgrom, P. (1998): "Game Theory and the Spectrum Auctions", *European Economic Review* **42**, 771-778.

[119] Milgrom, P., and J. Roberts (1982): "Limit Pricing and Entry under Incomplete Information", *Econometrica* **50**, 443-460.

[120] Milgrom, P., and R. J. Weber (1982): "A Theory of Auctions and Competitive Bidding", *Econometrica* **50**, 1089-1122.

[121] Milnor, J. W. (1965): *Topology from the Differentiable Viewpoint*, Charlottesville, Va, University Press of Virginia.

[122] Morris, S., R. Rob, and H. S. Shin (1995): "p-Dominance and Belief Potential", *Econometrica* **63**, 145-157.

[123] Moulin, H. (1979): "Dominance Solvable Voting Schemes", *Econometrica* **47**, 1337-1351.

[124] Moulin, H. (1984): "Dominance Solvability and Cournot Stability", *Mathematical Social Sciences* **7**, 83-102.

[125] Moulin, H. (1986): *Game Theory for the Social Sciences*, New York, New York University Press, second edition.

[126] Myerson, R. (1978): "Refinements of the Nash Equilibrium Concept", *International Journal of Game Theory* **7**, 73-80.

[127] Myerson, R. (1981): "Optimal Auction Design", *Mathematics of Operations Research* **6**, 58-73.

[128] Myerson, R. (1991): *Game Theory—Analysis of Conflict*, Cambridge, Mass., Harvard University Press.

[129] Nagel, R. (1995): "Unraveling in Guessing Games: An Experimental Study", *American Economic Review* **85**, 1313-1326.

[130] Nash, J. F. (1950): *Non-Cooperative Games*, PhD Dissertation at Princeton University.

[131] Nash, J. F. (1950): "Equilibrium Points in N-Person Games", *Proceedings of the National Academy of Sciences* **36**, 48-49.

[132] Nash, J. F. (1951): "Non-Cooperative Games", *Annals of Mathematics* **54**, 286-295.

[133] von Neumann, J. (1928): "Zur Theorie der Gesellschaftsspiele", *Mathematische Annalen* **100**, 295-320.

[134] von Neumann, J., and O. Morgenstern (1953): *Theory of Games and Economic Behavior*, Princeton, Princeton University Press, third edition (first edition 1944).

[135] Okada, A. (1981): "On Stability of Perfect Equilibrium Points", *International Journal of Game Theory* **10**, 67-73.

[136] Okada, A. (1987): "Complete Inflation and Perfect Recall in Extensive Games", *International Journal of Game Theory* **16**, 85-91.

[137] O'Neill, B. (1953): "Essential Sets and Fixed Points", *American Journal of Mathematics* **75**, 497-509.

[138] Osborne, M. J., and A. Rubinstein (1994): *A Course in Game Theory*, Cambridge, Mass., MIT Press.

[139] Owen, G. (1982): *Game Theory*, New York, Academic Press, second edition.

[140] Pearce, D. G. (1984): "Rationalizable Strategic Behavior and the Problem of Perfection", *Econometrica* **52**, 1029-1050.

[141] Piccione, M., and A. Rubinstein (1997): "On the Interpretation of Decision Problems with Imperfect Recall", *Games and Economic Behavior* **20**, 3-24.

[142] Png, I. (1983): "Strategic Behavior in Suit, Settlement, and Trial", *Bell Journal of Economics* **14**, 539-550.

[143] Pratt, J. (1964): "Risk Aversion in the Small and in the Large", *Econometrica* **32**, 122-136.

[144] Quiggin, J. (1982): "A Theory of Anticipated Utility", *Journal of Economic Behavior and Organization* **3**, 225-243.

[145] Ramsey, F. P. (1931): "Truth and Probability", in F. P. Ramsey, *The Foundations of Mathematics and Other Logical Essays*, New York, Harcourt, Brace; reprinted in H. E. Kyburg and H. E. Smokler (eds.), *Studies in Subjective Probability*, New York, John Wiley, 1964.

[146] Reny, P. J. (1992): "Backward Induction, Normal Form Perfection, and Explicable Equilibria", *Econometrica* **60**, 627-649.

[147] Reny, P. J. (1993): "Common Belief and the Theory of Games with Perfect Information", *Journal of Economic Theory* **59**, 257-274.

[148] Richter, M. (1971): "Rational Choice", in J. Chipman, L. Hurwicz, and H. Sonnenschein (eds.), *Preferences, Utility, and Demand*, New York, Harcourt Brace Jovanovich.

[149] Ritzberger, K. (1994): "The Theory of Normal Form Games from the Differentiable Viewpoint", *International Journal of Game Theory* **23**, 201-236.

[150] Ritzberger, K. (1996): "On Games under Expected Utility with Rank Dependent Probabilities", *Theory and Decision* **40**, 1-27.

[151] Ritzberger, K. (1999): "Recall in Extensive Form Games", *International Journal of Game Theory* **28**, 69-87.

[152] Ritzberger, K., and J. W. Weibull (1995): "Evolutionary Selection in Normal Form Games", *Econometrica* **63**, 1371-1399.

[153] Rosenthal, R. (1981): "Games of Perfect Information, Predatory Pricing, and the Chain-Store Paradox", *Journal of Economic Theory* **25**, 92-100.

[154] Rothschild, M., and J. Stiglitz (1970): "Increasing Risk I: A Definition", *Journal of Economic Theory* **2**, 225-243.

[155] Rubinstein, A. (1982): "Perfect Equilibrium in a Bargaining Model", *Econometrica* **50**, 97-109.

[156] Rubinstein, A., and A. Wolinsky (1985): "Equilibrium in a Market with Sequential Bargaining", *Econometrica* **53**, 1133-1150.

[157] Samuelson, L. (1992): "Dominated Strategies and Common Knowledge", *Games and Economic Behavior* **4**, 284-313.

[158] Samuelson, L. (1997): *Evolutionary Games and Equilibrium Selection*, Cambridge, Mass., MIT Press.

[159] Savage, L. J. (1954): *The Foundations of Statistics*, New York, John Wiley.

[160] Schelling, T. (1984): *Choice and Consequence*, Cambridge, Mass., Harvard University Press.

[161] Schmeidler, D. (1989): "Subjective Probability and Expected Utility without Additivity", *Econometrica* **57**, 571-587.

[162] Selten, R. (1965): "Spieltheoretische Behandlung eines Oligopolmodells mit Nachfrageträgheit", *Zeitschrift für die gesamte Staatswissenschaft* **121**, 301-324 and 667-689.

[163] Selten, R. (1975): "Reexamination of the Perfectness Concept for Equilibrium Points in Extensive Games", *International Journal of Game Theory* **4**, 25-55.

[164] Selten, R. (1978): "The Chain Store Paradox", *Theory and Decision* **9**, 127-159.

[165] Selten, R. (1999): "Multistage Game Models and Delay Supergames: Nobel Lecture, December 9, 1994", in R. Selten, *Game Theory and Economic Behaviour: Selected Essays*, Vol. 2, Cheltenham, U.K. and Northampton, Mass., Edward Elgar, 1999, 3-32.

[166] Shaked, A., and J. Sutton (1984): "Involuntary Unemployment as a Perfect Equilibrium in a Bargaining Model", *Econometrica* **52**, 1351-1364.

[167] Shapley, L. (1974): "A Note on the Lemke-Howson Algorithm", *Mathematical Programming Study* **1**, 175-189.

[168] Simon, L. K., and M. B. Stinchcombe (1995): "Equilibrium Refinement for Infinite Normal Form Games", *Econometrica* **63**, 1421-1443.

[169] Sion, M., and P. Wolfe (1957): "On a Game without a Value", in *Contributions to the Theory of Games*, vol. III, Princeton Annals of Mathematical Studies 39.

[170] Sion, M. (1958): "On General Minimax Theorems", *Pacific Journal of Mathematics* **8**, 171-176.

[171] Spence, A. M. (1973): "Job Market Signaling", *Quarterly Journal of Economics* **87**, 355-374.

[172] von Stackelberg, H. (1951): *Grundlagen der theoretischen Volkswirtschaftslehre*, Tübingen-Bern, second edition.

[173] Sutton, J. (1986): "Non-Cooperative Bargaining Theory: An Introduction", *Review of Economic Studies* **53**, 709-724.

[174] Tan, T., and S. R. d. C. Werlang (1988): "The Bayesian Foundations of Solution Concepts of Games", *Journal of Economic Theory* **45**, 370-391.

[175] Thompson, F. B. (1952): "Equivalence of Games in Extensive Form", RAND Research Memorandum No. 759.

[176] Vega-Redondo, F. (1996): *Evolution, Games, and Economic Behaviour*, Oxford, Oxford University Press.

[177] Vickrey, W. (1961): "Counterspeculation, Auctions, and Competitive Sealed Tenders", *Journal of Finance* **16**, 8-37.

[178] Wakker, P. (1994): "Separating Marginal Utility and Probabilistic Risk Aversion", *Theory and Decision* **36**, 1-44.

[179] Weibull, J. W. (1995): *Evolutionary Game Theory*, Cambridge, Mass., MIT Press.

[180] Wilson, R. (1969): "Competitive Bidding with Disparate Information", *Management Science* **15**, 446-448.

[181] Wilson, R. (1971): "Computing Equilibria of n-Person Games", *Siam Journal of Applied Mathematics* **21**, 80-87.

[182] Wilson, R. (1977): "A Bidding Model of Perfect Competition", *Review of Economic Studies* **44**, 511-518.

[183] Wu Wen-Tsün and Jiang Jia-He (1962): "Essential Equilibrium Points of n-Person Non-Cooperative Games", *Scientia Sinica* **11**, 1307-1322.

[184] Yaari, M. E. (1987): "The Dual Theory of Choice under Risk", *Econometrica* **55**, 95-115.

[185] Zermelo, E. (1913): "Über eine Anwendung der Mengenlehre auf die Theorie des Schachspiels", *Proceedings of the Fifth International Congress of Mathematicians* **2**, 501-504.

Name Index

Subject Index